WORKSHOP TECHNOLOGY

VIRENDER NARULA

WORKSHOP TECHNOLOGY

(Including Workshop Practice Tools and Exercises)

By

VIRENDER NARULA

M.Tech. (Manufacturing and Automation), B.E. (Mechanical)
Associate Prof. and Head
Department of Mechanical Engineering
Shri Ram College of Engineering & Management
Palwal Sohna Road, Gurgaon

S.K. KATARIA & SONS®

Publisher of Engineering Books

4885/109, Prakash Mahal, Dr. Subhash Bhargav Lane,
Opposite Delhi Medical Association, Daryaganj, New Delhi-110002 (INDIA)
Phone: +91-11-43551243; Mobile: +91-9871775858, 7678425209
e-mail: katariabook@gmail.com, katariabooks@yahoo.com
Website : www.skkatariaandsons.com
(e-books, subscriptions and bulk buying also available on our website)

Published by :
S.K. Kataria & Sons®
4885/109, Prakash Mahal, Dr. Subhash Bhargav Lane,
Opposite Delhi Medical Association, Daryaganj, New Delhi-110002
Phone: +91-11-43551243; Mobile: +91-9871775858, 7678425209
e-mail: katariabook@gmail.com, katariabooks@yahoo.com

First Edition: January 2007, Third Edition: August, 2012, Reprint: 2026

Price: ₹ 495/-

Laser Typesetting by : Rohan Graphics *Printed at : L.B. Enterprises*

Preface

I feel highly delighted to present this book on **"Workshop Technololgy".** The book is basically further elaboration of my lecture notes which I prepared during fifteen years academic experience. Some how Industrial experience has also helped me a lot to shape this book. The book is basically written strictly as per the syllabus of Indian Universities. Some more topics have been added in some chapters to make the concepts more clear. I felt the need of writing this book because no other book covers the syllabus according to the needs of first year engineering students. I have genuinely tried to present the book as the Introduction to manufacturing processes with a touch of science & mechanism behind these processes. The Workshop Practice tools have been described alongwith their diagram. The workshop practice exercises are also there in each shop as it helps students to think why things are made in that way. Book starts with Basic Introduction to manufacturing processes. Materials properties and testing has been discussed in more details then required as per syllabus because I felt that any engineering student must be through with the Material properties. Science behind forming processes has also been explained in detail. Large number of diagrams have been drawn in every chapter to make the concept more clear. Pictorial views are also drawn frequently to make easy understanding of concerned topic. Previous university paper problems are mentioned after every chapter.

Appendices containing valuable information for a engineering student like SI units, conversion factors, Dimension farmulae, mathematical farmulae have been added at the end of the book.

Overall it is a genuine effort to bring a book which will be quite useful and indespensable tool for first year engineering students. Any comments and suggestions from Teachers as well as students are most welcome.

Virender Narula
(Author)

Acknowledgement

I express my sincere gratitude to Dr. Stya Paul, honourable chairman, Apeejay Education society and management of Apeejay Education society for their constant inspiration and encouragement. I feel greatly indebted to Dr. Sarbjit Singh, Dr. U.S. Triar, Prof. P.V. Gupta & Prof. Harihar Singh for their constant support & valuable guidance. I wish to express warm regards to my friend Prof. Shashi Chawla for motivating me to initiate this work.

I wish to thank the publisher and entire staff of S.K. Kataria & Sons for showing sustained interest during entire work.

It is pleasure to express heart felt thanks to my family members who have borne long hours of inconvenience during the preparation of manuscript.

Author

Dedicated to

My Parents

CONTENTS

CHAPTER 1

INTRODUCTION TO WORKSHOP TECHNOLOGY

INSIDE THIS CHAPTER

1.1 INTRODUCTION

Everyday we use hundreds of items like furniture, clothing, telephones, cars, planes, railways, TVs, VCRs and so on. These products are manufactured in factories using some manufacturing processes. Even items of basic needs of people like house, clothings and food items, directly or indirectly, depend upon the manufacturing processes.

Manufacturing Engineering is a branch of Engineering, strictly speaking, Mechanical engineering. In all the branches of Engineering, manufacturing is perhaps most important because it involves the "Conversion of raw material into the final product". It is of interest of not only to mechanical engineers, but also to every discipline of engineering. The various products like plant machinery, equipment, tools etc. required for chemical, civil, electrical, information technology and textile industries are made using some manufacturing processes. We may say manufacturing is quite significant for development of industries and plays a great role in making a country a developed country.

We may conclude from the above discussion that manufacturing is a very basic subject which is related to basic needs of people as well as living standard of the people of a country. Knowledge of manufacturing process is thus essential for every engineer.

"Successful people don't do different things they just do the same things differently and make the difference."

1.2 CLASSIFICATION OF MANUFACTURING PROCESSES

The manufacturing processes can be broadly classified into two categories:

1. Primary Manufacturing Processes (Fig. 1.1).
   (*a*) Casting Processes
   (*b*) Forming Processes.
2. Secondary Manufacturing Processes (Fig. 1.5).
   (*a*) Fabrication Processes
   (*b*) Material Removal Processes.

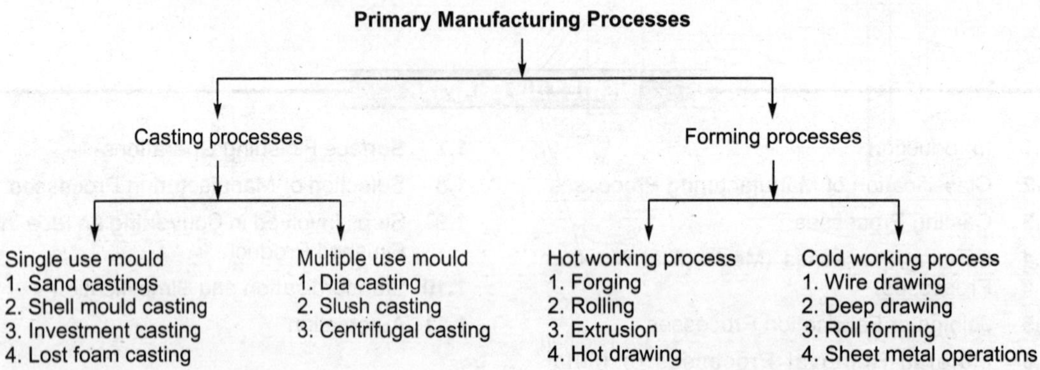

Fig. 1.1 Classification of primary manufacturing processes.

1.3 CASTING PROCESSES

These are the only processes where liquid metal is used. Casting is the one of the oldest known processes and also first step in manufacturing of most products. In casting processes a solid material (usually a metal) is first melted and then poured into a cavity or mould which resembles closely the shape of final object to be made. The cavity is made with the help of pattern. The molten metal is allowed to solidify and removed from cavity after solidification. Solidified object is called casting. In casting processes simple of complex shape can be made in single step.

Casting processes are universally used for manufacturing a wide variety of products. Extremely large, heavy metal objects can be made with casting process. Casting of even 200 ton can be made. Casting processes are best option when it is difficult or economically impossible to produce by any other process.

There are number of casting processes available which may be broadly classified as sand casting process and contemporary or special casting process.

1. **Sand Casting Process.** In sand casting process sand is used as primary material for making moulds. Sand casting process is used for most commercial metals. The process is equally suitable for production of very small batch as well as on a very large scale. These are used for producing large size objects.

2. **Contemporary or Special Casting Process.** Whenever the size of casting is not large and production is on a large scale, specialised casting process offers cost saving. Some of the specialised casting processes are:

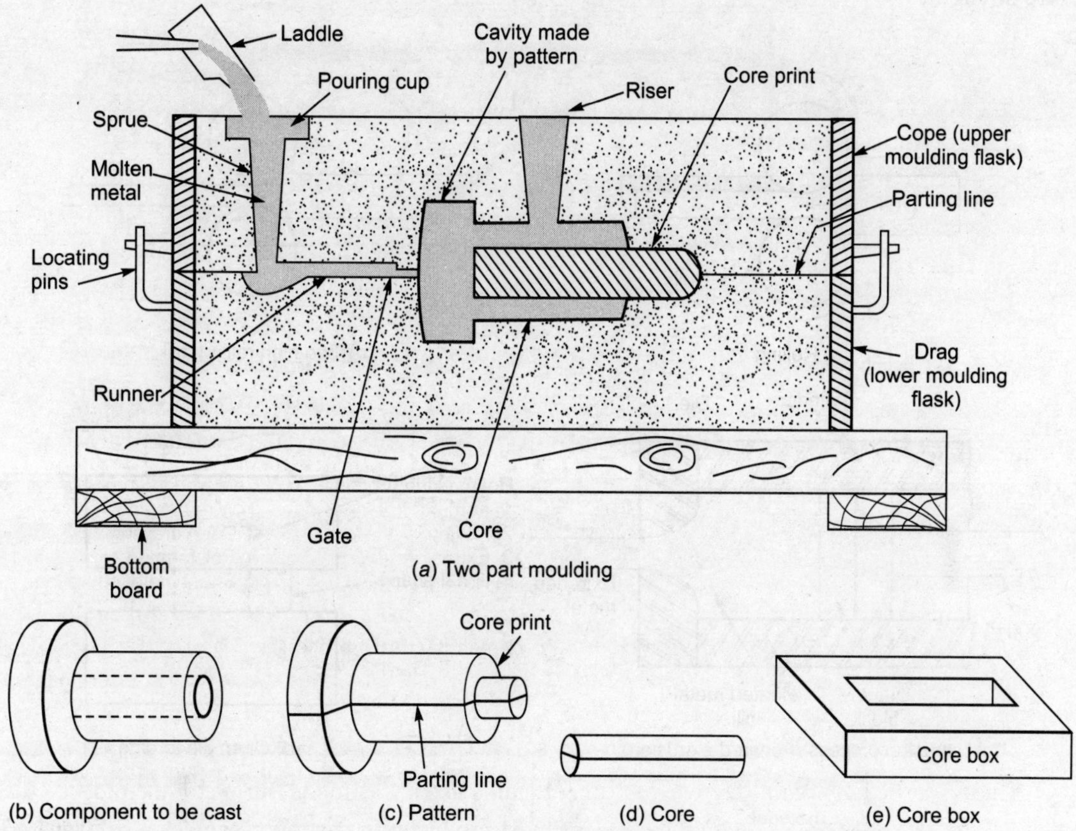

Fig. 1.2 Casting processes. (a) Two-part mould and its various components. (b) Component of be cast. (c) Pattern corresponding to that components. (d) Core for making hole in the component. (e) Core box for making core.

1. Shell mould casting
2. Precision investment casting
3. Permanent mould casting
4. Die casting
5. Centrifugal casting.

Some of the typical examples of casting include: pulleys, flywheels, machine tool beds, turbine blades, engine blocks, cast iron pipes etc.

1.4 FORMING PROCESSES (MECHANICAL WORKING PROCESSES)

Forming processes or mechanical working processes are primary manufacturing processes. In these processes metals are deformed to get desired shape and size. The necessary deformation can be achieved by application of large amount of mechanical force only or by heating the metal and then applying comparatively less force. The desired shape and size is obtained by means of set of tools called dies which may be open or closed dies. Fig. 1.3 (*d*) and 1.3 (*e*) shows a open and closed die respectively. The stresses induced during the process are greater than yield strength but less than fracture strength.

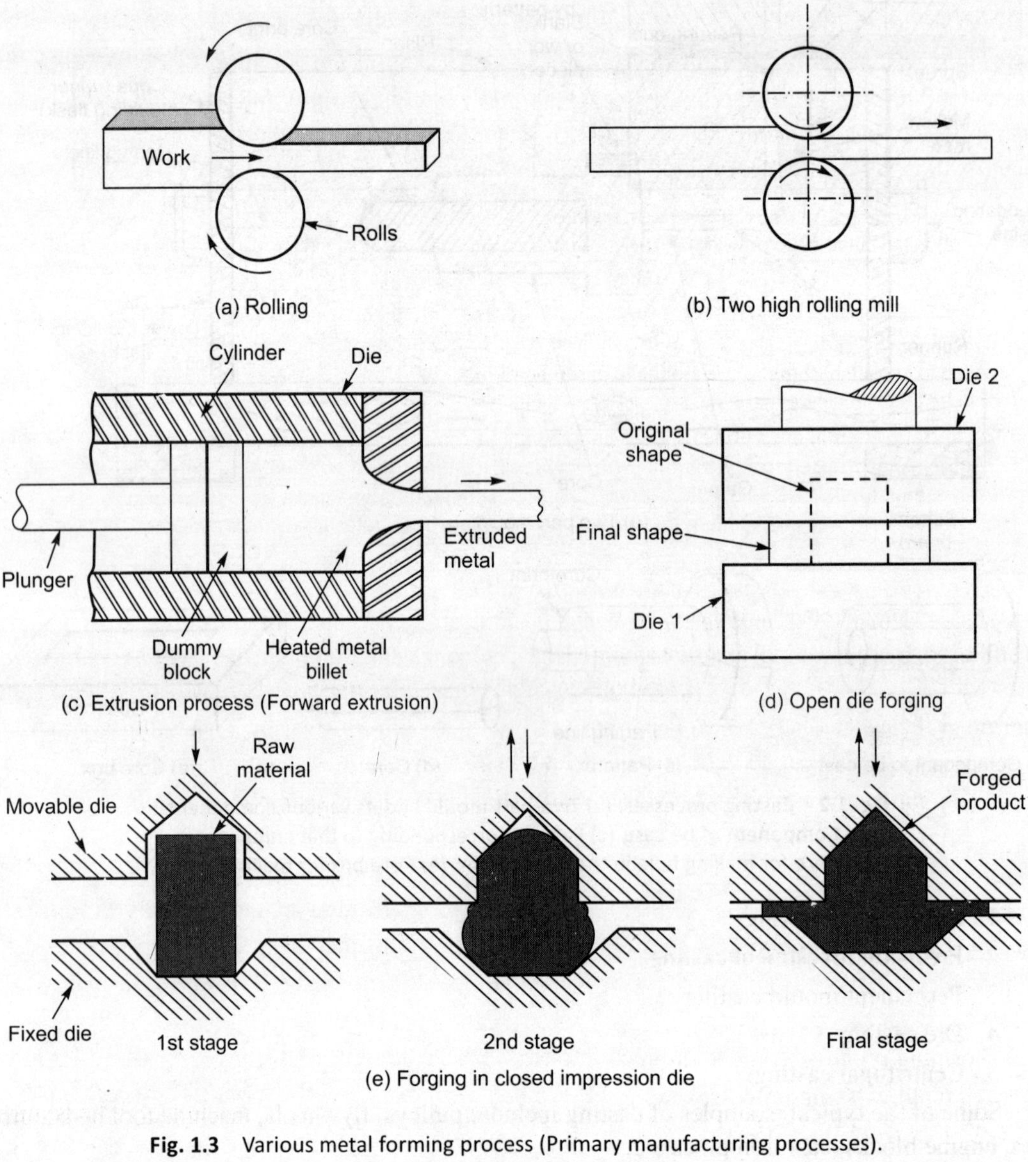

Fig. 1.3 Various metal forming process (Primary manufacturing processes).

Forming processes are very economical processes as they involve minimum amount of material wastage and result in faster production. The materials processed by forming processes develop superior mechanical properties like strength, hardness and toughness etc. Ductility is quite desirable property of materials in forming processes.

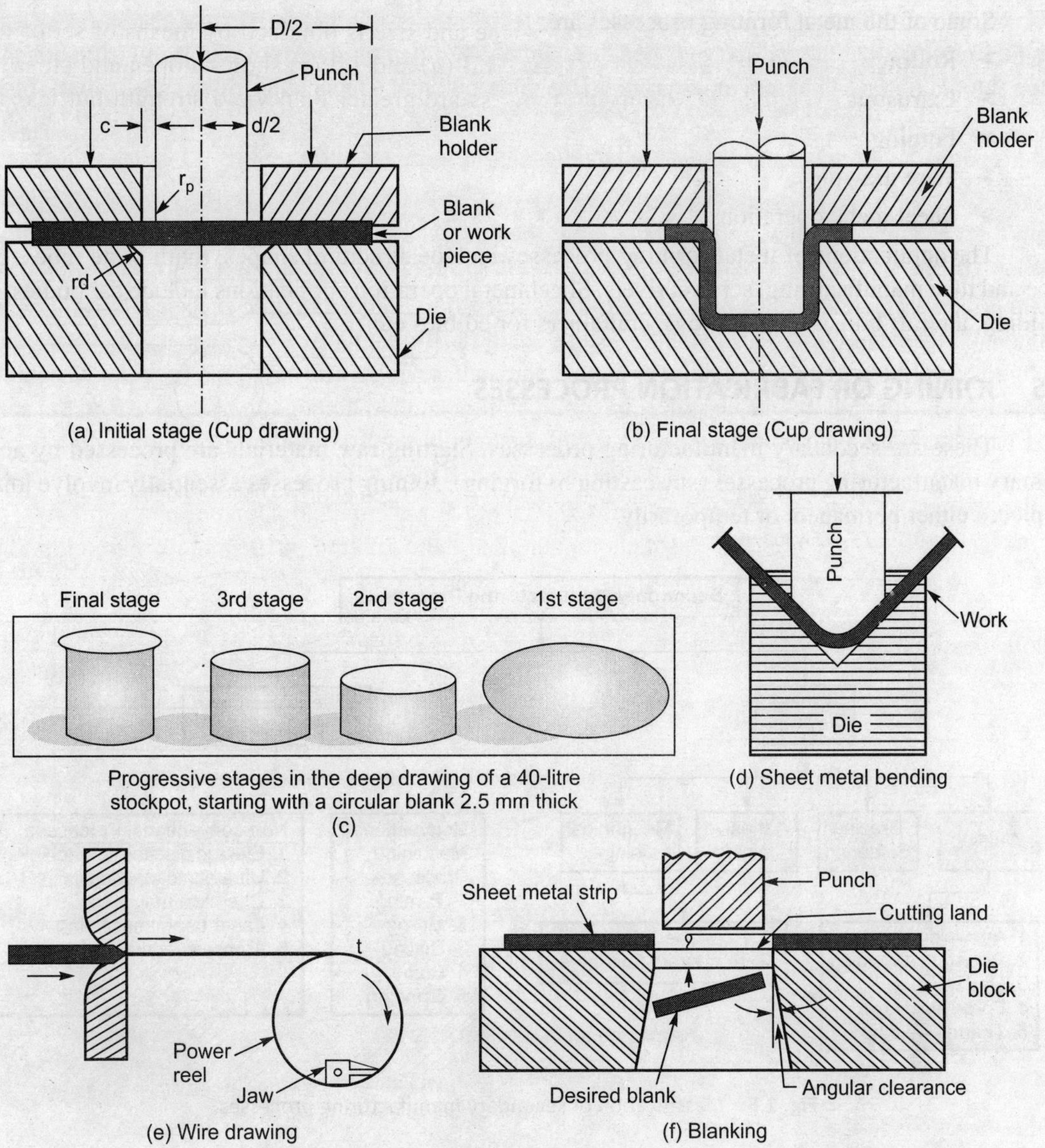

Fig. 1.4 Sheet metal processes.

Forming processes can be grouped into two broad categories:

1. Cold working processes
2. Hot working processes

If the working temperature is lower than recrystallisation temperature then process is termed as cold working process.

When the working temperature is above recrystallisation temperature then process is termed as hot working process.

Some of the metal forming processes are:

- Rolling
- Extrusion
- Forging
- Wire drawing
- Sheet metal operations.

The applications of metal forming processes include structural shapes, tooth paste tubes, cans, pipe and tube manufacturing, screws, rivets. Sheet metal operation applications include car bodies, ship building, aircraft bodies, boilers, toys, containers for edibles etc.

1.5 JOINING OR FABRICATION PROCESSES

These are secondary manufacturing processes. Starting raw materials are processed by any of primary manufacturing processes (say casting or forging). Joining processes essentially involve joining of pieces either permanent or temporarily.

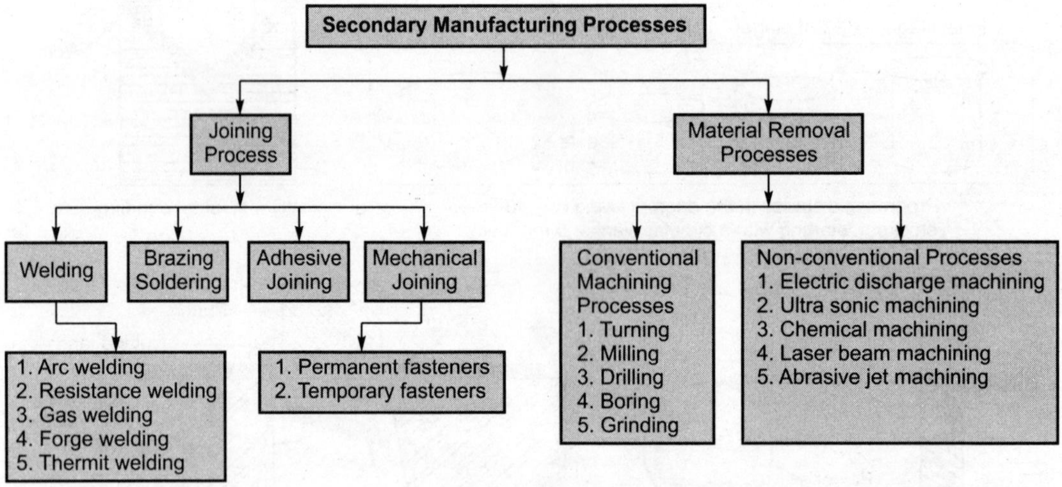

Fig. 1.5 Classification of secondary manufacturing processes.

1. **Permanent Joining Processes.** Permanent joining is achieved by means of riveting, adhesive joining, welding, soldering or brazing. Rivetting finds its application in joining sheet metal components (like boilers, aircraft and ship bodies). Adhesive joining helps in joining awkward shape parts or joining metals to non-metals (examples are automobile brake-shoe lining).

   In welding processes joint is obtained by applying either heat or pressure or by applying both, heat and pressure. Welding process is usually used to join similar metals. Joining material may or may not be used. Soldering and brazing are used to join dissimilar metals by means of a filler material. Figure 1.6 shows various welding processes.

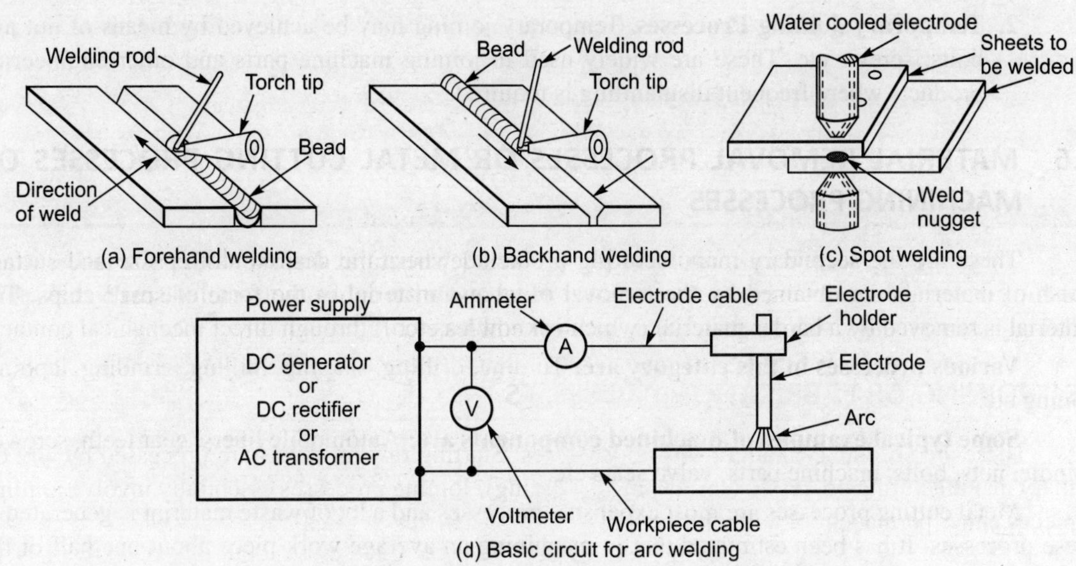

Fig. 1.6 Welding-joining processes (Secondary manufacturing processes).

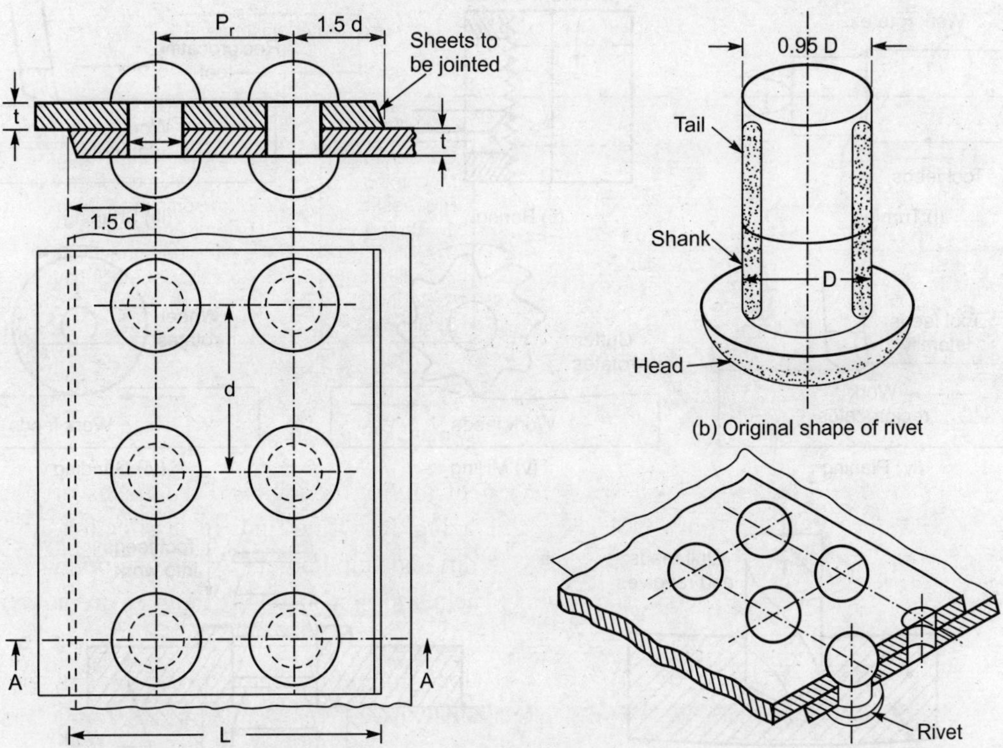

Fig. 1.7 Riveting - Mechanical joining processes (Secondary manufacturing processes).

2. **Temporary Joining Processes.** Temporary joining may be achieved by means of nut and bolts, screws etc. These are widely used in joining machine parts and other engineering products where frequent dismantling is required.

1.6 MATERIAL REMOVAL PROCESSES OR METAL CUTTING PROCESSES OR MACHINING PROCESSES

These are the secondary manufacturing processes where the desired shape, size and surface finish of materials are obtained by the removal of excess material in the form of small chips. The material is removed by a harder material (which is known as tool) through direct mechanical contact.

Various processes in this category are: Turning, drilling, shaping, milling, grinding, lapping, honing etc.

Some typical examples of machined components are: Automobile liners, gear teeth, screwed spindle, nuts, bolts, machine parts, valve seats etc.

Metal cutting processes are most expensive processes and a lot of waste material is generated in these processes. It has been estimated that in machining an average work piece about one half of the material is removed from work piece as chips. Still these are perhaps most versatile processes as they deliver good dimensional accuracy and good surface finish.

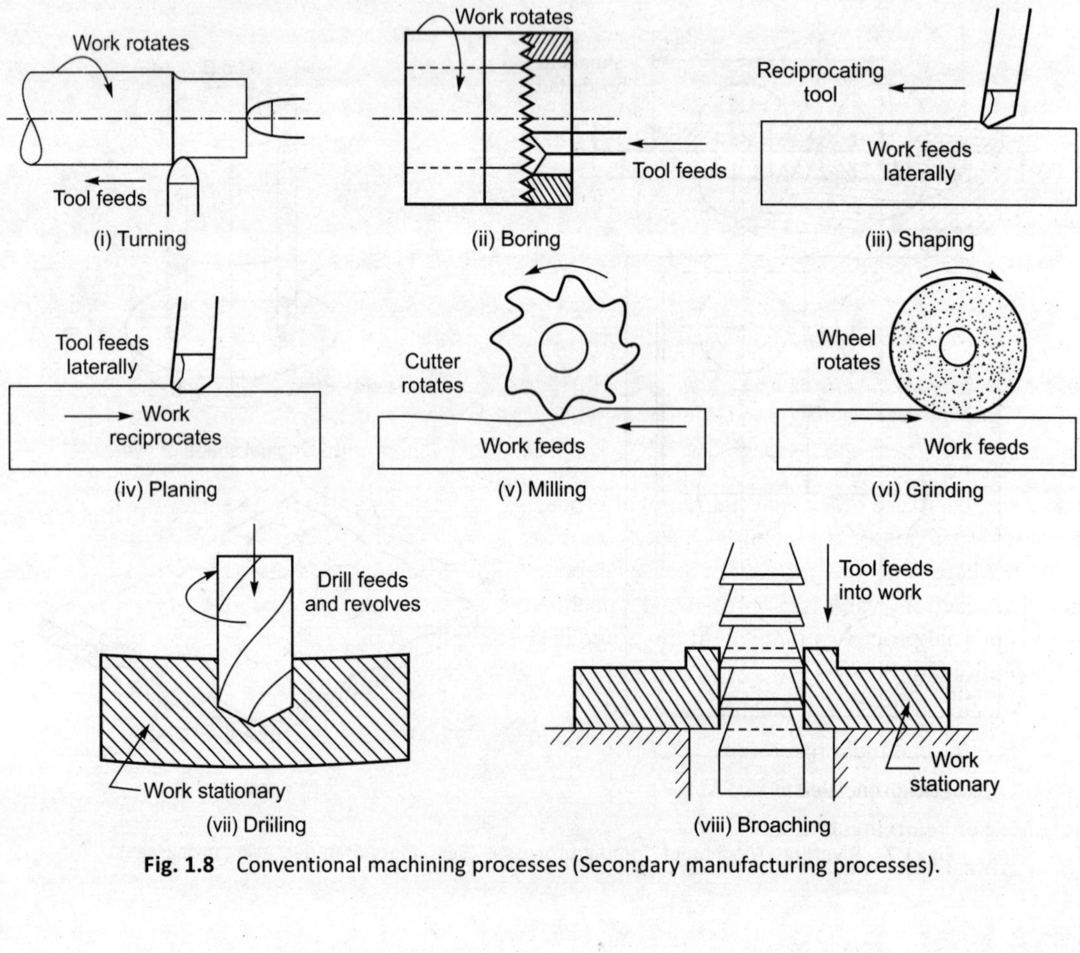

Fig. 1.8 Conventional machining processes (Secondary manufacturing processes).

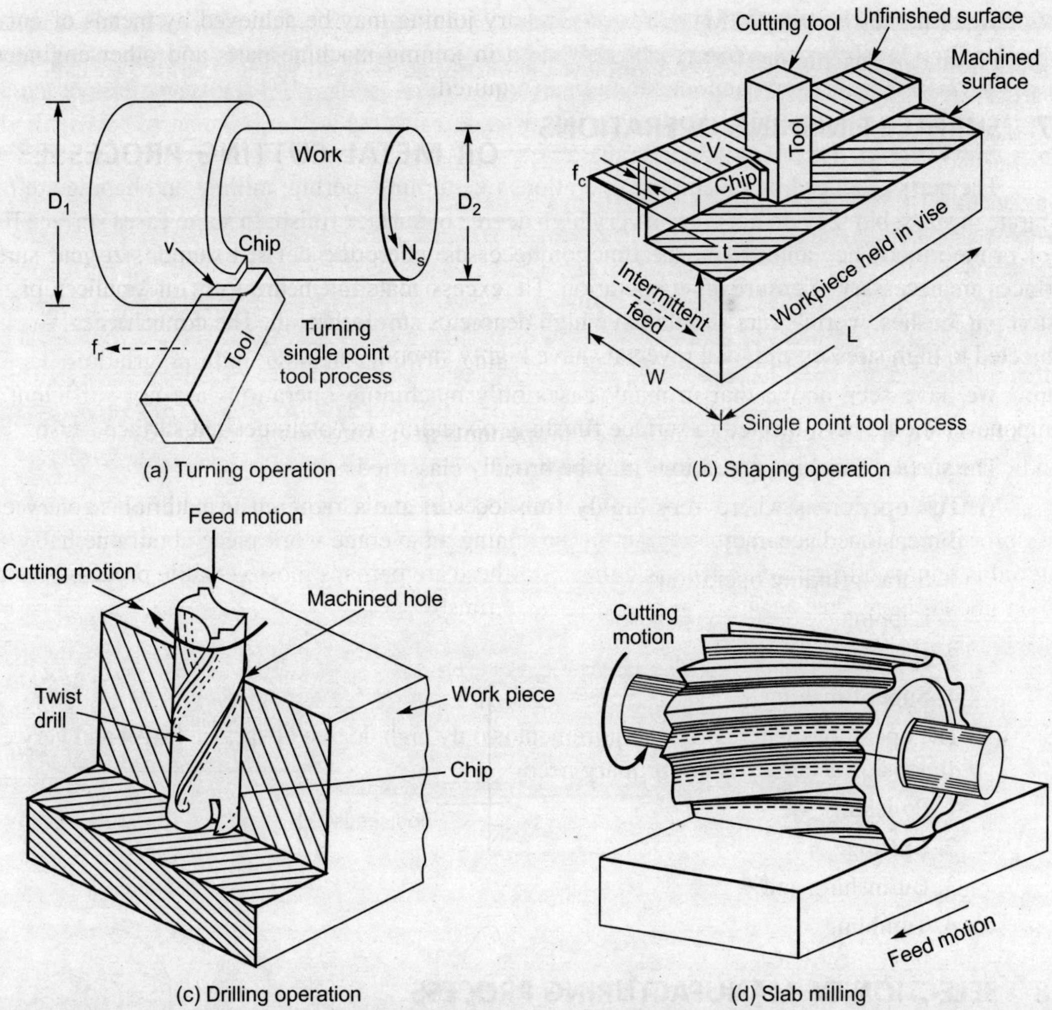

Fig. 1.9 Metal removal processes. Figures (a) and (b) present material removal processes using single point cutting tool whereas Figures (c) and (d) show material removing processes with multipoint cutting tools.

All the processes discussed in the above paragraph are strictly speaking conventional machining processes. There are other material removal processes which are termed as unconventional machining processes. Conventional machining processes make use of shearing process as the basis of material removal whereas unconventional processes depend on other factors such as electrolytic displacement, chemical reaction, vaporisation of metal, mechanical erosion etc. for removal of metal. Discussing unconventional processes in detail at this stage is beyond the scope of the book. Some of the common unconventional processes are:

Electrical discharge machining (EDM)
Electro chemical machining (ECM)
Ultra sonic machining (USM)
Laser beam machining (LBM)
Abrasive jet machining (AJM)

Water jet machining (WJM)

Plasma arc machining (PAM).

1.7 SURFACE FINISHING OPERATIONS

The parts produced by machining operations like turning, boring, milling and haping are fairly accurate in sizes, but they do not carry a very high degree of surface finish. In some cases surface finish is of prime impartance and affect, the functioning of the component. For example, in gear smooth surfaces are necessary to ensure quiet operation. The components like bearings cylinder liners, pistons, piston pin, bushes, worm-gears should have high degree of surface finish. The components which are subjected to high stresses and load reversals have highly smooth surfaces.

We have seen above that in many cases only machining operations are not sufficient and components should be subjected to surface finishing operations to obtain desired surface finish.

The surface finishing operations may be broadly classified as

(*i*) The operations where very highly finished surface is required in addition to very close dimensioned accuracy.

1. Some grinding operations
2. Lapping
3. Honning and,
4. Super finishing.

(*ii*) The operations where prime requirement is only high degree of surface finish and very close dimension accuracy is not primary need.

1. Polishing
2. Buffing
3. Burnishing and
4. Tumbling.

1.8 SELECTION OF MANUFACTURING PROCESS

Throughout the world market today there is continuous struggle for cheaper products with better quality. The product should also have functional and aesthetic appeal. The engineers should choose the process of choice which would require the lowest manufacturing cost and less time to deliver the product of desired quality. If the process is costly or time consuming he may change the design slightly to suit the particular manufacturing process.

Sometimes the engineer may choose the manufacturing process depending upon the major advantage gained by a particular manufacturing process, for example :

(*i*) Metal removal process produces smooth surface and dimensional accuracy not obtained in any other way.

(*ii*) Hot working aids in developing the ultimate fibre strength and toughness in steel.

(*iii*) Welding provides a convenient method or joining of fabricating wrought or cast products into more complex structures.

(*iv*) Sheet metal operation produces light weight sheet metal parts economically and in less time.

(*v*) Casting process produces the most complicated shapes in single step; as a result other operations such as machining, forging, and welding may be minimised or eliminated. Also for very large and heavy metal object casting is only the practical or feasible process.

Thus the engineer may select from number of manufacturing processes which are most suited to need of this work (influence of quantity).

The quantity of particular product to be produced has also an important influence upon the selection of manufacturing process. For example, if a part is to be cast, a pattern is required. The cost of pattern must be charged equally to all the parts produced. As the required quantity of parts becomes larger the share of the pattern cost for each part becomes smaller. When the required quantity of parts is sufficiently large, different manufacturing method may be used.

1.9 STEPS INVOLVED IN CONVERTING AN IDEA INTO FINISHED PRODUCT

Figure 1.10 shows steps required to convert an idea into a finished product. Design, material selection, tool selection and selection of manufacturing process plays most significant role in the manufacturing of finish product. These elements are all closely related, each having its effect on the other, so each of these elements must be carefully considered, planned and co-ordinated before manufacturing starts.

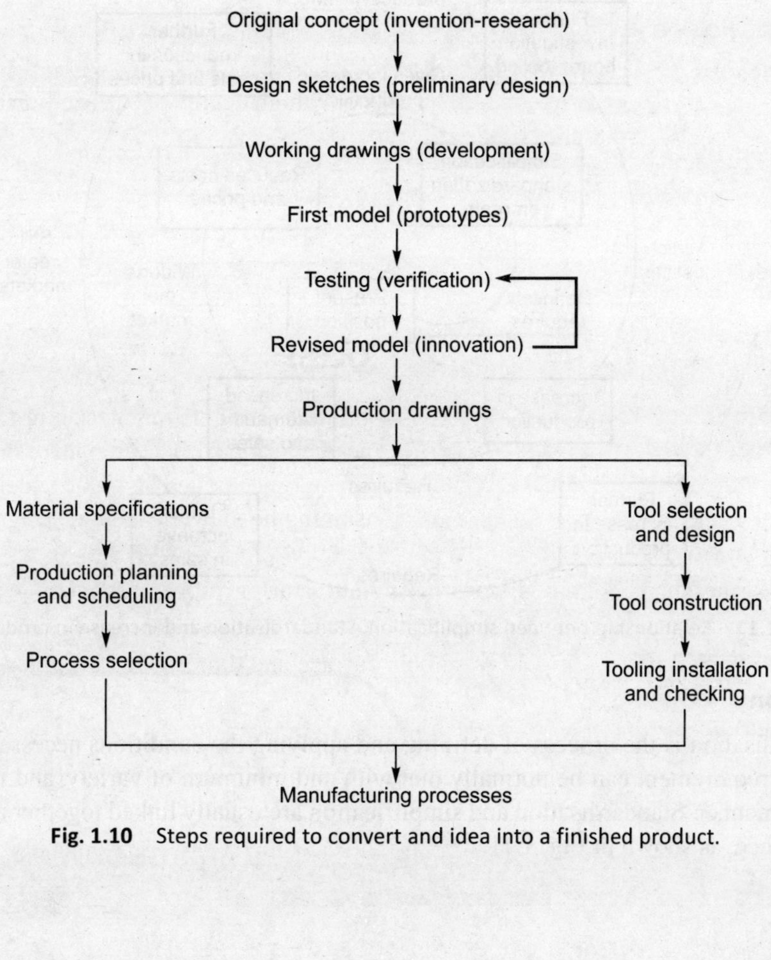

Fig. 1.10 Steps required to convert and idea into a finished product.

Design engineers prepare the original idea of product through invention research and then he prepare design sketches. Design engineers are a vital factor in manufacturing process. Design engineer should design for manufacturing, that is, the product should be designed in such a way that it can be manufactured and assembled economically.

Manufacturing engineers bring the ideas and designs into reality by proper selection of material, machine, tool and manufacturing process.

1.10 STANDARDISATION AND SIMPLIFICATION

In the industrial market different types of products model and components exist. Variety of products, components, models lead to a demand for different type of materials and methods for manufacturing. At various stages of manufacturing different materials, fabrication methods, tools, equipment, jig and fixtures, machines are used.

Simplification and standardisation are functions which aim at defining a limited variety of different types so that basic requirements are fulfilled and the efficiency of plant is increased. These are the roots of any economic analysis of product design.

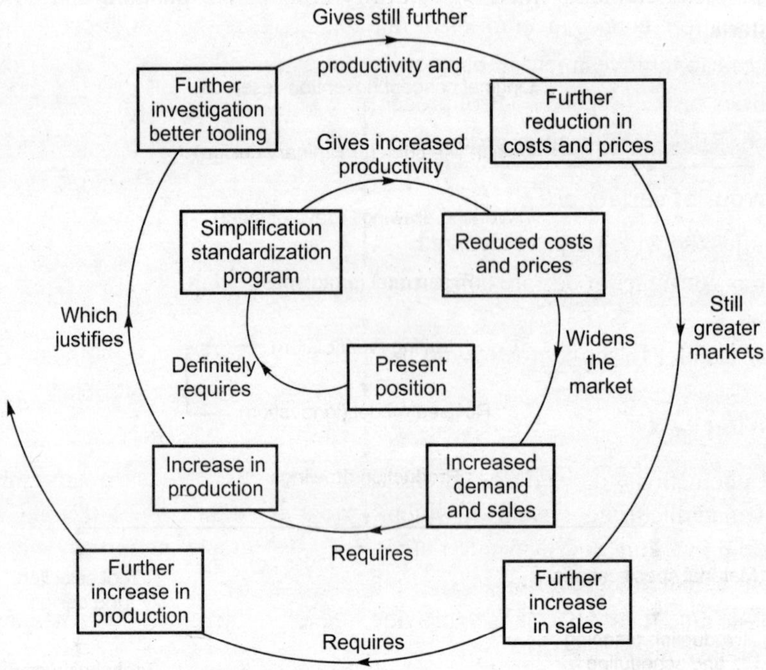

Fig. 1.11 Relationship between simplification, standardisation and increase in production.

Standardisation

Standardisation is the process of defining and applying the conditions necessary to ensure that given range of requirement can be normally met with and minimum of variety, and in a reproducible and economic manner. Standardisation and simplification are usually linked together and developed as a logical sequence, as shown in Fig. 1.11.

From a wide range of requirements it is first necessary to sort out the essential features, define them and work out in a scientific manner the minimum variety required to meet these essential features. This is the process of standardisation and it is mainly an engineering process.

Standardisation cover wide field of activity which may be

(*i*) Dimensions and tolerances of components.

(*ii*) Rating of machines in units of energy, temperature, current speed etc.

(*iii*) Methods of testing characteristic, installation and performance.

One of the important aim of standardisation is interchangeability of components or assemblies.

Simplification

Simplification is the process carried out with a view of reducing the variety of products and materials that are produced or purchased. It is both an economic and engineering process. The points in favour of simplification are:

(*i*) Simplification enables production department to improve planning, achieve high rates of product and machine utilization.

(*ii*) Simplification reduces inventories of materials and finished products.

(*iii*) Reduces capital investment in plant and equipment.

(*iv*) Simplifies inspection and control procedures.

(*v*) Reduces sales price through production simplification.

The points in favour of variety are

(*i*) It satisfies the wide range of demand.

(*ii*) As far as consumer goods are concerned, psychological effect of variety of products creates demand.

(*iii*) Variety enables better contracts with markets to study its tastes and requirements.

1.11 AUTOMATION

The word automation is derived from Greek and it means self-moving or self-thinking. Automation refers to type of manufacturing system in which various manufacturing functions such as material processing, material handling and inspection are carried out by automation machines (self operating machines) without human intervention.

(*i*) The basic aim of automation is to provide higher productivity, lower manufacturing cost and more precise manufacturing.

(*ii*) The need of automation was felt to bridge the gap between specialised machine tools and general purpose machine tools. Specialised machine tools are used in mass production. These are highly specialised but inflexible because of their longer set up time while changing over to new job. Flexibility is essential now a days because variety of products are to be manufactured. General purpose machine tools are highly flexible but are not suitable for mass production. Hence automation is needed. Automation is done in following cases:

(*a*) Loading and unloading of parts

(*b*) Automatic tool changing

(*c*) Automatic production line

(*d*) For production of complex parts
(*e*) For frequently changing design
(*f*) To reduce cycle time.

The advantages of automation are

1. Reduction in direct labour cost as less operator skill is required
2. High quality of product
3. Increased plant efficiency
4. Reduction in process inventory
5. Reduced material handling cost and time
6. Reduced machine breakdown
7. Increased flexibility
8. Increased manufacturing control
9. Increased safety

VIVA VOCE QUESTIONS

1. Why is it necessary for all engineers to know about manufacturing processes?
2. How would you classify the manufacturing processes?
3. Explain clearly with examples and neat sketches primary manufacturing processes and secondary manufacturing processes.
4. Explain the traditional steps required to convert an idea into a finished product.
5. What are the factors responsible for selection of manufacturing process?
6. Explain why standardisation, simplification and automation is required for industries.

UNIVERSITY PROBLEMS

1. What are the objectives of study of workshop technology?
2. What are primary and secondary operations? Explain.
3. Write short notes on the following:
   (*i*) Simplification (*ii*) Standardisation (*iii*) Automation
4. Explain various types of manufacturing processes with examples.
5. Explain how raw material is converted into product/services in workshop.
6. Give the various classifications of manufacturing processes, when and why are secondary processes used? Explain any one of the secondary processes with sketch.
7. Distinguish between primary and secondary shaping processes.
8. State the difference between metal shaping and metal machining process. Also discuss application and importance of these processes.
9. Explain how manufacturing processes are classified.

"The men who try to do something and fail are infinitely better than those who try to do nothing and succeed."

CHAPTER 2

INDUSTRIAL SAFETY AND ACCIDENTS

INSIDE THIS CHAPTER

2.1 INTRODUCTION

In all the chapters of the book we have discussed about the engineering materials and manufacturing processes of production. These products are basically meant for human needs and comfort. Therefore safety of human life is of prime importance.

An industry involves a large number of machines and other equipment which carries moving parts and dangerous projections. These may cause fatal accidents if proper measures are not taken for safeguarding against them. Electrical equipments, explosive and inflammable materials used in industries may also cause dangerous accidents for employees. So every employee must be made to learn the words 'safety first'. In industries approach should be to make each and every employee safety conscious so that safety measure becomes part and parcel of his regular working. It is an established fact that best materials, methods of manufacturing and techniques are not going to serve any purpose unless safe working habits are developed and practiced.

2.2 ACCIDENT

An accident may be defined as happening or disaster that results in some sort of injury to man, damage to machine or tools and equipment and hence a loss to company as a whole. The injury may be minor, may cause disability or even death and accordingly accident is termed as non-repartable or reportable.

"One can never go back, one should not ever try to go back, the essence of life is going forward, life is really a one way street."

2.2.1 Cost of Accidents

An accident can be very costly to the injured employees as well as to the employer. There are both direct and indirect costs associated with the accident. Indirect costs are immeasurable. Direct costs include:

(*i*) Wages of the employee during the period of his absence from duty.

(*ii*) The amount of compensation paid on account of death, permanent injury or temporary injury caused due to accident.

(*iii*) Cost of medical care and hospitalisation of injured employee.

Indirect cost of an accident

(*a*) The expenditure of recruiting and training a person as a replacement to the injured worker.

(*b*) Cost of machine, equipment and other infrastructure facilities that might have been damaged in the accident.

(*c*) The cost of time lost by other employees while assisting injured employee, investigating the cause of accident.

(*d*) Cost of delay in production due to accident.

(*e*) Loss of the production time and quality because of inexperience and lack of skill of new employee.

(*f*) Loss because of the low morale of the employees on account of fear of accident.

2.2.2 Causes of Accidents

Experts claim that almost 98% of the accidents can be easily avoided if safety measures and care were taken well in time. There is a quite famous slogan which states that "Accidents do not just happen but are caused due to failure of one element or other". Accidents may be caused due to

1. Human causes
2. Technical causes
3. Other causes.

1. **Human causes.** It has been observed that human element is most prominent of all which fail. There are four accidents caused by human error to a accident by technical error. The accidents occur because of lack of safety, awareness and carelessness. The causes of accidents include:

(*i*) Age and health of operator is not suitable to job assigned to him.

(*ii*) Lack of experience and interest for given job.

(*iii*) Day - dreaming, forgetfulness and inattentiveness.

(*iv*) Carelessness and recklessness.

(*v*) Emotional instability like: jealousy, revengefulness etc.

(*vi*) Fatigue.

(*vii*) Unnecessary exposures to risk.

(*viii*) Using intoxicants (like alcohol or tobacco) while working in industries.

(*ix*) The person working under any kind of stress (e.g., family problem, threat of losing a job) are more prone to accidents than others.

(*x*) Non-observance of safety measures.

(*xi*) Non-use of safety devices.

(*xii*) Working at unsafe speeds and improper use of tools.

2. **Technical causes.** Technical causes include unsafe working conditions. These conditions may be because of mechanical factors or environment factors.

   Mechanical factors include:

   (*i*) Defective machines.

   (*ii*) Inadequate safety devices.

   (*iii*) Improper machine guards.

   (*iv*) Improper material handling.

   (*v*) Wrong and faulty layout of plant.

   (*vi*) Improper design of machines.

   (*vii*) Broken safety guards.

   (*viii*) Improper maintenance of machines, equipment and infrastructural facilities.

   (*ix*) Leaking acid valves.

   (*x*) Untested boilers and pressure vessels.

3. **Environmental factors include.** Improper physical and atmospheric surrounding conditions of work place.

   (*i*) Improper ventilation of working place.

   (*ii*) Presence of dust, fumes and smokes and lack of ventilation.

   (*iii*) High noise levels have annoying and unpleasing effects and may cause accident because operator may not listen warning signals.

   (*iv*) Defective and inadequate lighting arrangements causing eye strain, glares and shadows.

   (*v*) Too high temperature to cause headache and sweating.

   (*vi*) Too low temperature to cause shivering.

   (*vii*) Too high humidity to cause uncomfort and fatigue.

   (*viii*) Bad odour or flash coming from nearby machinery or process.

2.2.3 Sources of the Accidents

Power driven machines are always a common source of accidents. However, the moving parts may well be protected. Following are the common sources of accident in machines:

1. Revolving shaft, spindles, bar, mandrels and tools like drills, boring bars and chucks.
2. Protruding nails.
3. Projecting fasteners like keyheads, pins, screws, bolts and nuts on revolving parts e.g. the rotating bolt on the driving plate of a lathe is a danger spot.
4. Driving belts are also source of accident. While moving belt from one pulley to other pulley for changing speed, hand may be entrapped between the belt and pulley. There is a danger of projecting part of steel belt, lacing, catching the hand and tearing the flesh.
5. Revolving cutting tools like milling cutters, circular saw blades, circular shears etc.
6. Material handling equipment such as crane hoisting devices and conveyer may be the sources of accidents.

7. Loose clothes are also a common source of accident. Loose clothes at any time may be caught in the moving parts of machines.
8. Other sources of accidents include boilers, electrical bus bar rooms, explosive and chemicals used in industries.

In coming article 2.3 various safety measures are discussed which include safety measures while working on machine, designing a layout, using material handling equipment, preventing electric shocks and five accidents. If these safety measures are not property taken then they become the sources of the accidents.

2.3 SAFETY MEASURES

The term safety measures means the provision of devices, equipment and other methods which ensure safety. The examples may be proper design and layout of industry, presence of safety guards, the fire extinguishers etc.

2.3.1 Safety Measure while Designing the Layout

Safety can be assured through proper location, design and layout of industry. Occurrence of large number of fire hazards, accidents and occupational diseases can be prevented by proper layout of building, and equipment such as providing adequate ventilation, proper lighting arrangement, sufficient working area to operator, clear pathways for movement of materials and parts, proper location of personal facilities like canteen, lunch rooms, dispensary, fire fighting services, etc. One common example of safety may be the direction of opening of doors in cinema theatres. The doors are designed in such a manner that they always open towards outside so that in case of emergency people sitting inside theatre can quickly rush out and door does not struck in their way.

Several codes and standards of industrial safety, health and hygiene, fire prevention etc. have been prescribed by central and state government and agencies and they should be taken care in an industry. In planning the layout of a new plant and its building a number of important features should be considered and incorporated for its safety.

2.3.2 Safety Measures Needed while Working on Machines

1. Every moving part of the machine should be properly guarded.
2. Work piece must be properly clamped to the table of the machine. If the clamping has been done carelessly the work piece may move when the force of cut comes on it.
3. Some machines are driven by belt. When these belts are moved from one pulley to another (for the purpose of changing speed of machine) the belt should not be touched by hand, it should be manipulated by a bar of metal or spanner.
4. Loose and improper clothing (like necktie) at any time may be caught in belt or moving parts which are not adequately covered. For this reason overalls without loose ends are to be recommended for workers.
5. The machines should have effective and properly working guards in proper place. No attempt should be made to work on machine unless all guards are correctly positioned and functioning properly.
6. No tools should be left on the way of movement of slides, rams, carriage etc.
7. Repair work on machines should be avoided during running of machines.

2.3.3 Safety Measures while Using Material Handling Equipment (Like Hoists and Conveyers) and String Materials

1. The hoisting devices used in industries must be of good mechanical design and made of sound material of appropriate tensile strength. These should be properly maintained and thoroughly examined by competent person atleast once in every six months for repair, maintenance etc.
2. Each crane should carry proper covers for its gears and other moving parts to prevent access to these parts during operations.
3. Hoisting devices specially cranes should be allowed to be operated by person properly trained and physically fit. The operator should be entirely trained for standard signals without which he should not operate the equipment.
4. The control panel in the crane should be provided with thick rubber matting to prevent the chances of electric shock to the operator.
5. All the housing devices should carry limit switches to prevent the load block from over travelling accidently.
6. Passage provided over and under the conveyer should be properly guarded.
7. Riding on the conveyer should be strictly prohibited.
8. While using the bucket conveyer it should be ensured that they are fully enclosed in a housing and sample, if required, should not be taken from moving conveyer.
9. While storing the material it should be ensured that the electrical panels, and fire extinguisher have easy accessibility, also the pathways, entries and exits should be kept clear.
10. Use of racks and bins enables more storage capacity, easy movement of material from one place to another and ensures better safety.

2.3.4 Safety Measures to Prevent Electric Shocks

1. All electrical equipment, wiring application and control should be insulated and periodically tested.
2. All metallic parts and covers of electrical machinery should be properly earthed.
3. Underground cables and wires should be safe from mechanical damage and exposure to excessive moisture.
4. The floor under the electrical panel should be covered with rubber insulation sheet.
5. Broken fittings like switches, plugs etc. should be immediately replaced.
6. Repair work should be started only after properly switching off power.
7. Inflammable materials should never be allowed to be kept near electrical machines, equipment, live wire, control panel, main switch etc.
8. The danger signs, safety boards and fire extinguishers used for electrical fires should be clearly marked and displayed at proper location.

2.3.5 Safety Measures Taken to Prevent Fire Accidents

1. All the personnel must be fire conscious and should know the location of nearest fire alarm, fire station and telephone number.
2. Adequate warning signals should be provided.

3. Explosives should be stored at adequate safe places.
4. Whenever equipment or processes (which give rise to fire accidents) are used, a suitable fire extinguisher should be placed within easy reach.
5. The doors should open outwards.
6. Fire exits and escape routes should be clearly marked and kept free from obstructions.
7. The fire extinguisher should be properly fixed at a convenient height.
8. Alongwith every fire extinguisher an instruction plate should be displayed which provides the method of using fire extinguisher and type of fire for which the fire extinguisher is suitable.
9. Training for fire-fighting should be given to all employees.

Types of Fires

Fires can be classified in three categories.

1. Class A Fires
2. Class B Fires
3. Class C Fires

1. **Class A Fires.** The fire in this class involve the fires fed by ordinary combustible materials such as paper, wood, clothes etc. Water filled extinguishers are most effective for class A fires.
2. **Class B Fires.** Fires caused by inflammable liquids such as petrol, oils, gas and fats. To extinguish these type of fires carbon dioxide type fire extinguishers are used. To extinguish fire supply of oxygen to burning materials is cut. Water must not be used for this type of fires.
3. **Class C Fires.** Fires caused by live electrical equipment comes under class C fires. Dry power extinguisher, vaporising liquid type extinguisher or carbon dioxide type fire extinguisher is used for class C fires. The extinguishing agent must be non-conductor of electricity. Water must never be used for these type of fires.

2.4 SAFETY METHODS

As we know that in modern machine tools a large number of driven elements move at high speeds. To protect the operator from these dangerous elements and to avoid the machine accidents various safety methods have been developed. Now a days various management techniques have come which not only assure the safety of machines and process but also make the system mistake proof.

Safety methods can be classified as:

1. Safety by construction
2. Safety by design
3. Safety by ergonomics consideration.
4. Safety by using guards.
   (*i*) Fixed guards
   (*ii*) Automatic guards
   (*iii*) Interlocking guards
   (*iv*) Trip guards

1. **Safety by Construction.** While designing a new machine all its dangerous parts should be adequately covered with suitable guards in order to eliminate chances of accident. Design of any dangerous part of machine may be modified keeping in view its effect on operation of other parts and the appearance of machine. For example, gear boxes of machine tools are fully enclosed and cast iron guards are provided for motor, belts and pulleys.
2. **Safety by Design.** The basic principle of this method is that machine should be designed in such a manner that dangerous parts are always beyond the reach of operator. The principle is, however, neither very practical because while designing a machine a lot of other factors are also considered like aesthetic aspects, (how the machine should look like), ergonomics aspects, vibration problems etc. So it is not always possible to keep all the dangerous parts away from reach of operator. It is always advisable that all the dangerous parts of the machine should invariably be guarded and efforts should be made to make them enclosed in housing of the machine.
3. **Safety by Ergonomics Consideration.** The basic principle in ergonomics designing is that men are not meant for machine rather machines are meant for men, so machine should be designed according to the comfort and physical dimensions of human beings in that region. All the control panel and levers should be within easy reach of the operator so that operator could perform various operations of machines in a comfortable manner and in case of emergency he could switch off the machine immediately.
4. **Safety by Using Guards.** There are various types of guards provided on machine to protect the operator against chips, splashing coolant, and also to eliminate the possibility of operators figures, hands, hair or clothes being caught by rapidly moving parts. Guards should be designed in such a manner that
   (*a*) They should be in proper position during the machining operations and can easily be removed during servicing.
   (*b*) They should not create obstacle for operator during operations.
   (*c*) They should allow the operator to watch the operation.

There are many types of guards used. Some of them are mentioned below:

(*i*) **Fixed Guards.** Fixed guards are either integral part of machine or are so tightly fastened to the machine that they cannot be easily removed. The fixed guards should be placed in such a way that any access to the dangerous parts of machine is prevented from all the directions. The material used for fixed guards can be cast irons and steel sheet. Cast iron is not used where guards are subjected to shocks. Expanded metal with small meshes may be used to fabricate the guard as it gives clear view of machining operations and also prevent entry of operator's fingers or hand through meshes.

(*ii*) **Automatic Guards.** The operation of automatic guard is actuated by moving part of machine and both are linked in such a way that guard will cover the part before the operation of machine starts. The guard does not permit the access to this area until machine stops. As enough time is required for activation by moving part, this guards is not suitable for quick acting and fast running machine. It is usually used in heavy and slow acting machines.

(*iii*) **Interlocking Guards.** In interlock guard machines cannot be started to work unless guard covers the dangerous part, similarly guard cannot be removed until and unless machine is totally stopped.

(*iv*) **Trip Guards :** With this type of guards always tripping device is associated, which quickly brings the machine to rest as soon as operator reaches within the reach of dangerous part.

2.5 FIRST AID

Even after following all safety measures and precautions accidents may occur in industries. First aid means providing the preliminary treatment given to injured person before he may be taken to hospital. First aid box is also meant for small cuts, wounds or minor injuries which may happen to employees at any time. It is mandatory to have first aid box in industries and industry should employ full time, at least one person who has successfully completed Red Cross first aid course. The training of first aid may also be given to supervisors or managers.

A first aid box contains the following items:

Items	**Nos.**
Rolled Bandages—10 cm wide	12
Rolled Bandages—5 cm wide	12
Large size sterlised dressings	12
Medium size sterlised dressings	12
Large size burn dressings	12
Packet of sterlised cotton wool	2
Adhesive plaster	2 Rollers
Eye drops	1 bottle
Bottle of salvolative wiui instruction of dose and methods of administration	1
Bottle containing 2% alcoholic solution	1
Bottle containing $KMNO_4$	1
Pair of scissors	1
Packet of safety pins	12 (1 Packet)

VIVA VOCE QUESTIONS

1. What do you mean by an accident. How can these accidents be prevented?
2. What are the various causes and sources for the occurrence of accidents?
3. "Accidents are caused not only due to unsafe acts but also due to unsafe conditions"— comment.
4. What common safety measures are taken while working on machines?
5. What common precautions are necessary to prevent fires and electrical hazard?
6. What is meant by first aid? What are the contents of a first aid box?

UNIVERSITY PROBLEMS

1. What is an accident, what are the common causes of an accident?
2. Write short note on first aid.
3. What is an accident? Give the various classifications of accidents and write their preventions.
4. What is first aid ? What are the main contents of first aid box? Mention their uses.

5. Write short note on factory act regulation.
6. What is an accident? What are the types of accidents? How can accidents be prevented in a Machine Shop.
7. Justify the statement "Accident does not happen, it is caused." Explain the various causes of accidents.
8. What are accidents? Give various classifications of accidents and write their preventions.
9. What is first aid? What are the main contents of first aid box? Mention there uses.
10. Why first aid station is required in industry? Explain the contents of a first aid box alongwith their applications.
11. What is an accident? Classify accidents and suggest methods for prevention of accidents in a machine shop.
12. What is safety? Why is it necessary and how is it pay? List the common safety measured required to be taken in welding shop.
13. Enlist various types of accidents that occur in industry. Also discuss various causes and sources of accidents responsible for such accidents.
14. Write note on 'First Aid' in view of Industrial Safety.
15. What do you understand by Industrial Safety? What are its objectives?
16. What is an accident? Explain in brief the causes of accident.

"You cannot turn back the clock. But you can wind it up again."
"When one door of happiness closes another opens, but often we look so long at the closed door that we do not see the one which is opened for us."

CHAPTER 3

ENGINEERING MATERIALS

INSIDE THIS CHAPTER

3.1 INTRODUCTION

Materials are one of the key elements of manufacturing system. Manufacturing of a component starts with selection of raw material.

Design engineer designs a component to perform some function once the design is established and drawings are made. Next significant task for design engineer is to determine the material for the manufacturing of component. The designed component will not function properly and would have short life span if material selected cannot carry the working load.

Now a days material selected should be strong (having good strength), durable and light in weight. Cost of material is also a significant factor which should not be overlooked. Technology of materials is continuously changing. So engineer must have knowledge of innovative and economical materials those are being used. He should also have understanding of various material systems, their advantages, disadvantages, so that he could select right material for a given job.

In addition to these engineer must be familiar with the effects which the manufacturing process and heat treatment have on the properties of material, for example if a low carbon steel is cold worked (cold working is a primary manufacturing process which comes under the category of metal forming process), it may develop the mechanical properties equivalent to those of heat treated higher carbon steel or low alloy steel. Similarly heat treatment also affects the mechanical properties of the materials.

"Human heart should possess the properties like toughness, malleability, ductility and elasticity and not the hardness and brittleness".

We may conclude from above discussion that three basic and most significant factors which influence the function of any component are designing of part, material and manufacturing process used. Discussion regarding design of the component is beyond the scope of this book and will be discussed in machine design books. In chapters covering manufacturing processes we will see how they affect the properties of materials and in present chapter engineering materials and their properties will be discussed.

One example which explains the significance of material and its properties is the mishap of Titanic ship. Perhaps engineers had not studied in detail the behaviour of material (which was a steel alloy) of ship at a very low temperature. If they knew that at low temperature brittle cracks might develop in that steel alloy, they would have probably changed the alloy and perhaps mishap could have been avoided.

3.2 CLASSIFICATION OF EXISTING MATERIALS IN THIS UNIVERSE

In this universe, every thing in our world is made from elements. They can be broadly classified as organic and inorganic substances.

1. **Organic Substances.** Organic substances contain carbon (and usually hydrogen) as a key part of their structure. All living elements and plants are organic.
2. **Inorganic Materials.** Inorganic materials are those substances not derived from living things, for example sand, rocks, metals, water etc. The engineering materials discussed in this text are inorganic solids. The classification of existing materials in this universe is shown in Fig. 3.1.

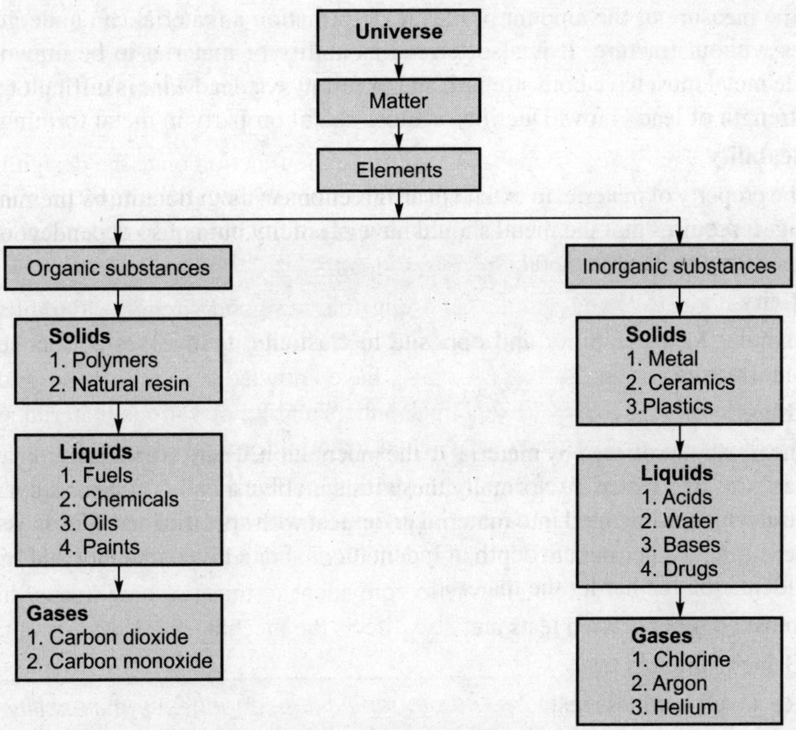

Fig. 3.1 Classification of existing Materials in this Universe.

3.3 STATIC PROPERTIES

When the forces that are applied to a material are constant they are said to be static. Standardised tests like tensile tests and hardness test have been developed to determine the static properties of materials. These results can be used to select materials provided that the material service conditions and test conditions are similar. Even these conditions differ the result of these tests can be used to qualitatively rate and compare materials. Static properties of materials are described below:

(*a*) **Elasticity**

It is the ability of a material to regain its original shape and size after removal of deforming force.

(*b*) **Elastic Limit**

Elastic limit is defined as the greatest stress up to which material deforms elastically without any permanent deformation. It is expressed as newton per unit area (N/m^2, N/cm^2, or N/mm^2). For example if the elastic limit of material is 25 KN per square centimeter (25 KN/cm^2), it means a bar with cross-sectional area 1 cm × 1 cm would return to its original length when load of 25 KN is removed. If the intensity of loading were exceeded the bar would be permanently deformed.

(*c*) **Brittleness**

It is the property of breaking of a material without much permanent deformation. Brittleness is the property opposite to ductility. Brittle material does not mean that it has less strength. It simply means that it lacks significant ductility. Cast iron is a brittle material.

(*d*) **Ductility**

It is the measure of the amount of plastic deformation a material can undergo under tensile forces without fracture. It is also termed as ability of material to be drawn into wires. A ductile metal must have both strength and plasticity e.g. lead wire is difficult to draw because the strength of lead is low. Ductility is quite useful property in metal forming processes.

(*e*) **Malleability**

It is the property of material to extend in all directions without fracture by pressing, hammering, rolling. It requires that the metal should have plasticity but not so dependent on strength e.g. lead is a very malleable metal.

(*f*) **Plasticity**

It is similar to malleability and opposite to elasticity. It involves permanent deformation without fracture.

(*g*) **Hardness**

It is the resistance offered by material to the indentation. There are indentation tests to measure the hardness of a material. Normally these tests involve a ball, a cone or a pyramid of harder material which is indented into material under test with specific load. Hardness is determined by measuring the permanent depth of indentations. For a given indentor and test load smaller the indentation the harder the material.

The most commonly used tests are:

(*i*) Brinell hardness test

(*ii*) Rockwell hardness test

(*iii*) Vickers hardness test.

(*h*) **Strength**

The strength of the material is its ability to resist the application of force without fracture. The force may be tensile, compressive, shearing, bending or twisting. Lets see what is meant by tensile stresses, compressive stresses and shear stresses.

Tensile and Compressive Stresses. Tensile stresses or compressive stresses are termed as direct stresses. These stresses are normal to plane on which they act.

$$\text{Direct stress } (\sigma) = \frac{P \rightarrow \text{Load}}{A \rightarrow \text{Cross-section area}}$$

$$\text{Units} = \text{Nmm}^2 \text{ or N/m}^2.$$

The rope attached to a crank is subjected to tensile stresses and the leg of a table is example of compressive stress.

Tensile strength of a material is determined by tensile testing machine.

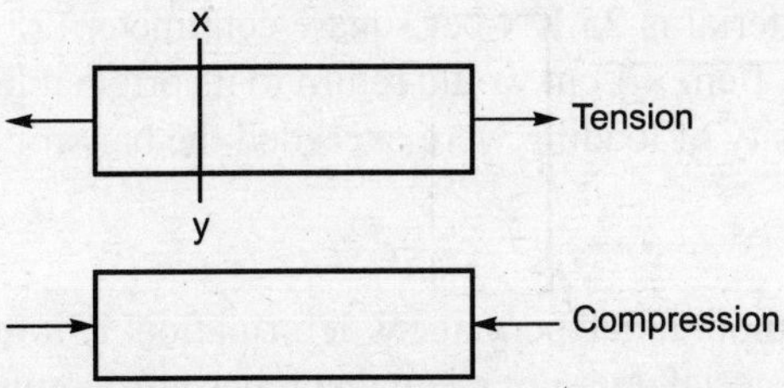

Fig. 3.2(a) Compressive and tensile forces.

Fig. 3.2(b) Position of body before and after shearing force is applied.

Shear Stress. If the applied force P consists of two equal and opposite parallel forces not in same line then there is tendency for one part of a body to slide over from other part across any section. Shear stresses are tangential to plane over which they act.

$$\text{Average shear stress} = \tau = \frac{P}{A}$$

Rivets, keys, couplings are subjected to shear stresses. Nuts and bolts are subjected to shear as well as tensile stresses. The Fig. 3.3 shows various components subjected to tensile, compressive, bending and shear stresses.

(a) Tensile stresses (b) Compressive stresses

Fig. 3.3 Contd.

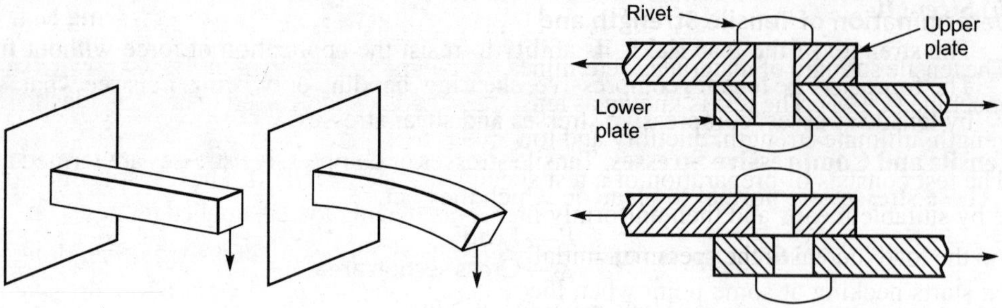

(c) Bending stresses (d) Shear stresses

Fig. 3.3 Various components subjected to tensile, compressive, bending and shear stresses. Arrowheads represent the direction of force applied.

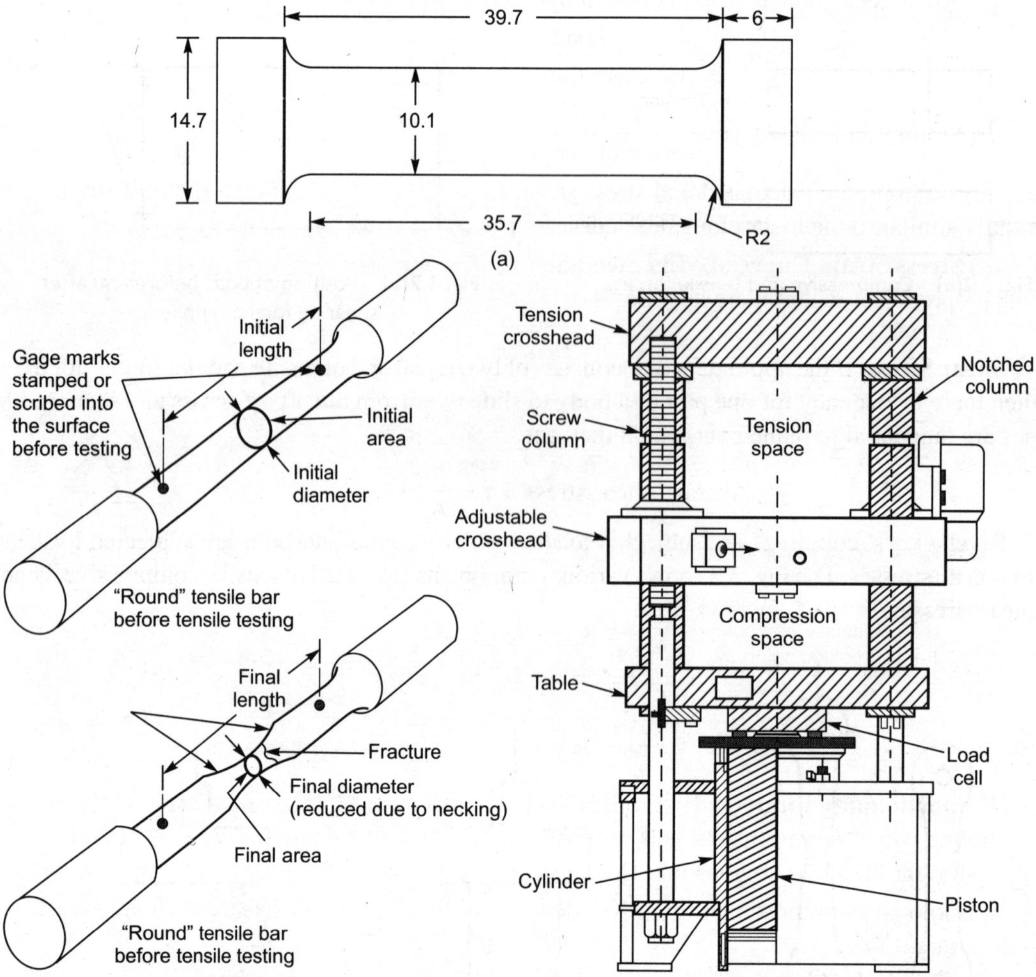

Fig. 3.4 (*a*) The dimensions of specimen for tensile testing on universal testing machine. (*b*) The diameter of specimen before and after tensile listing. It is clearly shown that diameter of specimen reduces before fracture due to necking (*c*) Universal testing machine.

3.3.1 Determination of Tensile Strength and Other Properties on Universal Testing Machine

The tensile strength of material is determined by loading it in a testing machine known as universal testing machine (UTM). The test is known as tensile test and is widely used to indicate elastic strength, yield strength, ultimate strength, ductility and toughness of a metal.

The test consists of preparation of a test specimen as per standard. The test specimen is held in machine by suitable means and then uniformly increasing tensile load is applied on it.

As the load increases the specimen initially gets elastically elongated. On further elongation the specimen starts necking at some point when the material goes beyond the elastic range. The reduced cross-section of the specimen would further be reduced under the force of the load and finally develop fracture when test is completed. During the test a record is maintained of the load and corresponding elongation.

(*i*) Stress is calculated as load divided by cross-section area

$$\text{Stress} = \frac{\text{Load}}{\text{Cross-sectional area}}$$

(*ii*) Strain is calculated as $= \dfrac{\text{Length}}{\text{Original Length}}$

Stress, strain are calculated and stress strain curve is drawn. The shape of stress strain curve is essentially similar to the load- elongation curve.

(*a*) **Stress Strain Curve of Mild Steel:** Stress strain curve of mild steel is shown in Fig. 3.5. The various properties of material are as follows.

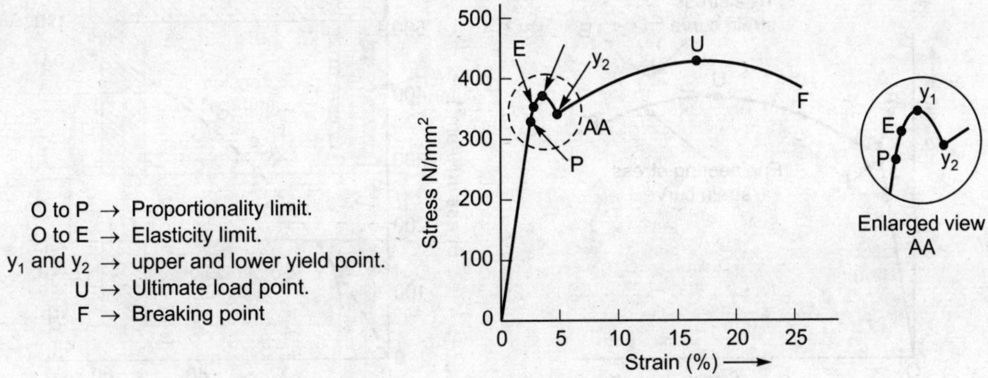

Fig. 3.5 Engineering stress strain curve for mild steel.

Proportionality Limit (O-P). From zero load point O to P the stress stain curve is geometrically a straight line which represents stress is proportional to strain. Beyond point P curve slightly deviates from the straight line. It is obvious that Hooke's law holds good up to point P.

It is known as proportional limit. It is defined as the stress at which stress strain curve deviates from the straight line.

Elastic Limit (P-E). From point P to E Hooke's law is not obeyed although the metal is still in elastic state *i.e.* test piece will come to original length upon unloading, but the load extension relation is non-linear. After point E plastic flow or yielding starts, hence E is known as elastic limit for the material. For mild steel it occurs at 0.12% of strain. If the material is stressed beyond this point plastic deformation will occur *i.e.* strain which is not recoverable if the load is removed.

Yield Point (y_1 and y_2). Beyond point E load increases a little after which there is momentary decrease of load as the extension increases. This upper point y_1 just after the elastic limit is known as yield point or more precisely as upper yield point. The lower load point after y_1 is known as lower yield point y_2. Upper and lower yield points can be obtained only in a carefully controlled test otherwise the region from E and y_1 to y_2 merges into one curve almost horizontal showing apparently an increase of strain from 0.12% to 2% at constant stress.

Ultimate Load Point (U). After y_2 point curve is smooth and much flatter. The stress strain curve continues to rise up to the point U. The strain in the region from (y_1, y_2) to **U** is 100 times more than in region from O to y_1, y_2 and is partly elastic (recoverable) but mainly plastic (permanent strain). The stress corresponding to point U is called ultimate strength of material.

Breaking Point (F). Beyond point U, a neck forms, the apparent stress decreases and material breaks at F. The stress corresponding to F is called breaking (fraction) strength of the material.

(*b*) **True Stress Strain Curve.** The Fig. 3.6 shows true stress strain diagram. Here it should be pointed out that true stress in the necked region (from point D to E) is much higher than shown in Fig. 3.5 and in fact reaches its greatest value at the (Point E) breaking load. It is because as the neck starts forming, [Fig. 3.4 (*b*)] cross-sectional area is decreased and so stress is increased. Stress = $\frac{\text{Load}}{\text{Area}}$ and when area decreases stress increases.

However, it is an established practice to draw stress-strain curve on the basis of original cross-sectional area of specimen which is known as engineering stress-strain curve and shown in Fig. 3.5.

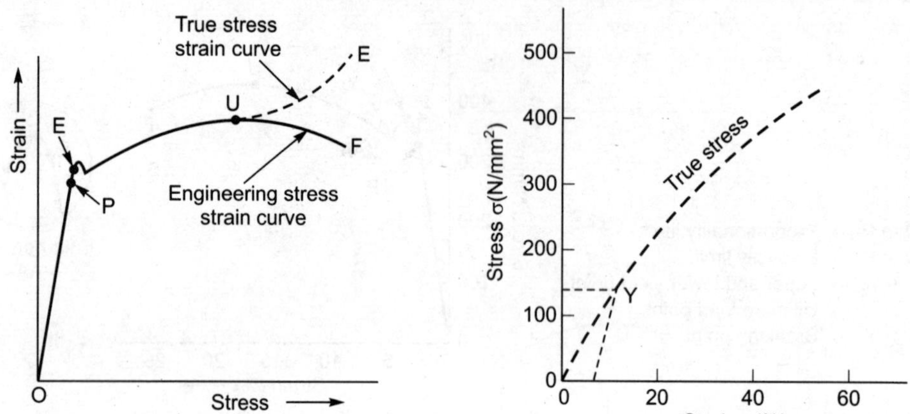

(a) True stress strain curve and Engineering stress strain curve

(b) True stress strain curve for copper

Fig. 3.6 Stress strain curves for mild steel and copper.

(*c*) **Stress Strain Curve for Brittle Material.** Figure 3.7 shows stress strain curve for a brittle material. As it can be observed from figure that material fractures with very little or no plastic strain. If the percentage elongation is less than 5% material is considered to be brittle.

(*d*) **Proof Stress.** It should be noted that most of the materials like some alloy steel and light alloys of aluminium and magnesium do not possess any definite proportional limit, elastic limit, or yield point in tensile test. In such cases yield stress is defined as follows.

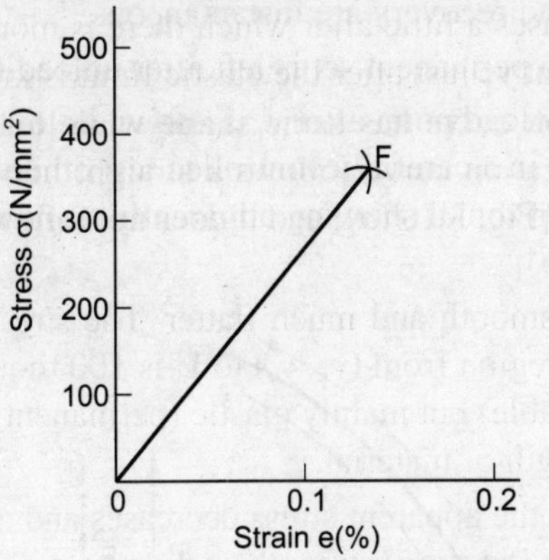

Fig. 3.7 Stress strain curve for cast iron. Some other brittle materials like glasses, ceramics, some metal and some plastic shows same type of stress strain curve.

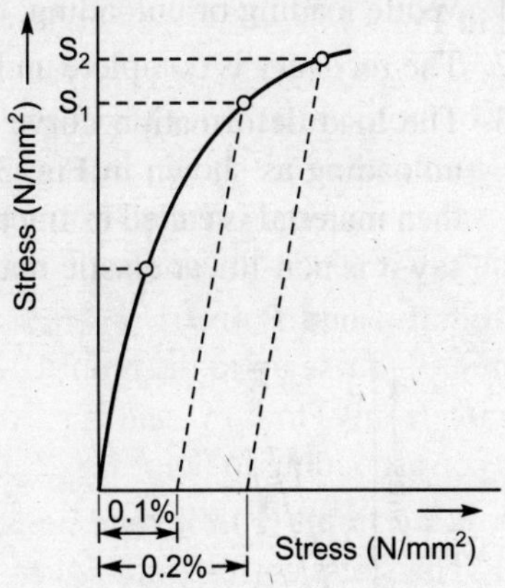

Fig. 3.8 Stress strain curve for material not having well defined yield point. Offset method is used for determining yield strength. S_1, is the 0.1% offset yield strength; S_2 is 0.2% offset yield strength.

A line parallel to the tangent of stress strain curve at the origin is drawn from the point representing 0.2 strain. This line intersects the stress strain curve at point *y*. The stress at this point is taken as yield stress (Fig. 3.8).

(*e*) **Resilience.** Resilience is the ability of material to absorb energy when deformed elastically and return it when unloaded is called resilience. The cross-hatched area O to I in Fig. 3.9 indicates the modulus of resilience for structural steel and area OYG represent the modulus of resilience for high carbon spring steel. Because of higher yield strength spring steel has greater resilience.

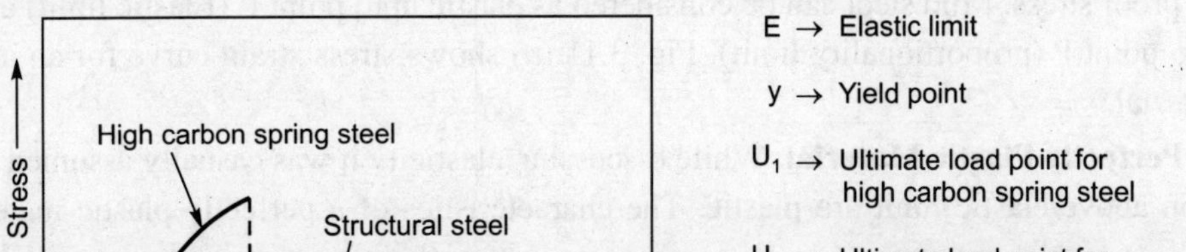

Fig. 3.9 Compares the resilience of high carbon spring steel and structural steel. As area OYG is greater than area OEI, it clearly indicates that high carbon spring steel has greater resilience.

(*f*) **Perfectly Elastic Material.** In the beginning of chapter, loose definition of elasticity has been given. Now we shall discuss elasticity and plasticity formally. A material can be called perfectly elastic, if and only if:

1. While loading or unloading, the deformation and recovery are instantaneous.
2. The recovery is complete and immediate i.e. no permanent set is left after unloading.
3. The load deformation curve or load-elongation curve has same shape while loading and unloading as shown in Fig. 3.10. If load elongation curve follows a straight line equation then material is called as linear elastic material [Fig. 3.10 (*a*)] and if does not follow then we say it is non-linear elastic material [Fig. 3.10 (*b*)].

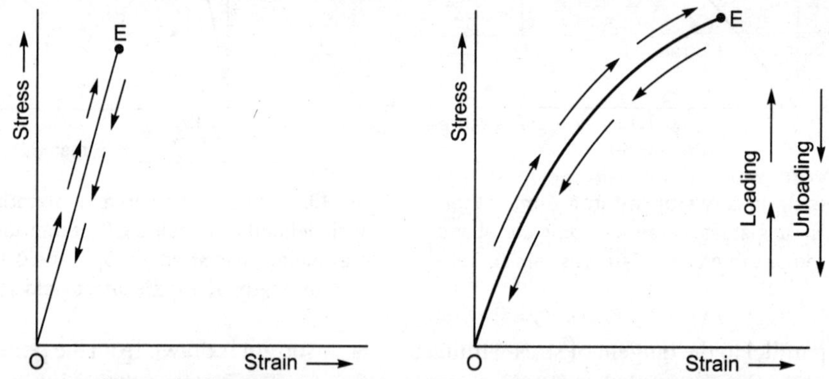

Fig. 3.10 Ideal case of linear plastic and nonlinear elastic material. These are ideal cases as loading and unloading follows same curve.

Actually there is no material which satisfies above three requirements for perfect elasticity. However, for practical purposes most of steels and other structural material satisfy these requirement up to their proof stress. Mild steel can be considered as elastic upto point E (Elastic limit) and linearly elastic upto point P (proportionality limit). Fig. 3.11 (*a*) shows stress strain curve for an ideal elastic plastic material.

(*g*) **Perfectly Plastic Material.** While discussing elasticity it was casually assumed that all the deformation above elastic limit are plastic. The characteristics of a perfectly plastic material are as follows:

(*i*) A perfectly plastic material shows an unlimited amount of deformation or strain at constant stress.

(*ii*) After removal of load plastic strain cannot be recovered.

(*iii*) Plastic deformations could be instantaneous.

Stress strain curve for ideal plastic material is shown in Fig. 3.11(*b*). If the elastic deformation is negligibly small, the rigid plastic idealisation is valid and for a linear strain hardening material Fig. 3.11 (*c*) could be reasonable approximation.

It should be kept in mind that like perfect elasticity perfect plasticity is also only an ideal concept and no material in nature is perfectly plastic. Plastic deformation are almost always time dependent. Perfect plastic idealisation can be used only when time is very small for deformation. In general all non-recoverable deformation are termed as plastic and all recoverable deformations are called elastic irrespective of the time of deformation.

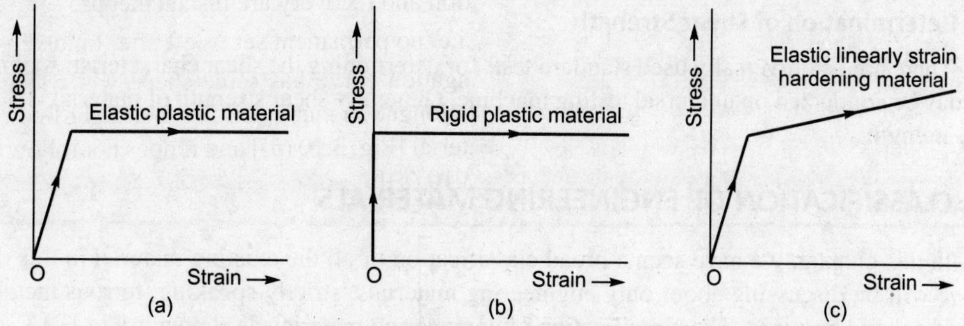

Fig. 3.11 Represents three ideal cases of plastic materials.

(*h*) **Creep:** While discussing perfectly elastic or plastic materials we assumed that material shows no time dependent flow. However, metal under high stress or high temperature shows this phenomenon. It can be defined as slow and progressive deformation of a material under constant stress. If a specimen is tested under the conditions of constant load and constant temperature and a graph is plotted for deformation (elongation) versus time then it would be similar to as shown in Fig. 3.14. Although the rate of elongation is small it is of great importance in design of equipment such as steam turbine, power plants and high temperature pressure vessels that operate under high loads and temperature for a long period of time.

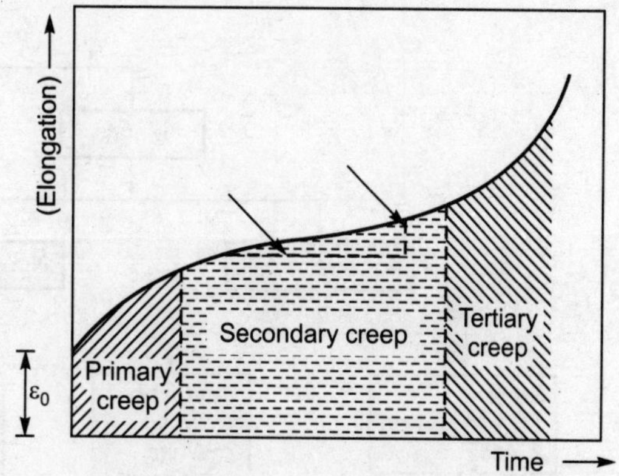

Fig. 3.12 Creep curve of a specimen at a fixed temperature and fixed load.

3.3.2 Determination of Compressive Strength

While discussing definition of strength it was mentioned that material may be subjected to tensile, compressive or shear strength or combination of these stresses. Tensile strength and other significant properties are measured using universal testing machine. Compressive strength can also be determined using the same machine [Fig. 3.4 (*c*)].

When a material is subjected to compressive loading the relationship between stress and strain are similar to those for a tension test. Up to a certain value of stress the material behaves elastically. Beyond this value plastic flow occurs. In general, however, a compression test is more difficult to conduct than standard tensile test. Test specimens must have larger cross-sectional areas to resist bending or buckling. As deformation proceeds the material strengthens by strain hardening and the cross-section of specimen increases, producing a substantial increase in required load.

Compressive strength of brittle material is easily determined for brittle material that will fracture when sufficient load is applied. The compressive strength of cast iron, a relatively brittle material is three to four times its tensile strength whereas for a more ductile material such as steel prediction of compressive strength cannot be made so easily.

Friction between the testing machine surface and ends of test specimen tends to alter the results if not properly considered. The selection of compression testing is determined largely by the type of application for which the material is intended.

3.3.3 Determination of Shear Strength

There are no universally used standard tests for determining the shear characteristics of material. Tests may be conducted on universal testing machine. Generally shear strength of material is 50% of its tensile strength.

3.4 CLASSIFICATION OF ENGINEERING MATERIALS

In this chapter, we have seen a broad classification of all the existing material in this universe. Here we will be discussing about only engineering materials, strictly speaking, ferrous metals which include steel and cast iron. The classification of engineering materials is shown in Fig. 3.13.

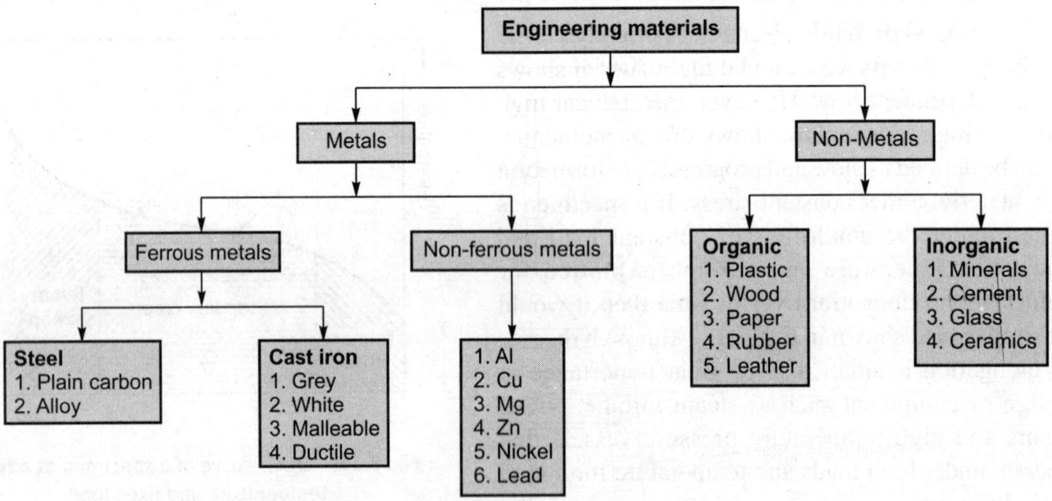

Fig. 3.13 Classification of engineering materials.

3.5 EXTRACTION OF METALS (IRON)

Before discussing about the metals lets see how and in which form these metals are found and how they are extracted. The majority of metals are found in the form of metal bearing substances called minerals. The minerals from which metals can be extracted profitably are called ores. Gold, silver, copper, platinum and bismuth, however, are found in free state because of their low reactivity.

Aluminium is most abundant metal in the earth's crust. It is followed by iron, calcium, sodium, potassium, magnesium and titanium. All other metals are present in much smaller extent.

Extraction of metals from their ores and refining them for use is called metallurgy. Metallurgical process basically consists of three steps:

1. Primary treatment
2. Reduction
3. Refining

All the processes are not necessarily used for all metals. Depending upon the chemical composition of the ore and impurities associated with it different methods are to be choosen. Let's now discuss the extraction of iron.

Most of the iron ore used today is iron oxide mixed with silicon dioxide. Two things are required to obtain iron from iron ore. Firstly a suitable agent to reduce iron oxide to iron and secondly a suitable solvent to dissolve silicon dioxide. Reducing agent CO is obtained by burning coke in an insufficient supply of oxygen. Silicon dioxide is separated using molten calcium oxide.

To achieve both these purposes iron ore lime stone ($CaCO_3$) and coke are put together in a big blast furnace. Lime stone get associated because of intense heat in the furnace and following reaction takes place

$$CaCO_3 \longrightarrow CaO + CO_2$$

Calcium oxide reacts with silicon dioxide to form liquid calcium silicate.

$$CaO + SiO_2 \longrightarrow CaSiO_3$$

Iron oxide thus freed from silicon oxide gets reduced as shown in following reaction:

$$2C + O_2 \longrightarrow 2CO$$

$$Fe_2O_3 + 3CO \longrightarrow 2Fe + 3CO_2$$

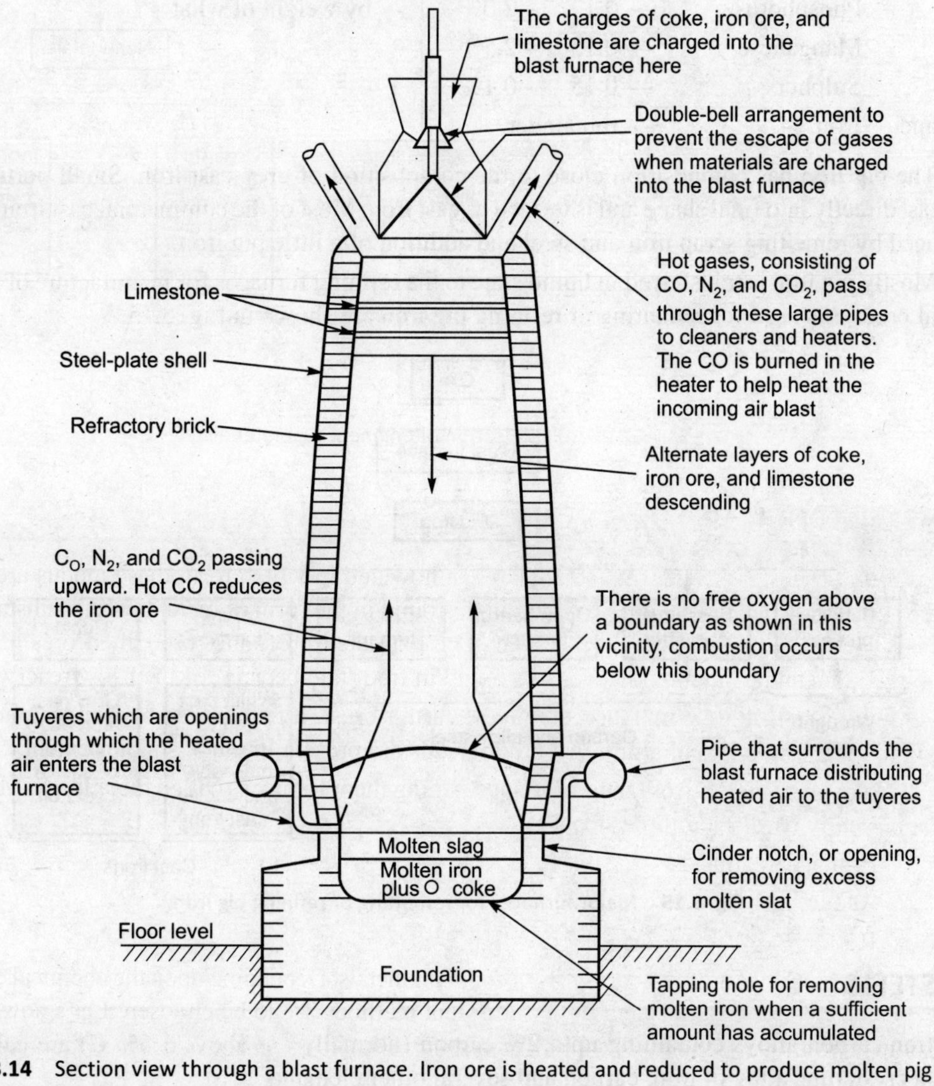

Fig. 3.14 Section view through a blast furnace. Iron ore is heated and reduced to produce molten pig iron.

It is interesting to note that formation of calcium silicate not only removes the unwanted silicon dioxide but also keeps freshly prepared iron away from oxygen. Calcium silicate is lighter than molten metal and hence floats on freshly prepared molten metal otherwise iron would get converted into oxide.

The chemical process described above is carried out in big furnace called blast furnace (Fig. 3.14) and the product obtained from blast furnace is "Pig iron".

3.6 PIG IRON

Pig iron is the principal raw material for all ferrous products. As we have seen above that pig iron is product of blast furnace and obtained by melting iron ore with coke and lime stone. Pig iron roughly has following composition:

	Carbon	— 3.0	– 4.5%	by weight of what
	Silicon	— 1	– 3%	
	Phosphorus	— 0.1	– 2.0%	
	Manganese	— 0.15	– 2.5%	
	Sulphur	— 0.15	– 0.1%	
and	Iron	— remainder.		

The pig iron has composition close to the composition of grey cast iron. Small portion of pig iron is cast directly into final shape and is termed as cast iron. Most of the commercial cast iron however, is produced by remelting scrap iron and steel and addition of a little pig iron.

Mostly pig iron is transferred in liquid state to the refining furnaces for manufacture of steel. The principal processes used for remelting or refining pig iron are shown in Fig. 3.15.

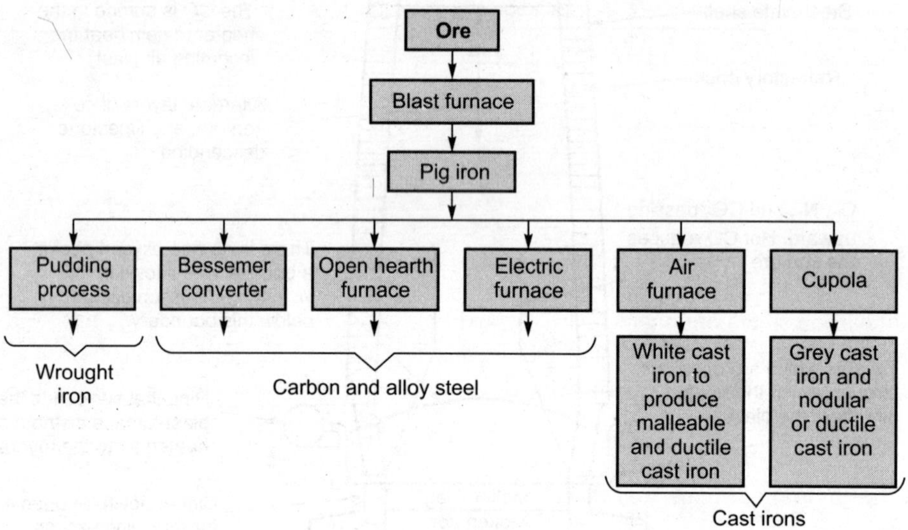

Fig. 3.15 Major furnaces for remelting or refining pig iron.

3.7 STEELS

Iron carbon alloys containing upto 2% carbon (normally above 1.5% C) are called steel. Steel is a crystalline alloy of iron, carbon and several other elements

Steels find maximum consumption in engineering applications among metallic materials. This is due to the wide range and combination of physical and machanical properties which steel can possess. Any combination of hardness, strength and ductility can be obtained in steels by controlling (1) carbon content, (2) alloying element, (3) heat treatment. Carbon is an important constituent because of its ability to increase the hardness and strength of steel. Steels can be easily processed by all types of manufacturing processes such as machining, forming, cold rolling, hot working, castings, welding etc. For a ferrous metal to be termed as steel there must be no free graphite in its composition. All the carbon is present in chemical combination with the iron. Some of the drawbacks of steel are poor scaling and corrosion resistance, high specific gravity and low electrical conductivity.

Classification of Steels

Steel is classified according to alloying elements it contains. Carbon is the most important element and, therefore, all steels are classified according to carbon content. Steels are broadly classified as

1. Plane carbon steel
2. Alloy steel.

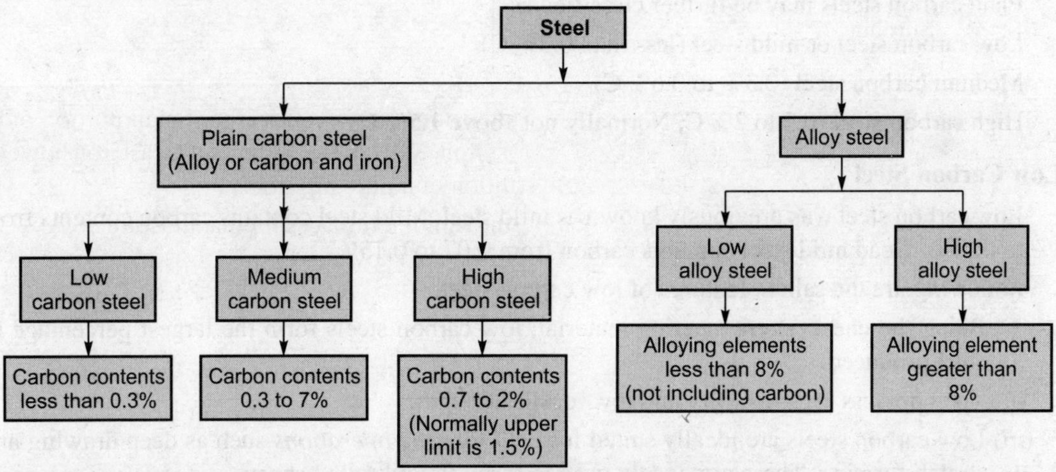

Fig. 3.16 Classification of steel.

3.7.1 Plain Carbon Steel

Plain carbon steel contains primarily iron and carbon. Besides carbon these alloys also contain small amount of manganese, sulphur, phosphorus and silicon. These are generally considered as impurities and hence need to be controlled. However, in certain conditions some of these, such as silicon, are treated as alloying elements. The properties of steel are influenced significantly by an increase in carbon contents which is shown in Fig. 3.17.

As the carbon contents increase:

1. Tensile strength is increased
2. Hardness is increased
3. Ductility decreases
4. Weldability decreases
5. Toughness decreases

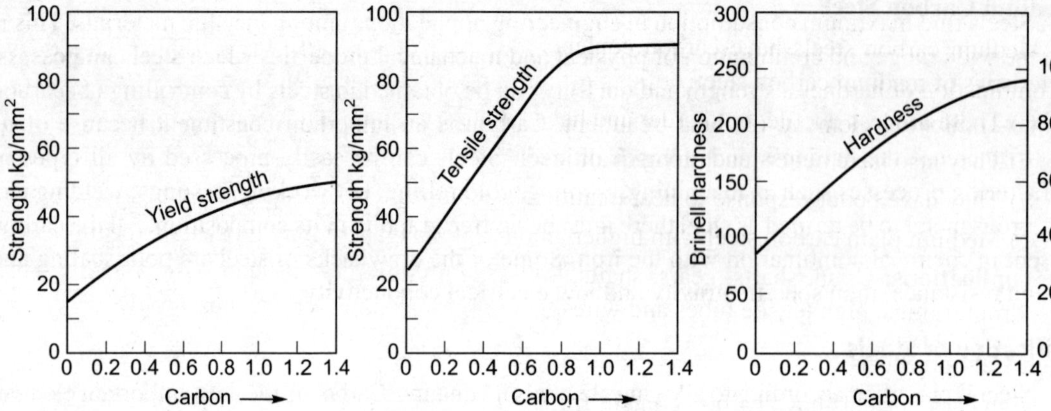

Fig. 3.17 Influence of increasing carbon contents on strength and hardness of steel.

A point, however, to be noted is that in all steels (even in alloy steels) the maximum hardness that can be obtained after the heat treatment process is basically a function of carbon content.

Plain carbon steels may be further classified as

Low carbon steel or mild steel (less than 0.3% C)

Medium carbon steel (0.3% to 0.6% C)

High carbon steel (0.7 to 2% C, Normally not above 1.5% C)

(1) Low Carbon Steel

Low carbon steel was previously known as mild steel. Mild steel contains carbon contents from 0.15% to 0.25%. Dead mild steel contains carbon from 0.07 to 0.15%.

Following are the salient features of low carbon steel:

(*i*) Being the cheapest engineering material, low carbon steels form the largest percentage of steel produced.

(*ii*) They possess high ductility and low tensile strength.

(*iii*) Low carbon steels are ideally suited for *cold forming operations* such as deep drawing and stretch forming. These are widely used in form of cold rolled sheets.

(*iv*) These are also used for operations such as spinning, cold bending, swaging etc.

(*v*) Surface of low carbon steel can be hardened by a process called case hardening, as they are not responsive to normal heat treatment.

In recent years efforts have made for the development of high strength low carbon steels without an appreciable loss of formability. Dual phase steel have thus been developed. These have the characteristic like

(*a*) The absence of discontinuous yield point

(*b*) A high work-hardening or strain-hardening rate

(*c*) A high tensile strength

(*d*) A large uniform elongation

(*e*) A low ratio of yield strength to tensile strength.

Applications. Wire rod, thin sheets, solid drawn tubes, nails, nuts, bolts, screws, washers, boiler plates, light and having structural members, drop forgings, automobile body sheets, sheet metal products.

Low carbon steel is available in the form of sheets, plates, squares, round and wires.

(2) Medium Carbon Steel

Medium carbon steels have carbon contents varying from 0.3% to 0.7%. Following are the characteristics of medium carbon steel.

(*i*) These are less ductile, but harder and have greater strength than low carbon steels.

(*ii*) They also have better machining qualities.

(*iii*) They have better response to heat treatment as compared to low carbon steel.

(*iv*) Medium plain carbon steel with higher amount of manganese may be used as spring steels.

Applications: Crank shafts, gears, shafts, bolts, connecting rods, machine components, rails, springs, dropforgings, high tensile tubes and wires, wire ropes, hammers, agricultural tools.

(3) High Carbon Steels

High carbon steels have carbon content from 0.7 to 2% C. Normally not above 1.5% C. They have higher strength and are harder than other plain carbon steels. They also quickly respond to heat treatment.

Applications. Cold chisels, press dies, punches, axes, wood working tools, razors, hand files, drills, gauges, metal cutting tools, shear blades, milling cutters, taps, hacksaw blades, railroads.

3.7.2 Alloy Steel

An alloy steel may be defined as a carbon steel to which one or more elements are added to get some beneficial effects. The commonly added elements include: silicon, manganese, nickel, chromium, molybdenum, tungsten, vanadium, copper, boron, aluminium, etc.

Due to the presence of these elements physical, mechanical, chemical and magnetic properties are improved. Alloy steels possess the following improvement in properties as compared to plain carbon steel:

1. Alloy steels can have high hardness, strength and toughness.
2. Ductility is improved without lowering of tensile strength.
3. They possess much higher hardenability, which has a great significance in heat treatment.
4. Alloy steels have higher corrosion and oxidation resistance.
5. They retain physical properties at extremely high or low temperatures.
6. They possess high hardness (red hardness) at temperatures upto 600°C due to the presence of alloy carbides.
7. Alloying elements improve machinability.
8. Alloying elements improve electrical and magnetic properties.

Alloy steels can be classified as low alloy steels or high alloy steels. The steels containing alloying element below 8% (not including carbon) are called low carbon steels and steel containing alloying element above 8% are high alloy steel.

Alloy steels may be low carbon alloy steels like creep resisting steel, which has following composition.

Molybdenum (0.4 – 0.6%)

Vanadium (0.25 – 1.0%) and

Chromium (upto 6%)

Example of medium carbon alloy steel is spring steel which contains manganese and silicon as main alloying elements. Chromium, nickel, molybdenum, and vanadium may be added in place of silicon.

High speed steel is example of high carbon steel alloys. It contains

Tungsten – 18%

Chromium – 4%

Vanadium – 1%

Carbon – 0.8%

The other typical examples of alloy steels are rail steels, concrete reinforcing steel, ball bearing steel, marging steels, dual phase steels.

(1) High Speed Steel

High speed steels are used as cutting tool for cutting metals at much higher speeds. Before discussing the composition of high speed steels let's see what are the properties required for cutting tool:

(*a*) **Wear and Abrasion Resistance.** A property that is achieved by the presence of hard alloy carbide.

(*b*) **Toughness.** Obtained by lower carbon contents.

(*c*) **Red Hardness.** The ability to retain hardness at elevated temperatures. Red hardness is imparted by secondary hardening during tempering.

(*d*) **Non Deforming Properties.** It is the resistance to distortion or dimensional changes during heat treatment, a property that is achieved by high hardenability that permits air cooling during hardening.

(e) **Minimum Cost.** The steel should be no more expensive than is needed to meet the particular requirements.

The above required properties are the characteristics of high speed steel. They cut at twenty times the speed of plain carbon steel tool without losing their sharpness even when they are hot enough to glow with dull red colour in dark room. These steels retain their hardness at temperature up 600°C due to formation of alloy carbide (high temperature is developed during the cutting process). Typical examples of this group are 18 : 4 : 1 which means.

Tungsten – 18% Chromium – 4% Vanadium – 1% Carbon – 0.8%	Tungsten based high speed steel

and 6 : 5 : 4 : 2 which means

Tungsten – 6% Molybdenum – 5% Chromium – 4% Vanadium – 2%	Molybdenum based high speed steel

Tungsten provides for toughness, wear resistance and cutting ability, chromium serves to increase hardenability and vanadium is for grain refinement. Molybdenum replaces the part of tungsten or totally replaces the tungsten in molybdenum based high speed steel. Molybdenum performs similar functions as tungsten.

3.8 EFFECTS OF ALLOYING ELEMENTS ON STEELS

Generally alloying elements are added to steel in small percentage (normally less than 5%) to improve strength or hardenability, or in much higher amount (often upto 20%) to produce special properties such as corrosion resistance, stability at high or low temperature etc. Figure 4.6 shows the following influences of various alloying elements on hardness.

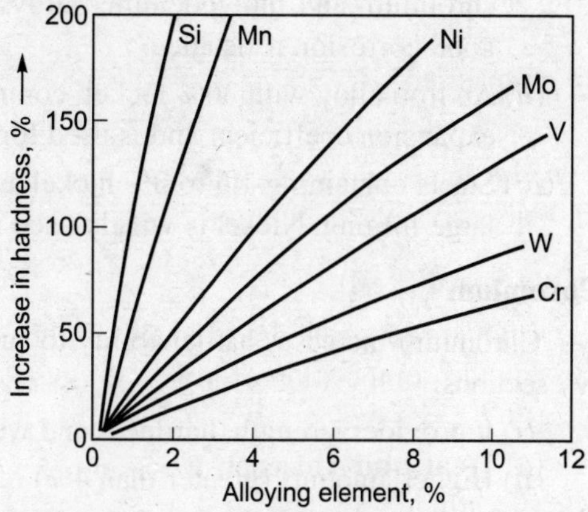

Fig. 3.18 Effects of various alloying elements when dissolved in ferrite, manganese, silicon and nickel have a greater influence on hardness and strength.

(*a*) Alloying elements may be added during steel of manufacturing to remove dissolved oxygen from molten metal. Manganese, silicon and aluminium are used for this deoxidation.

(*b*) Aluminium, and to less extent vanadium and titanium are added to control austanite grain size.

(*c*) Manganese, silicon, nickel and copper add strength by forming solid solution in ferrite.

(*d*) Chromium, vanadium, molybdenum and tungsten add strength by forming carbides.

(*e*) Nickel and copper are added in small amounts to improve corrosion resistance.

1. Manganese

(*i*) Manganese is least expensive and most common of all alloying elements. It is always present in steels.

(*ii*) Manganese combines with sulphur to produce soft manganese sulphide, thereby preventing the formation of iron sulphide which causes brittleness in steels.

(*iii*) It acts as deoxidiser as well as purifier and increases the strength and hardness of steel.

(*iv*) When managnese is added in large amount (11 to 14%) an austenitic alloy is produced which has high hardness, good ductility, high strain, hardening capacity and excellent wear resistance. These properties make it ideal for impact resisting tools and similar applications.

2. Sulphur

Sulphur is always present in steel as inclusion of iron sulphide (FeS) and manganese sulphide.

(*i*) It is usually not desirable and its contents should be kept as low as possible, preferably below 0.1% because higher amount of sulphur reacts with iron to produce (Fes) iron sulphide which causes brittleness and makes it unsuitable for hot rolling and forging operations.

(*ii*) However, sulphur is not harmful if it combines with manganese to form manganese sulphide. Steels with relatively high manganese sulphide have excellent machining properties and are known as free cutting steels. Some free-cutting steels contain as much as 0.08 to 0.15% sulphur in combination with increased manganese.

3. Nickel

Nickel is added to refine grain structure, increase strength toughness and resistance to corrosion. It has no effect upon ductility.

(*i*) It is generally used (2% to 5%) often in combination with other alloying elements such as chromium and molybdenum. Alloys with low carbon and 12 to 20% nickel possess very good corrosion resistance.

(*ii*) An iron alloy with 36% nickel, commonly known as invar, has approximately zero thermal expansion coefficient and is used for sensitive measuring devices.

(*iii*) Steels containing 1.5 to 3% nickel are used for locomotive boilers, bolts, railway axles and large forging. Nickel is widely used in stainless steels.

4. Chromium

Chromium increases hardenability to such extent that chromium steels are readily hardened in heavy sections:

(*i*) It provides strength, hardness and wear resistance.

(*ii*) Higher amounts (greater than 4%) of chromium are added to get corrosion resistance.

(*iii*) Chromium is relatively less expensive and most commonly found in steels. It is the principal alloying element in stainless steels. Chromium is always present in tool steels. It is widely used in structural steels and heat resisting steels.

5. Silicon

(*i*) One of the important use of silicon is as a de-oxidiser in molten metal and for its ability to resist oxidation in steel. For this purpose amount added varies between 0.1 to 0.3%.

(*ii*) It dissolves in ferrite increasing strength and hardness without lowering ductility but above 2.5% it develops poor machinability and prone to decarburisation.

(*iii*) Silicon is extensively used in electrical and magnetic application like transformers, motor, and generators.

(*iv*) It is present in almost all steels. It is purposely added in spring steels, chisels and punches to increase their toughness.

6. Tungsten

Tungsten increases the strength by forming hard carbides.

(*i*) It increases the wear resistance, toughness and cutting abilities.

(*ii*) It is also used as primary alloying element in tool steels. Cutting tools of these alloys remain hard and sharp even after they have become heated to dull red colour during machining. These alloys are known as high speed steels. Typical examples are tungsten based high speed steel. (18 : 4 : 1)

Tungsten – 18%. Chromium – 4%, Vanadium – 1%, Carbon – 0.8%.

7. Molybdenum

(*i*) Molybdenum increases the strength and impact resistance particularly at high temperature.

(*ii*) It forms carbides which are extremely stable at high temperature and are wear resistant.

(*iii*) It is cheaper than tungsten and is replacing the tungsten to a great extent in tool steels.

(*iv*) In molybdenum based high speed steel part or all of the tungsten is replaced by molybdenum (6 : 5 : 4 : 2)— carbon — 0.8% tungsten 6%, Molybdenum 5%, chromium 4%, vanadium 1%.

8. Vanadium

Vanadium is another alloying element which forms strong carbides that are stable at elevated temperatures.

(*i*) It increases the hardness at high temperatures.

(*ii*) It refines the grains. Vanadium steels have much finer grain structure than the steels without vanadium.

(*iii*) It increases the strength properties like elastic limit, yield point and impact strength with almost no loss of ductility.

9. Aluminium

It is primarily used as deoxidiser in steels.

Aluminium (0.95 to 1.30%) forms hard nitrides when heated in contact with nitrogen resulting in extremely hard steel.

10. Copper

About 0.1 to 0.4% copper is added to get increased atmospheric corrosion resistance. Low carbon steel sheets and structural steels often contain a copper addition to enhance corrosion resistance but surface quality and hot working behaviour of metal tends to deteriorate somewhat.

11. Titanium

(*i*) It is the strongest carbide former among all alloying elements and because of this ability it does not quench harden.

(*ii*) It acts as de-oxidiser.

(*iii*) It is added to stainless steels to prevent precipitation of chromium carbide.

12. Boron

It is a very powerful hardening agent, 250 to 750 times as effective as nickel, 75 to 125 times as effective as molybdenum, about 100 times as powerful as chromium. It is even added as 0.001 to 0.005%. But the results diminish rapidly with increasing carbon contents.

3.9 CAST IRON

Cast iron is an engineering material which is used since fourteenth century. Before that time ferrous objects were made from wrought iron (which was obtained by the crude reduction of iron core). Steels occurred in the nineteenth century.

Definition

Cast iron may be defined as an alloy of iron and carbon containing more than two percent of carbon. It is the family of iron, carbon and silicon alloys with the carbon percentage greater than that can be accommodated in solid solution. A graphite or carbide phase is usually present in cast iron. In addition to above cast iron contains other elements such as sulphur, phosphorus and manganese.

Production of Cast Iron

Cast iron is produced by remelting the pig iron in any one of the following furnaces:

1. Melting in cupola furnace. The cupola furnace is widely used in foundries.
2. An electric or air furnace. It gives higher quality cast iron.
3. Duplex melting method. Using cupola for melting and air furnace or electric furnace for refining (duplex melting method).

Advantages

Cast iron possesses following advantages and characteristics over the steel and other materials.

1. It is the least expensive material. All the raw materials used for are relatively cheap like pig iron, cast iron scrap, steel scrap, lime steel, coke and iron ore.
2. Cast iron possesses good casting properties such as high fluidity, low shrinkage and ease of production. These properties enable it to be used widely for making casting of the parts having intricate shapes such as bodies of machines and other components.
3. These can be easily machined.
4. Cast iron possesses very high compressive strength (about 3 to 4 times its tensile strength) It is used in beds, slides, bases of machine tools, cylinders of engines.
5. An important characteristic of cast iron is its high damping capacity. It is the property which permits a material to absorb vibrational stresses, for example beds of machine tools should have high damping capacity to absorb vibration and wear resistance. The tensile strength of mild steel and grey cast iron is shown in stress strain curve Fig. 3.19.
6. Cast irons can provide a very wide span of metallic properties ranging from high yield strength to high ductility and toughness.
7. They provide high wear and abrasion resistance.
8. Alloyed cast iron possesses high corrosion and heat resistance.
9. Cast iron has lower melting temperature (1150°C to 1200°C) than steel (1380°C – 1500°C).

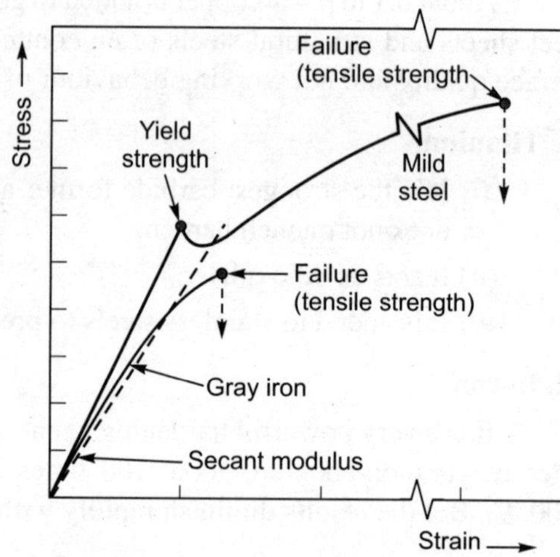

Fig. 3.19 The figure compares stress strain curves of mild steel and grey cast iron.

Applications

Machine bodies, bed, slides and bases of machine tools, the cylinders and beds of engines, pump liners, mill liners, grinding balls, ingot or pig moulds, industrial and consumer products such as elevator counter weights, industrial furnace counter weight, pump housing, housing for steam turbine, cylinder, pistons, and engine frames.

3.10 TYPES OF CAST IRON

There are five major types of cast iron. Grey iron is oldest type and still used. The classification of cast iron is shown in Fig. 3.20. Approximate ranges of carbon and silicon percentage in various types of cast irons are presented in Fig. 3.21.

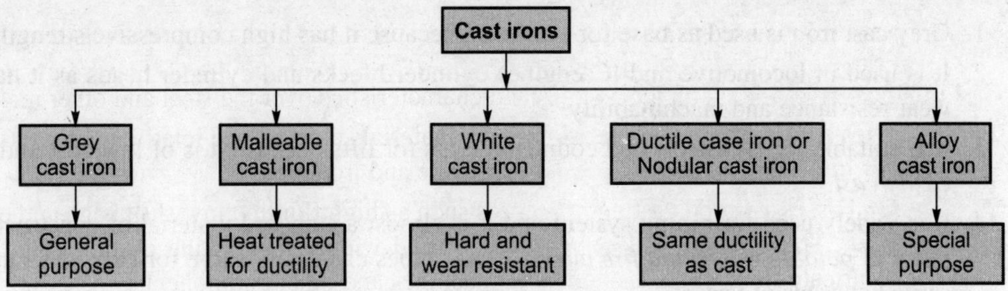

Fig. 3.20 Classification of cast iron.

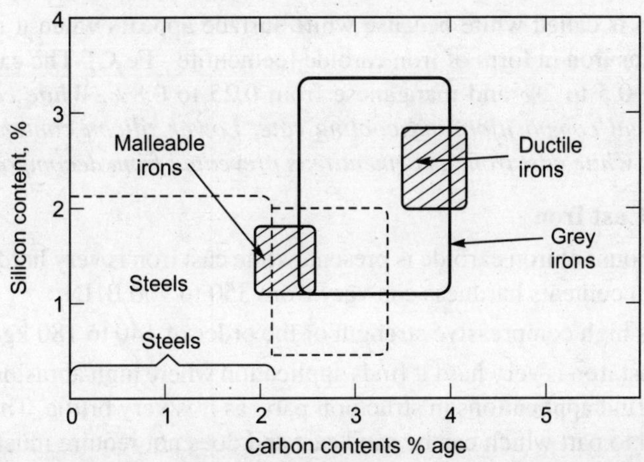

Fig. 3.21 Approximate range of carbon and silicon in various types of cast iron.

3.10.1 Grey Cast Iron

Grey cast iron is iron-carbon-silicon alloy with high carbon. The chemical composition of grey cast iron ranges from 2% to 4% of carbon contents with minimum 1% silicon. In microstructure graphite is really present in the form of the three dimensional rose like structures but metallographic sections show graphite as flakes. Grey cast iron has certain unique properties.

1. It is cheaper than steel.
2. The wear resistance of grey cast iron is very good as graphite acts as lubricant.
3. Compressive strength of grey cast iron is one of its good mechanical properties. Compressive strength can be three to five times tensile strength. Shear strength is about equal to tensile strength (from 1 to 1.5 times). Tensile strength has a range of 138 MPa to 414 MPa.
4. It is easy to machine as chip formation is promoted by graphite flakes. In addition flakes serve as lubricant for cutting tool.

5. It has excellent fluidity.
6. The damping capacity (ability to damp vibrations) of grey iron is superior to that of steel.
7. It has low ductility and very poor toughness. It is best to design grey cast iron parts in such a manner that they never receive shock loads. Fatigue strengths are about 40% of tensile strength.

Application of Grey Cast Iron

1. Grey cast iron is used as base for machines because it has high compressive strength.
2. It is used in locomotive and IC engines cylinder blocks and cylinder heads as it has good wear resistance and machinability.
3. It is suitable for flywheels and counterweights for lifts because it is of low cost and can be *easily cast.*
4. It is widely used for piping system and it is almost a standard material of construction for pipes of *potable water and fire mains.* These pipes can easily serve for fifty years and may serve hundreds of years.

3.10.2 White Cast Iron

White cast iron is called white because white surface appears when it is fractured. White cast iron essentially contains iron in form of iron carbide [cementite - Fe_2C]. The carbon is in range from 2 to 3.5%, silicon from 0.5 to 2% and manganese from 0.25 to 0.8%. *White cast iron is obtained by controlling the chemical composition and cooling rate. Lower silicon content and rapid cooling of molten alloy produces white cast iron, as cementite is prevented from decomposing to graphite.*

Properties of White Cast Iron

1. As large amount of iron carbide is present, white cast iron is very hard and brittle. Depending upon carbon contents hardness can vary from 350 to 500 BHN.
2. It possesses high compressive strength of the order of 140 to 180 kg/mm^2.
3. As white cast iron is very hard it finds application where high abrasion resistance is required. It does not find applications in structural parts as it is very brittle. The use of white cast iron is restricted to part which can be sand cast and does not require much machining as it is not easy to machine white cast iron.
4. Large quantity of white cast iron is used for making malleable cast iron.

Applications

Wearing plates, pump liners, mill liners, grinding balls etc.

3.10.3 Malleable Cast Iron

Malleable cast iron is made from white cast iron. The castings are first cast as white cast iron. Then by suitable heat treatment for an average 100 hours, *virtually all the combined* carbon in white cast iron is precipitated as nodules of graphite. These nodules are small, (fairly compact aggregations) and are much more desirable than flakes from a strength view point.

Malleable cast irons are iron-carbon-silicon alloys with silicon usually in range of 1 to 1.8%.

Malleable Cast Iron has following Properties

1. Malleable cast iron are more expensive than grey cast iron because of the cost of heat treatment involved. The cost may not be justified for one or two castings.

2. Most of the physical properties of malleable cast iron are similar to those of grey cast iron. There is, however, important difference in mechanical properties. The ductility of malleable cast iron is undoubtly the most important difference.
3. The tensile strength of ferrite malleable cast iron is around 370 MPa and has upto 20% elongation which means that malleable cast iron can bend without breaking.
4. Machinability of malleable cast iron is excellent below 300 HB. It is also extremely stable because of long annealing heat treatment.
5. Fatigue strength is as much as 60% of tensile strength. Compressive strength can be four times tensile strength. It also has considerable shock resistance.
6. Wear characteristics are still good because of temper carbon in micro structure.
7. The value of modulus of elasticity is around (172×10^3 MPa) for ferrite malleable iron and upto (186×10^3 MPa) for malleable iron. This puts the stiffness of malleable iron almost in range of steel which is quite significant design consideration.

Applications

Rail roads, automobile, crank shafts, chain links, brackets, *agriculture,* plumbing etc.

3.10.4 Ductile or Nodular Cast Iron

We have seen that malleable cast iron is produced by heat treating a white iron casting of suitable chemical composition and has improved machanical properties. The properties would be even better if iron is treated in molten stage.

Ductile cast iron or nodular cast iron or spheroidal graphite cast iron is produced by adding small amount of magnesium (0.3% to 1.2%), cerium or other modifiers to specially prepared iron in molten condition. These modifiers cause a large proportion of carbon to occur as spheroids of graphite rather than flakes. Ductility is obtained in ductile cast iron as a result of formation of spheroidal type of graphite. Ductite iron is a carbon-iron-silicon alloy with carbon (3% to 4%) and silicon (2 to 3%).

Properties of Ductile Cast Iron

1. Nodular iron or ductile iron is a major engineering material as it combines the engineering advantages of steels with the processing economics of cast iron.
2. Ductility of ferrite grade ductile iron is almost as good as that of mild steel.
3. Compressive strength is about twice its tensile strength.
4. Tensile strength ranges from 400 MPa to 700 MPa and elongation is between 10% to 18%.
5. Impact strength of *lower strength grades* is almost as good as mild steel. The fatigue strength is 40-50% of tensile strength.
6. The most important difference is magnetic and electrical properties of ductile cast iron. For example electrical resistivity of ductile iron is much lower.

Applications

Agriculture implements, industrial fan hubs, Crank shafts, and gears.

3.10.5 Alloy Cast Iron

The family of cast irons possesses a wide range of engineering properties in addition to manufacturing advantages such as low melting temperature, high castability, good machinability etc. but still cast irons lack two important engineering properties which are:

1. Shock and impact resistance combined with high wear and abrasion resistance.
2. Corrosion and heat resistance.

The above two shortcomings of cast iron can be overcome by use of alloyed cast iron. Two recent examples of alloy cast iron arc:

1. Acicular cast iron and spheroidal cast iron. Acicular iron contains nickel and molybdenum in its composition and is being used for casting crank shafts.
2. In spheroidal cast iron graphite content is converted from flaky to spheroidal form by alloying small amount of magnesium and cerium. This change increases the tensile strength to about 600 to 750 N/mm^2 and produces tough metal which can be twisted and bent.
3. When alloying elements such as nickel, molybdenum and chromium are added to various irons, it is possible to get chilled-iron-type structures in heavy sections.

In summary of our discussion of cast iron, there is an iron that can be used for almost any application. A wide range of properties are available and the property variations are obtained by process controls, composition control and thermal treatment of basic alloy of iron-carbon and silicon.

VIVA VOCE QUESTIONS

1. Explain the importance of materials in manufacturing processes.
2. How would you classify the materials in this world?
3. What are the properties of materials? How do material properties help while selecting a material for a particular application?
4. What is meant by static properties of a material? Explain some of the static properties of a material.
5. What are the tests which are performed on universal testing machine?
6. What are some tensile test properties that relate to elastic to plastic transition in a material?
7. What is the shape of an engineering stress-strain curve for a brittle and ductile material?
8. What is the difference between true stress strain curve and engineering stress strain curve?
9. How tensile test can be utilized to measure the ductility of material?
10. What is the shape of engineering stress strain curve for mild steel? What are the salient points in the curve?
11. Is a brittle material necessarily weak? Comment.
12. What do you mean by resilience?
13. Under which conditions material can be called perfectly elastic and perfectly plastic?
14. What do you understand by creep? What are the various stages of creep?
15. What is the difference between static and dynamic loading? Describe several different types of dynamic loading.
16. What is toughness of a material? How is it determined?
17. Under what conditions might a metal fracture when exposed to stresses that lie below its yield strength?
18. Explain the phenomena of fatigue. How the fatigue strength or endurance limit of a material is determined?
19. How would you classify the engineering materials?
20. How and in which form metals are found?
21. Explain the process of extraction of iron from its ore.
22. How does steel differ from pig iron?
23. How would you classify steels? Describe different types of plain carbon steel, their properties and applications.

24. Explain the various properties, which are affected in steel by increase of carbon contents.
25. Discuss the various applications of Alloy steel.
26. What is high speed steel and where is it used?
27. What is the most common alloying element added to steel?
28. What is the effect of alloying elements to the properties of steels?
29. What is the major difference between steels and cast iron?
30. Explain the various types of cast iron with their properties and applications.

UNIVERSITY PROBLEMS

1. How would you classify material properties?
2. Explain the following mechanical properties:
   (*i*) Ductility
   (*ii*) Malleability
   (*iii*) Plasticity.
3. Explain the following mechanical properties of engineering materials: Toughness, Stiffness, Resilience, Hardness and Strength.
4. Write short notes on following:
   (*a*) Grey Cast Iron
   (*b*) White Cast Iron
   (*c*) Malleable Cast Iron
   (*d*) Nodular Cast Iron
   (*e*) Alloy Cast Iron
5. What are plain carbon steel? How are they classified? Give their properties and applications.
6. What are the effects of different alloying elements on cast iron?
7. Explain the general properties applications and differences between mild steel, medium carbon steel and high carbon steel.
8. Discuss the general properties and area of applications of:
   (*i*) Mild steels
   (*ii*) High speed steels.
9. Describe various properties of engineering materials.

"Keep away from people who try to brittle your ambitions. Small people always do that, but the really great make you feel that you too can become great".

CHAPTER

4

BENCH WORK AND FITTING (FITTING SHOP)

INSIDE THIS CHAPTER

4.1 INTRODUCTION

Bench work and fitting plays a significant role in every workshop. Although now a days job can be finished to desired accuracy with the help of automatic machine, still they required some operations to be done manually by hand. The work carried out on the part for finishing and assembling the parts by hand tool, after the machining operations, is called Bench Work and fitting. As the bench work and fitting requires a large number of hand tools, one must have good working knowledge of all these tools to carry out the finishing and assembling in more effective and efficient manner.

Fitting bench work tools may be classified as:

(1) Folding and supporting tools

(2) Measuring and marking tools

(3) Striking tools

(4) Filing tools

(5) Cutting tools

(6) Drilling tools

(7) Threading tools.

Genious is one percent inspiration and ninty nine percent perspiration.

4.2 GEOMETRICAL PROPERTIES OF TOOLS

Before getting in details of these tools lets discuss the geometrical properties related to some of these tools.

1. **Flatness:** The extent to which all points on a surface lie in single plane. A perfectly flat surface represents a true plane. Flatness is an unrelated characteristic and should not be confused with parallelism or perpendicularity. Flatness can be measured with help of surface plate or try square.
2. **Straightness:** The extent to which a part feature such as line or axis is straight line. Straightness may include
   (*a*) Straightness of a line on a surface
   (*b*) Straightness of line on a single plane
   (*c*) Straightness of line which is axis of feature (say cylinder)
   Straightness can be checked with help of surface plate or straight edge.
3. **Parallelism:** The extent to which all point on surface, line or axis are always separated by uniform distance from reference surface, line or axis. Parallelism can be checked with help of dial indicator, seriber and surface gauge or universal surface gauge.
4. **Squareness (Perpendicularity):** The extent to which all points on surface, plane, line or axis are at 90° from reference surface, plane, line or axis. Squareness can be checked with help of try square, dial indicator or combination set.
5. **Concentricity:** The extent to which two or more cylindrical surfaces or cylindrical holes have a common axis. Concentricity can be checked with the help of dial indicator.
6. **Circularity (Roundness):** It is applicable for surface of revolution such as cylinder, circular hole, core or sphere. It is the extent to which all points the surface are at uniform distance from axis, The roundness is measured on a plane which is perpendicular to the axis of revolution (for example, the surfaces of the flat bases in case of cylinders).

   The circularity (roundness) is an unrelated characteristic and should not be confused with concentricity or cylindericity. The circularity can be checked with help of *V* block and dial indicator.
7. **Cylindericity:** Cylindericity may be assumed as combination of circularity, straightness and parallelism. The extent to which all points on the surface of revolution such as cylinder are at a uniform distance from axis of revolution.

   A component having circularity does not necessarily have cylindericity. The individual characteristics which comprise cylindericity are to be checked separately.

4.3 HOLDING AND SUPPORTING TOOLS

4.3.1 Vices

Vice is the most common tool and is used for clamping the workpiece during operations carried out in a fitting shop. The vice are made of cast iron or steel cast body. There are number of vices available like bench vice, hand vice, pipe vice etc. Out of these bench vice is most commonly used in general fitting work.

(*i*) **Bench Vice:** Bench vice is fitted to fitter's bench with the help of nuts and bolts. Main parts of bench vice are jaws, handle and base.

The bench vice carries fixed jaw and movable jaw. The fix jaw is integrated with the bench vice body and have internal square threads. The movable jaw is operated by screw having square threads. The holding faces of jaws are knurled or have serration to hold the workpiece firmly. The jaws of the vice are usually kept over hanginging the edge of the bench.

For gripping the workpiece in benchwise jaws are opened by with drawing the movable jaw with the help of handle. The work is kept between the jaws are tightened with the help of handle. Handle carried a collar inside, to prevent the coming out of movable jaw from vice.

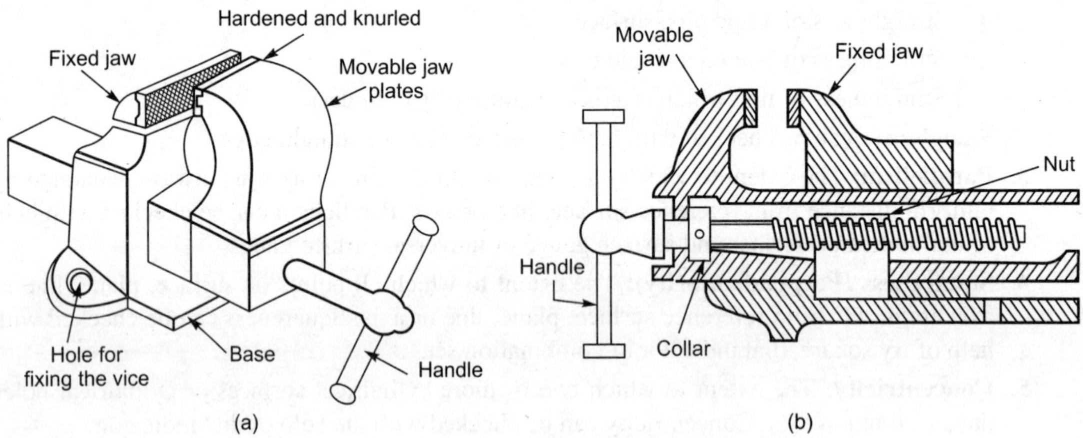

Fig. 4.1 Bench vice.

(*ii*) **Hand Vice:** For gripping very small object hand vice is used. The hand vices are made in various shape and sizes to accommodate workpieces. A commonly used hand vice is shown in Fig. 4.2.

It consists of two steel legs hinged together at the bottom. At the other end jaws are provided. A spring is provided between two legs to keep them separate. The jaws are opened and closed by means of wing nut.

(*iii*) **Pipe Vice:** As name suggests, pipe vice is used for holding pipes, shafts or circular jobs. It consists of movable jaw and fixed jaw. The movable jaw is attached to the vertical screw. The screw has square threads and handle to rotate the movable jaw for gripping the round workpiece. The shape of movable as well as fixed jaw is *V* shape to grip the workpiece at four points.

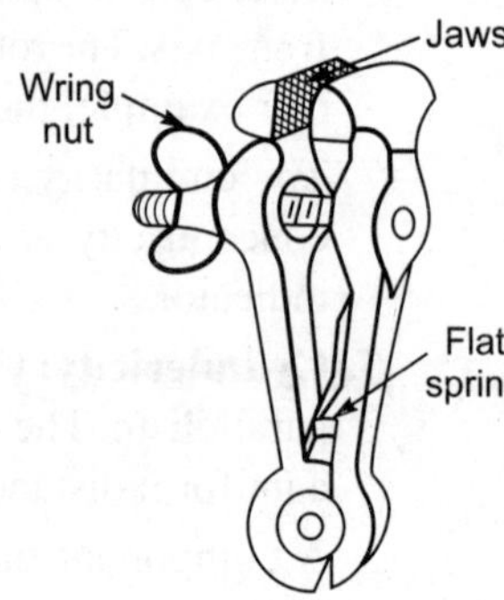

Fig. 4.2 Hand vice.

4.3.2 V Block

V-block is used for supporting the shaft and bushes in marking out and drilling. The *V* grooves provided in *V*-block grip the circular object longitudinally. *V*-block works in combination with *U* clamp to hold the round job as shown in Fig. 4.3. The small size *V* block is made of mild steel hardened and ground. *V* blocks are used in pairs to support two ends of a long shafts.

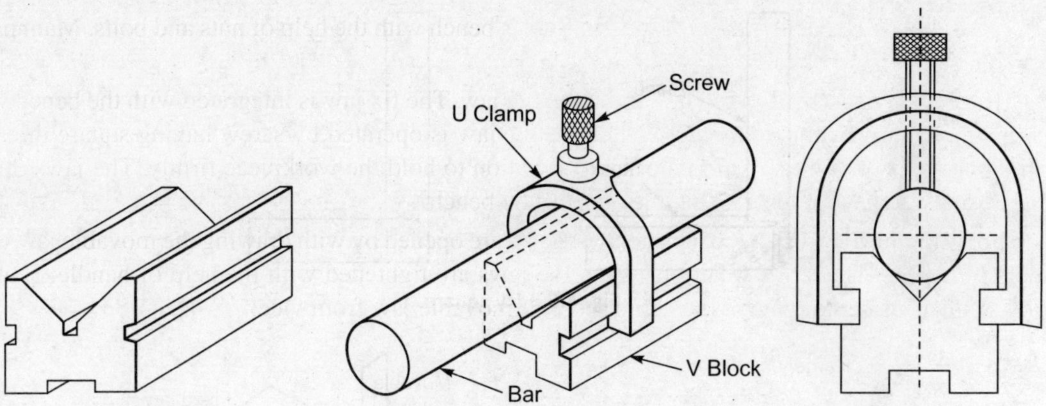

(a) V-Block (b) bar held in V-Block with help of U-Clamp for marking purpose

Fig. 4.3 V Block and its use.

4.3.3 Angle Plate

Angle plate is used for supporting the work piece surface at right angle to the surface of the table. The angle plate has two plane surfaces which are at right angle. It is provided with a number of slots in both surfaces to hold the piece by means of nut bolts or clamps.

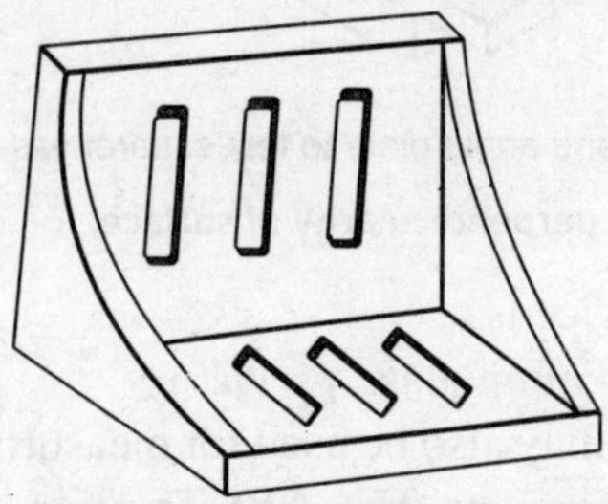

Fig. 4.4 Angle plate.

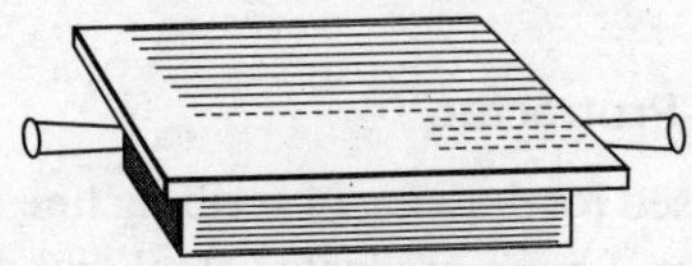

Fig. 4.5 Surface plate.

4.3.4 Surface Plate

It is used for testing the trueness and flatness of finished surface. (Fig. 4.5) Marking with the *V* block and angle plate is done by putting them on surface plate. Surface plate is having a rectangular or square face which is a perfect horizontal plane.

4.4 MARKING AND MEASURING TOOLS

4.4.1 Try Square

As the name suggests, it is used for testing the *squareness of two surfaces.* It consists of steel blade which is fitted in a stock (Fig. 4.6). For testing the squareness of surface, blade and stock are placed against the edges and then held up to the light. Any small deviation from squareness can be detected by the appearance of light between blade and workpiece edge. The blade is at 90° to the stock. The stock is made of steel or cast iron and has rectangular cross-section. Surface plate and angle plate are used to test the squareness of try square.

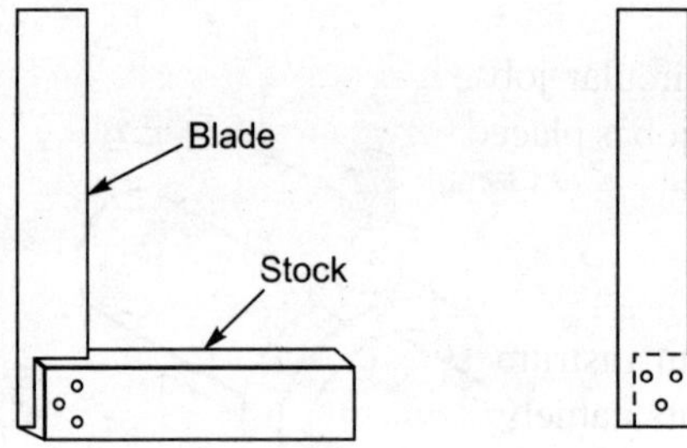

Fig. 4.6 Try square.

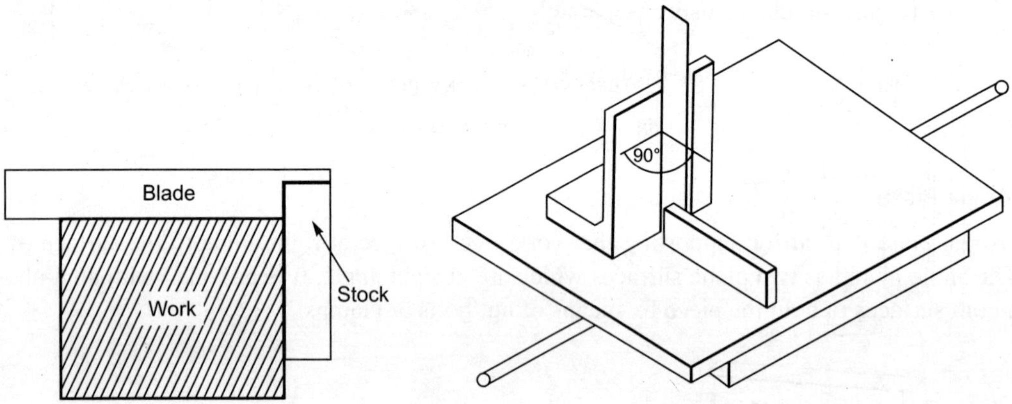

(a) Use of try square (b) Use of surface plate and angle plate to test squareness of try square

Fig. 4.7 Method of using try square for testing perpendicularity of surface.

4.4.2 Bevel Protractor

It is used for drawing or scribing lines at any angle. It may also be used for measuring inclination of any surface. I has a graduated steel dial divided in 360 divisions (Fig. 4.8). The dial can be rotated around the centre and can also be locked in specific position with help of lock screw. The angle is marked or measured by straight edge of the blade. Blade can also be slided along it length.

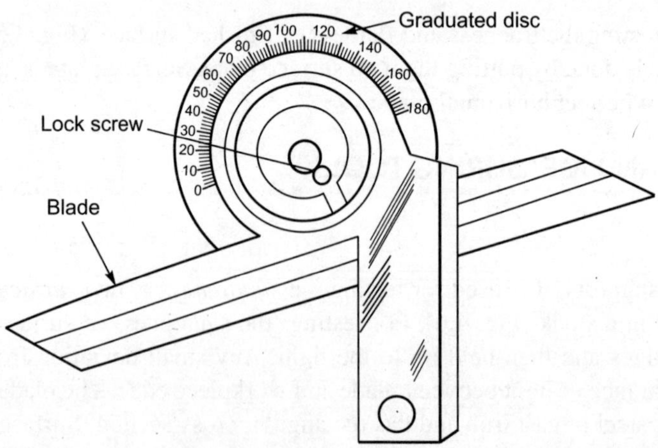

Fig. 4.8 Bevel protractor.

4.4.3 Centre Square

It is used to find the centre of circular jobs. It has a straight edge and two arms. The circular job is placed between the arms to find its centre.

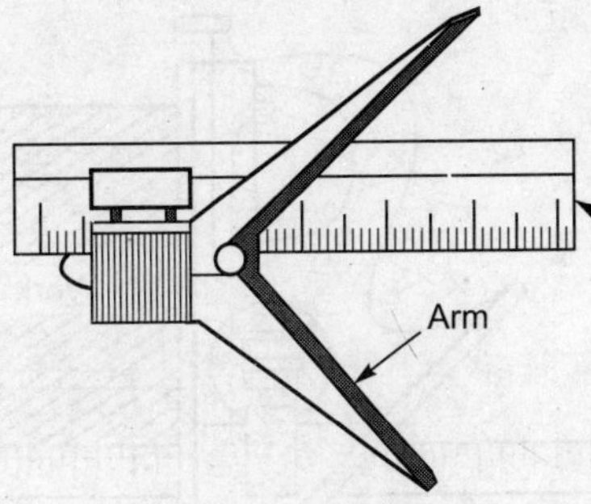

Fig. 4.9 Centre square.

4.4.4 Combination Set

Combination set is a very useful instrument and is combination of five different instruments namely steel rule, centre head, bevel protector, try square (square head) and bubble type spirit level (Fig. 4.10 and 4.11). All these attachments can be used simultaneously or separately or in combination, depending upon the requirement.

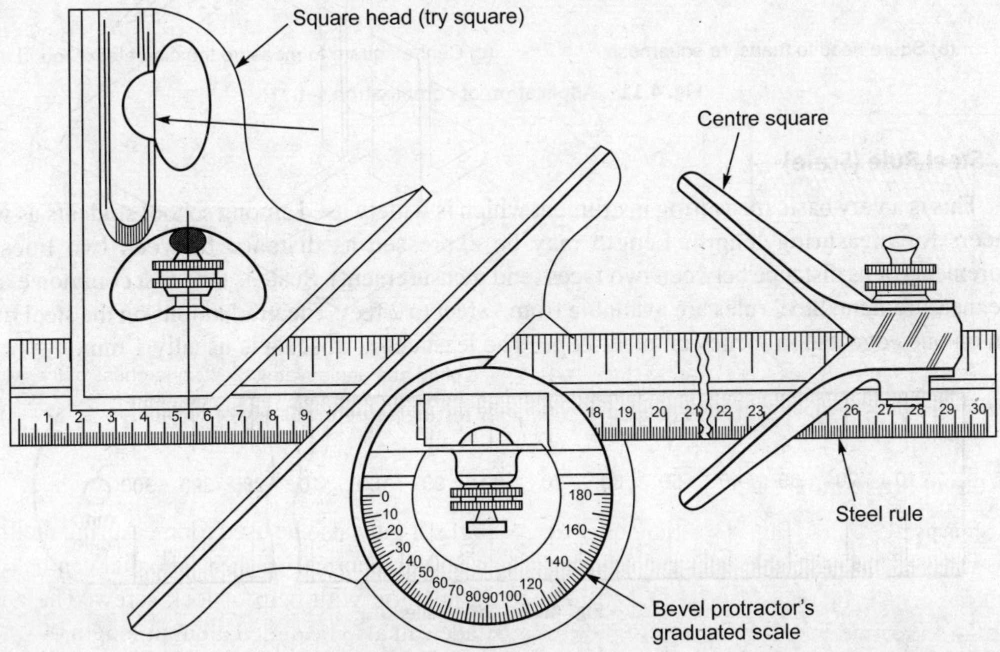

Fig. 4.10 Combination set.

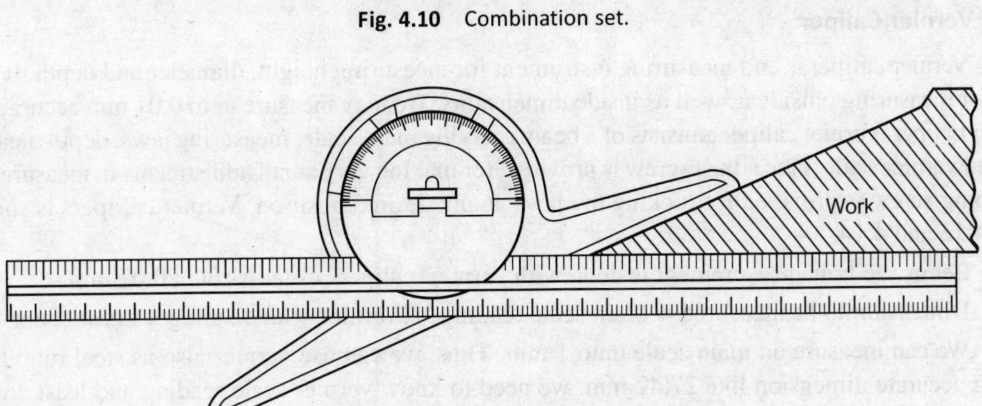

(a) Bevel protractor to measure angle

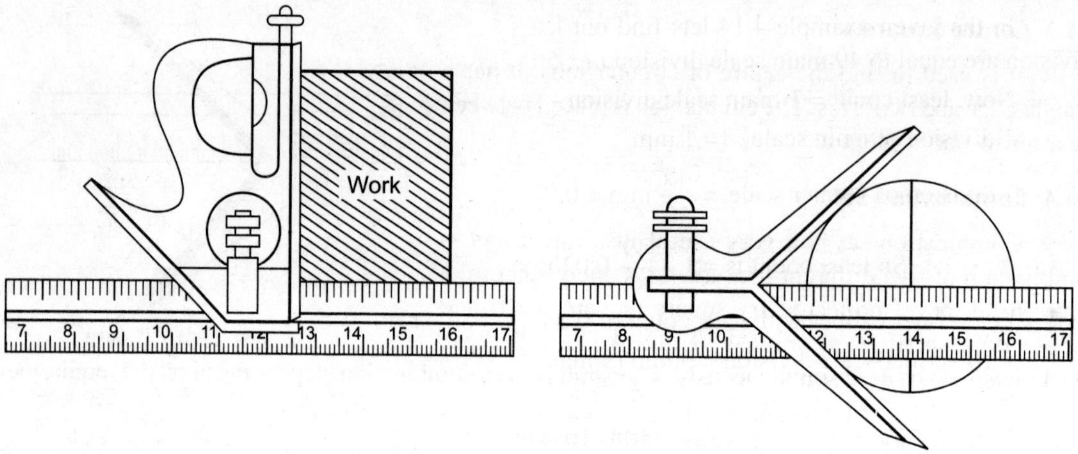

(b) Squre head to measure squarness (c) Centre square to measure the diameter of round object

Fig. 4.11 Application of combination set.

4.4.5 Steel Rule (Scale)

This is a very basic measuring instrument which is widely used among school students as well as engineers for measuring length. Length may be expressed as distance between two lines (line measurement) or as distance between two faces (end measurement). Scale is the most common example of line measurement. Steel rules are available from ½ feet to 2 feet. The graduations on the steel rule are in mm on one edge and in inches on other edge. The least count of scale is usually 1 mm.

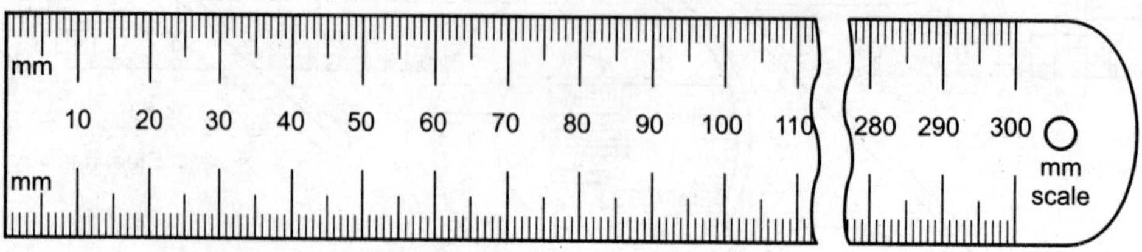

Fig. 4.12 Steel rule.

4.4.6 Vernier Caliper

Vernier caliper is end measuring instrument for measuring height, diameter and depth. It can be used for measuring outside as well as inside dimensions. We may measure upto 0.01 mm accuracy with vernier caliper. Vernier caliper consists of a beam carrying main scale, measuring jaws, depth measuring bar, and vernier scale. Fine adjust screw is provided for making very small adjustments in measurements. Clamping screw is provided for locking the jaws at any desired position. Vernier calilpers is shown in Figs. 4.13 and 4.14.

Let us see how measurement is done with vernier calipers of least count 0.02 mm.

Dimension to be measured = Main scale reading + (Vernier scale reading × Least count)

We can measure on main scale upto 1 mm. Thus, we can use vernier also as steel rule. But for getting accurate dimension like 27.42 mm, we need to know vernier scale reading and least count of vernier caliper (least count is minimum reading that an instrument can measure).

In the given example 4.14 lets find out least count of vernier calipers. Here 50 vernier scale division are equal to 49 main scale division *i.e.*, 50 vernier scale division = 49 mm.

Now, least count = 1 main scale division – 1 vernier scale division

1 division of main scale = 1 mm

$$1 \text{ division of vernier scale} = \frac{49}{50} \text{ mm} = 0.98 \text{ mm}$$

$$\text{So least count is} = 1 - \frac{49}{50}\ 0.02 \text{ mm}$$

For vernier caliper with least count of 0.1 mm

10 vernier scale division = 9 mm (9 main scale divisions)

$$\text{Least count} = 1 - \frac{9}{10} = \frac{1}{10} = 0.1 \text{ mm}$$

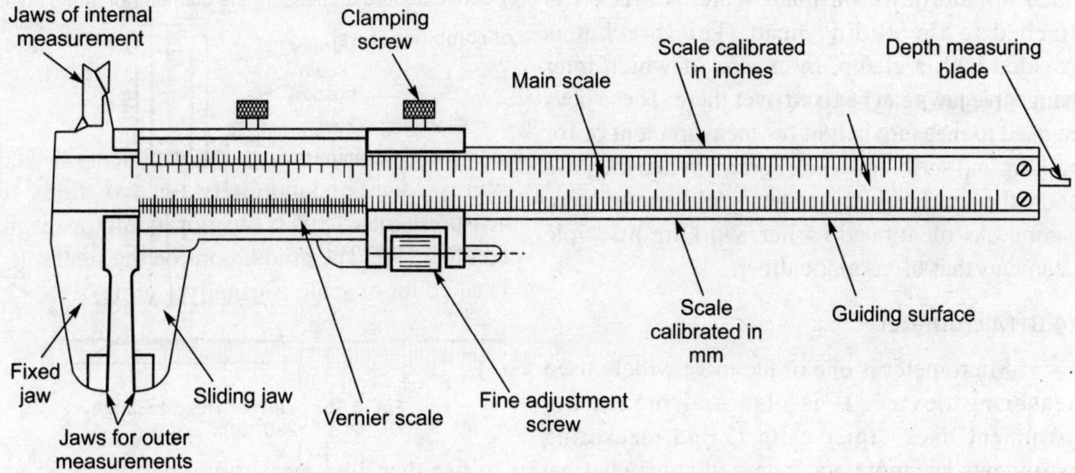

Fig. 4.13 Vernier caliper.

For reading vernier scale, we see that which division (on vernier scale) is exactly coinciding with a division on main scale, say 21st division of vernier scale is exactly coinsiding with the main scale division and reading on the main scale is slightly ahead of 27 mm. So total reading is 27 + 0.02 × 21 = 27.42 mm as shown in Fig. 4.14.

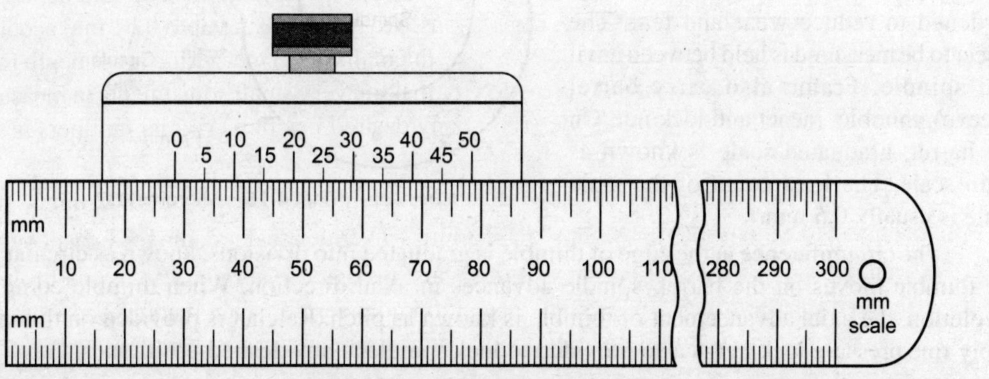

Fig. 4.14 Reading the vernier caliper.

Here the zero of vernier is between 27 and 28 mm and 21st reading of vernier coinsides with main scale reading, so total reading of vernier caliper is

$$27 + 0.02 \times 21 = 27.42 \text{ mm}$$

4.4.7 Vernier Height Gauge

Vernier height gauge works on the principle of vernier caliper. It is used for measuring the height of parts and for marking a work piece at a specific height (Fig. 4.15). It consists of a heavy base with accurately finished bottom surface. A vertical bar is mounted to the base which carries graduated main scale. A sliding head (which carries graduated vernier scale and fine adjustment screw) can be slided up and down on main scale. A bracket is attached to the sliding head. This bracket is provided with a clamp, by means of which inter changeable jaws can be fixed over there. These jaws are used to measure height for measurement or for marking and scribing on workpiece. When the jaws touch the work table, vernier scale shows zero reading. As mentioned earlier, working principle is same as that of vernier caliper.

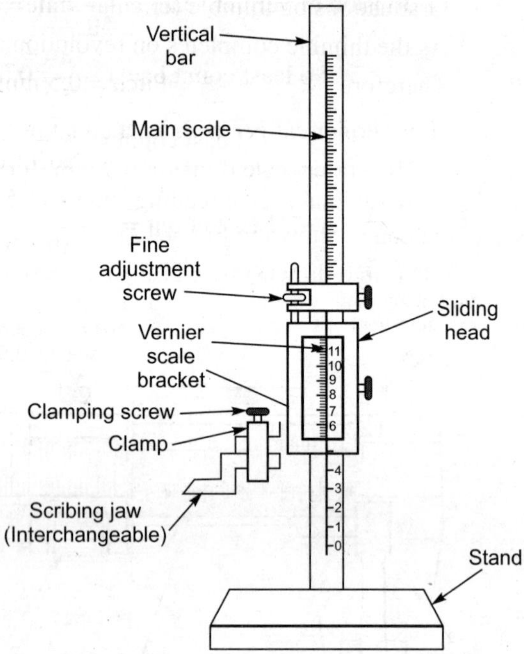

Fig. 4.15 Vernier Height Gauge.

4.4.8 Micrometer

Micrometer is one of the most widely used measuring device. It is also end measuring instrument like vernier caliper. End measuring instruments are more accurate and somewhat easier to use than line measuring instrument because, object to be measured is held in jaws and any difficulty in aligning the edges or line is avoided.

Micro meter is more accurate than vernier caliper. It consists of a hook type frame. An anvil is fixed at one end and on other end there is a movable spindle. The ends of both anvil and spindle are hardened to reduce wear and tear. The object to be measured is held between anvil and spindle. Frame also carry barrel (sleeve), thimble, rachet and lock nut. On the barrel, graduated scale is known as main scale. The least count of the main scale is usually 0.5 mm.

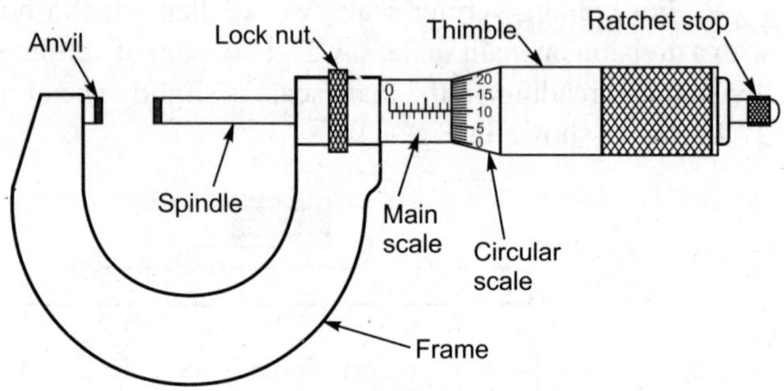

Fig. 4.16 Micrometer.

The circumference at the edge of thimble is graduated into divisions, known as circular scale. As the thimble moves on the barrel, spindle advances in axial direction. When thimble completes one revolution, the axial advancement of thimble is known as pitch. Ratchet is provided on thimble end to apply fine pressure lock nut is also provided to lock spindle at any desired position.

Lets see how dimensions are being measured on micrometer having least count of 0.01 mm.

Dimension to be measured

= Main scale reading + (Least count of micrometer × Thimble scale reading)

Least of micrometer is found as follows:

Least count of main scale = 0.5 mm

Graduations on thimble (circular scale) = 50

As the thimble completes on revolution axial advancement of spindle = 0.5 mm

Therefore Pitch = 0.5 mm

$$\text{Least count} = \frac{\text{Pitch}}{\text{No. of division on circular scale}} = \frac{0.5}{50} = 0.01$$

Now say main scale reading shows 11.5 mm and 31st division of circular scale coincides main scale division as shown in Fig. 4.17.

Measured dimension = 11.50 + 33 × 0.01 = 11.83

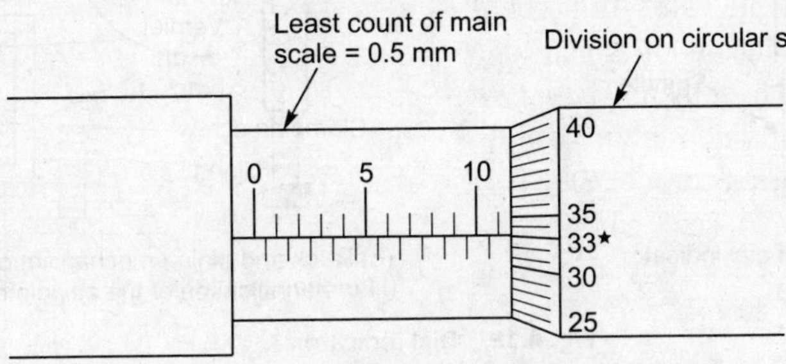

Fig. 4.17 Reading the micrometer.

Here least count of the main scale is 0.5 mm, divisions on circular scale = 50, pitch = 0.5 mm so least count is 0.01 mm. The main scale reading is 11.5 and 33rd division of circular scale is considing with main scale so reading of micrometer is

11.50 + 33 × 0.01 = 11.83

4.4.9 Dial Indicator

There are the precision instruments used for making the checking linear measurement. These are widely used for checking flatness of surface, parallelism of bars and rod and run out or ovality of circular jobs. They are generally used as comparators that is to compare a part to reference setting. They require less skill in their use than other precision instruments like vernier caliper, micrometer and gauges etc. Dial indicators usually have the accuracy of 0.01 mm.

Dial indicator is circular in shape with outer metallic casing and a graduated scale along its circumference (Fig. 4.18), much smaller graduated scale with pointer A_2, is also present to note number of revolution the pointer (A_1) has made.

The pointer A_1 and A_2 are moved by rack and pinion arrangement. The dial is divided into 100 equal divison. Each division represents a plunger movement of 0.01 mm. During measurement the end of plunger touched with the job. A very slight upward movement of the plunger is indicated by movement of pointer. The very small movement of plunger is magnified many times by use of gear and pinion arrangement as shown in Fig. 4.18. As the pointer A_1 makes one revolution A_2 moves through 1 graduation registering number of turns made by pointer A_1.

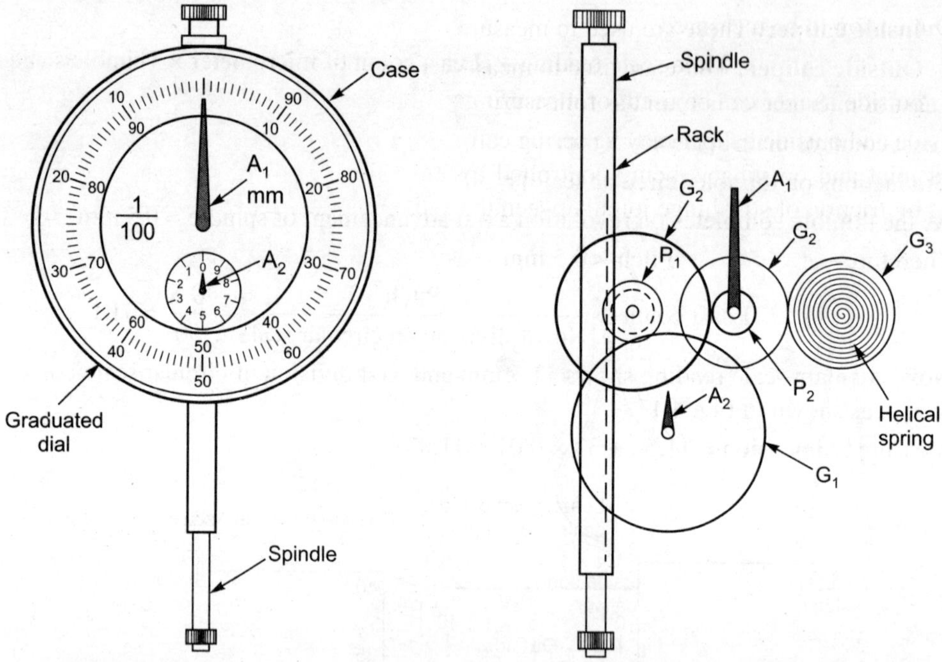

(a) Front view of dial indicator

(b) Rack and pinion mechanism of dial indicator for magnification of the spindle displacement

Fig. 4.18 Dial indicator.

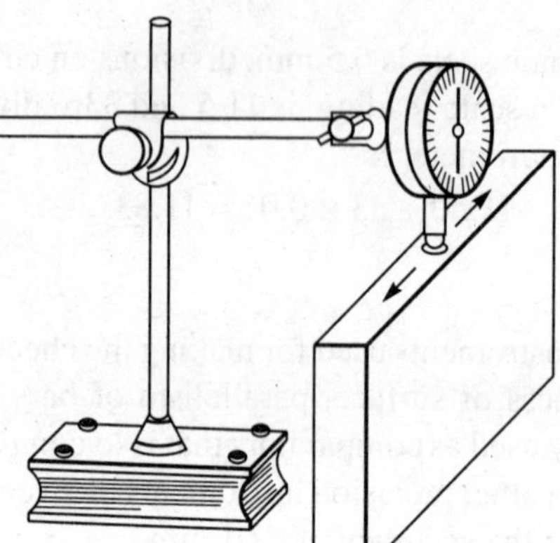

Fig. 4.19 Use of dial indicator for checking parallelism of the surface.

Please note here the end of plunger touches with surface and the plunger movement is magnified with gear pinion arrangement.

4.4.10 Calipers

Calipers are used for measuring inside or outside diameter of circular part and for measuring the length of bar. These are end measuring instruments and are more convenient to use as compared to steel rule. Depending upon the application of use and shape they are of two types (Fig. 4.20).

1. **Inside caliper:** These are used to measure inside diameter of circular hole in a part.
2. **Outside caliper:** These are used to measure inside diameter of circular hole both inside and outside caliper can be used for measuring length also.

Inside and outside caliper may be spring caliper or firm joint caliper. In spring caliper is spring controlled joint and opening is being controlled by nut and screw. In firm joint caliper opening is controlled by friction of stiff hinge joint between the caliper legs.

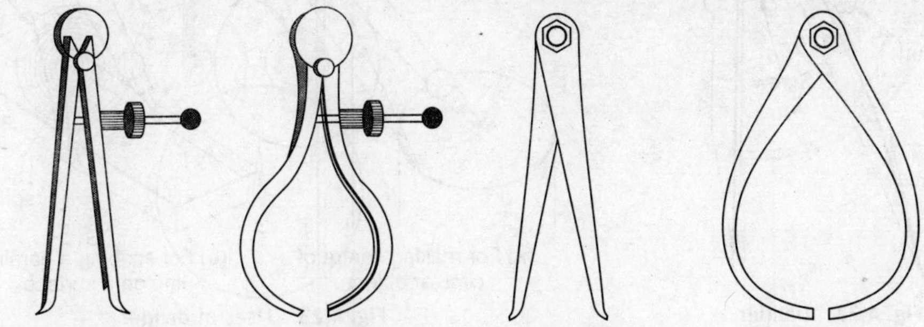

(a) Inside spring caliper (b) Outside spring caliper (c) Inside firm joint caliper (d) Outside firm joint caliper

Fig. 4.20 Outside and inside calipers.

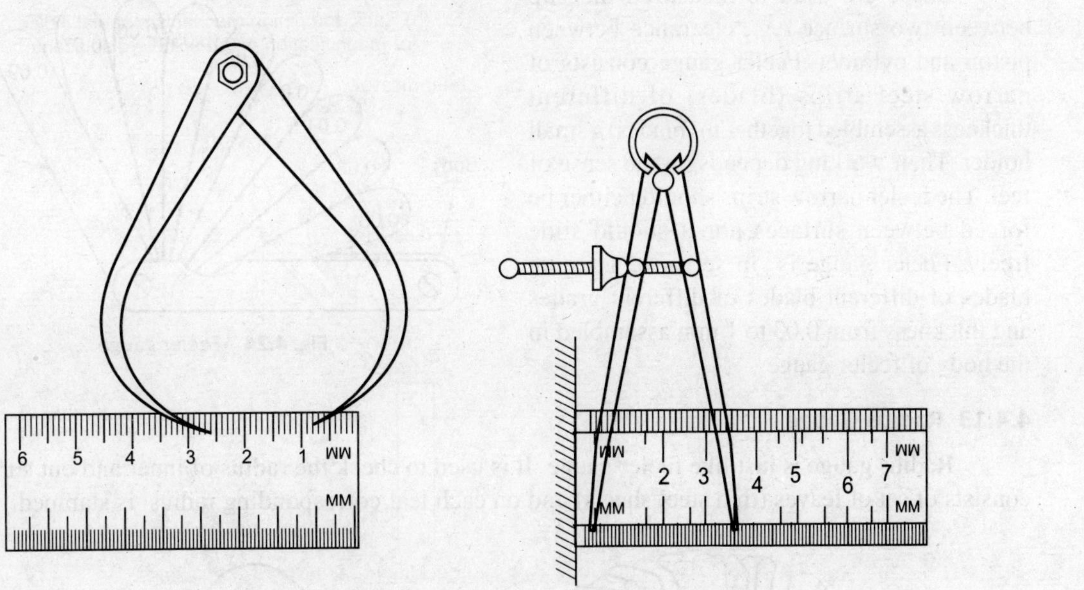

Fig. 4.21 Reading the outside and inside caliper with steel rule.

4.4.11 Dividers

These are similar in construction to the calipers but their legs are not bent. At the free end of two legs they have sharp point. These are either having hinged joint or spring controlled joints. There are used for transferring dimensions, measuring the distance between two points or lines, drawing circle or arc etc. The divider and its uses are shown in Figs. 4.22 and 4.23.

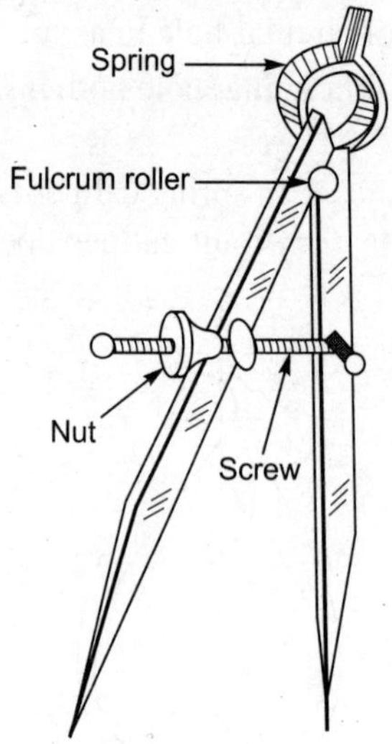

Fig. 4.22 Divider.

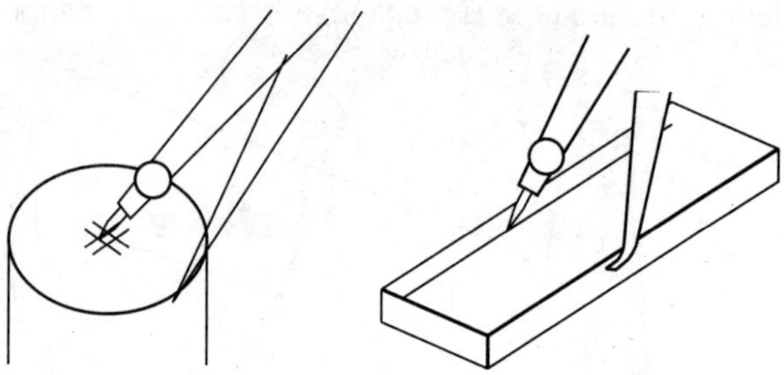

(a) For making centre of circular object

(b) For scribing a parallel line on a surface

Fig. 4.23 Uses of divider.

4.4.12 Feeler Gauge

These are used to measured the gap between two surface *e.g.*, clearance between piston and cylinder. Feeler gauge consists of narrow steel strips (blades) of different thickness assembled together in (blades) a small holder. Their working depends on the sense of feel. The feeler narrow strips should neither be forced between surfaces nor it should slide freely. Feeler gauge is in series of gauging blades of different blades of different grades and thickness from 0.03 to 1 mm assembled in the body of feeler gauge.

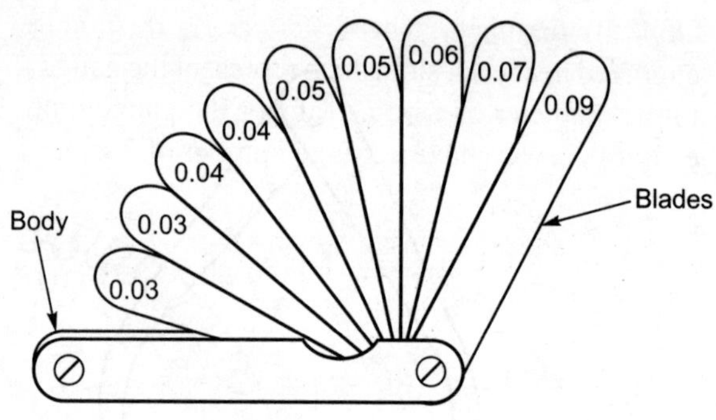

Fig. 4.24 Feeler gauge.

4.4.13 Radius Gauge

Radius gauge is just like feeler gauge. It is used to check the radius of inner and out surfaces. It consists of set of leaves (thin steel sheets) and on each leaf corresponding radius is stamped.

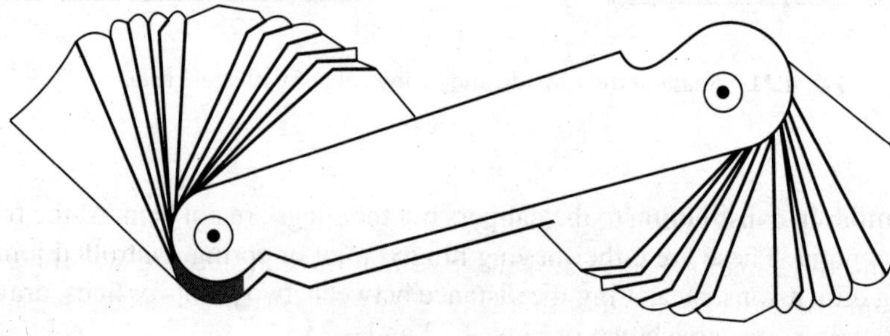

Fig. 4.25 Radius gauge.

4.4.14 Thread Gauge

It is also just like feeler gauge. It is used to determine the pitch of various threads by matching the teeth of the leaves (thin steel strips) with the threads of workpiece.

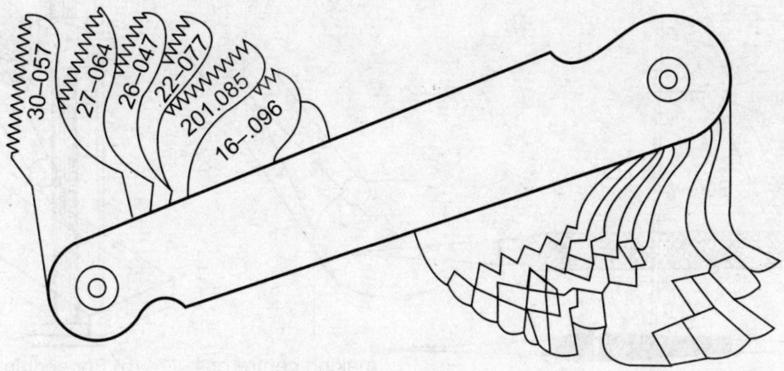

Fig. 4.26 Thread gauge.

4.4.15 Wire Gauge

It is used to check the diameter to wire. Wire gauge is flat circular steel sheet having slots all along its periphery. These slots are corresponding to different diameters of wires. The numbers are engraved near each slot which represent the gauge number of the wire. The gauge number varies inversely to the diameter of wire *i.e.* higher the gauge number the lower the diamer of wire and vice versa. For example, a wire having gauge number of 7 is of higher diameter than wire having gauge number of 29.

Fig. 4.27 Wire gauge.

4.4.16 Scriber and Surface Gauge

Scriber is used for marking lines. It is made of high carbon steel and is having hardened front edge. Scriber mounted on a vertical bar with a base is known as surface gauge. Surface gauge is also used for marking and drawing parallel lines.

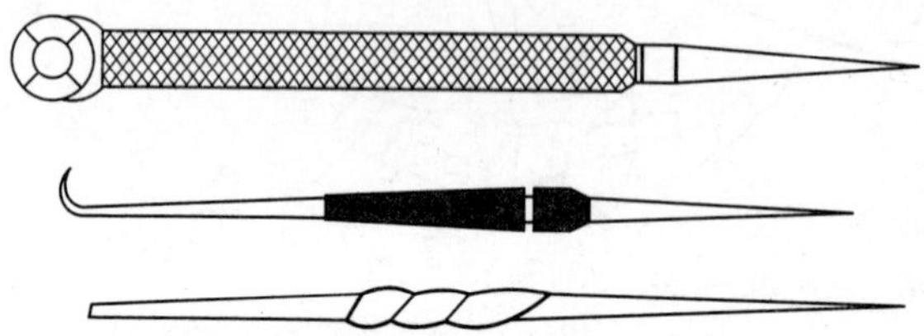

Fig. 4.28 Various types of scribers.

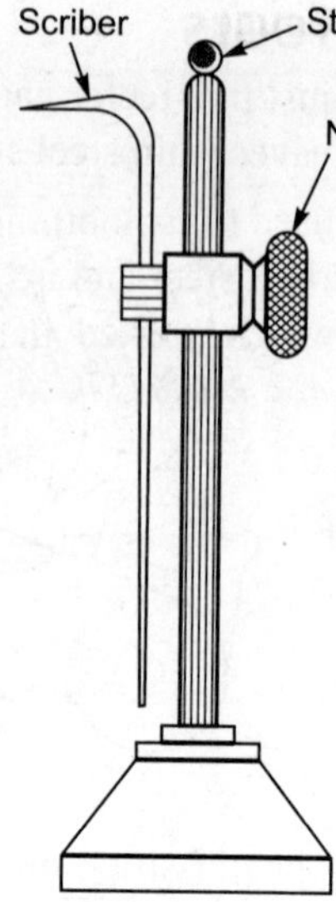

Fig. 4.29 Surface gauge.

4.4.17 Universal Surface Gauge

It is improved form of surfaces gauge. It has cast iron base, vertical bar (spindle), adjusting screw, scriber and pin. The base is having the *V*-groove at the bottom to make it suitable for use on round objects. Guide pins are provided at the end of the base which can be pressed down to project below base. These pins can be used for guiding during scribing. Spindle can be set at any angle and can be locked at any position with the help of adjusting screw.

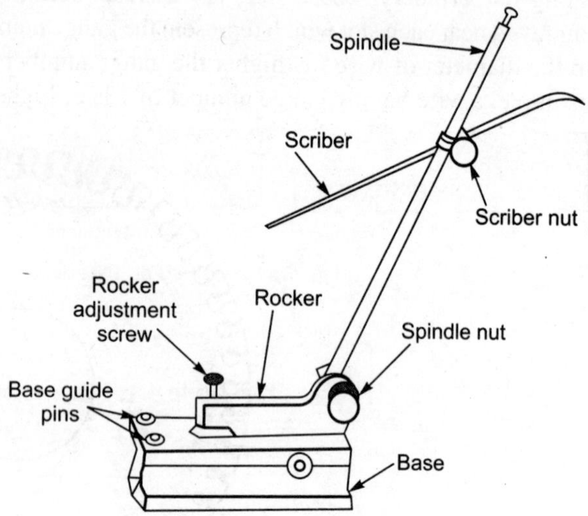

Fig. 4.30 Universal surface gauge.

4.4.18 Centre Punch

Centre punch is basic marking tool in bench work and fitting. These are used for marking dots on surfaces. Centre punches are also used for marking the centre point on surfaces before drilling operation. These are made of cast steel, being hardened and tempered.

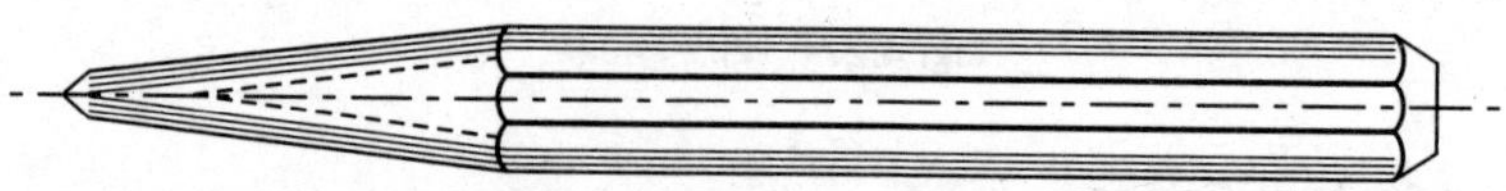

Fig. 4.31 Centre punch.

4.5 STRIKING TOOLS

Striking tools includes hammer and are discussed in detail in 5.4.12.

4.6 FILING TOOLS

(1) Files

Files are used for smoothing a rough surface or for reducing the thickness of metal surface file is made of high carbon steel and is hardened and tempered. The file consists of two major parts toothed blade and tang (which is fitted in a wooden handle). The different parts of files are shown in Fig. 4.32.

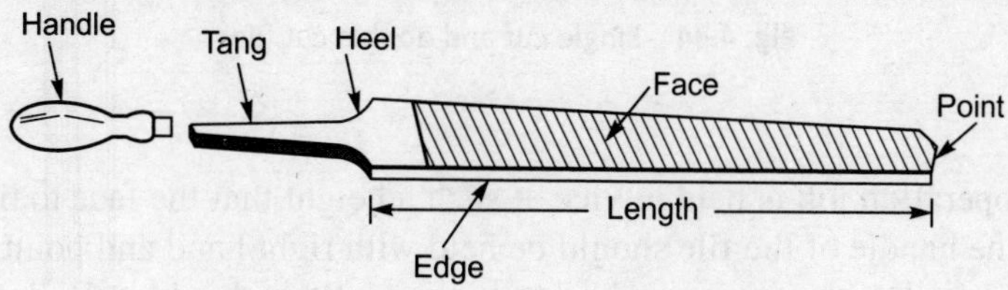

Fig. 4.32 Various parts of single cut file.

Files may be classified on the basis of cross-section, on the basis of grade and on the basis of type of teeth (Fig. 4.32, 4.33 and 4.34).

➢ The classification of files on the basis of cross-section is shown in Fig. 4.33. Depending upon the cross-section, these are used for various different applications.

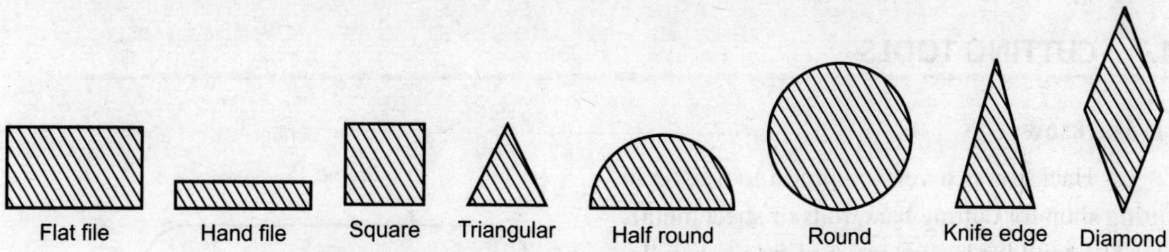

Fig. 4.33 Classified on files depending upon their cross-section.

➢ Depending upon the number of teeth per cm files are classified as:

Table 4.1. Files of different grades

Name of file	Pitch (teeth per cm)
Rough files	8 teeth per cm
Bastard file	12 teeth per cm
Second cut files	16 teeth per cm
Smooth files	20 to 24 teeth per cm
Dead smooth file	40 or more teeth per cm

➢ Depending upon the type of teeth files may be classified as single cut file double cut file (Fig. 4.34)

In single cut files, there is only one series of parallel grooves. The groove may be inclined at 45° to 60°, the file edge as shown in Fig. 4.34. In double cut file there are two series of grooves, one similar to those of single cut file and other and an angle of 70° to the edges of file blade. In both type of files all the teeth are having negative rake in sloping backward and thus perform cutting operation during forward stroke only.

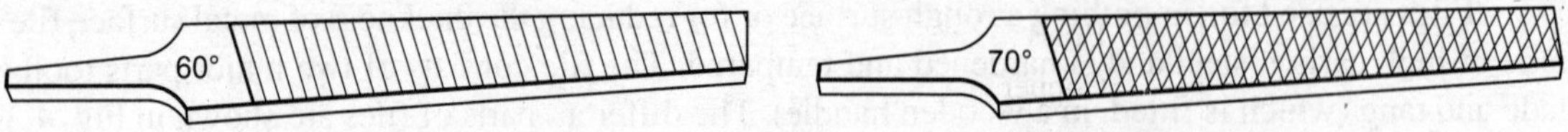

Fig. 4.34 Single cut and double cut files.

Method of Filing

During file operation job is held in vice at such a height that the face to be filed at level with operator's elbow. The handle of the file should be held with right hand and point of the file with left hand. The right hand applies the force required to cut and left hand guides the cut. It should be noted that file must remain horizontal during filing with force applied only in forward stroke.

Cross filing: This operation is used for removal of maximum amount of material in the shrotest possible time. The file must remain horizontal during cross filing. The rough files are used for this operation.

Draw filing: In the operation files removes very little material from surfaec. A fine cut file is used in draw filing. This operation is used for obtaining good surface finish and to remove file marks.

4.7 CUTTING TOOLS

(1) Hacksaw

Hacksaw is a very common tool used in fitting shop for cutting bars, rods or sheet metals etc. The hand hacksaw consist of frame, handle, blade and wing nut. Frame is made of metal and handle is usually made of wood. The frame may be fixed or adjustable type which is capable of accomodating baldes of different lengths. The blade is tightened by means of wing nut.

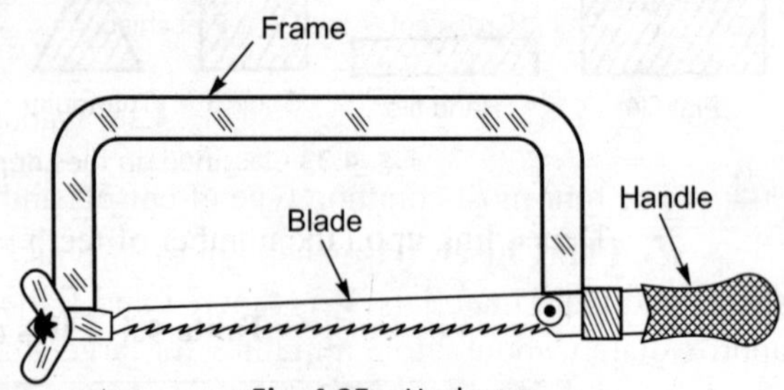

Fig. 4.35 Hacksaw.

Hacksaw blades are uaually forward stroke blade *i.e.*, they perform cutting only during forward stroke. The teeth of the blade may very from 5 teeth per cm to 12 teeth per centimeter. As the number of teeth increases fine cutting is achieved.

(2) Chisels

These are used for chipping away the material from work piece. The chisels are all over made from high carbon steel and have no wooden handle. The chisels are made from carbon steel bar having octagonal or hexagonal cross-section. The cutting edges of the chisel are hardened and tempered. The most commonly used cutting angle is 60°, but it may vary according to the type of material being cut. Approximate values of cutting angles are shown in Table 4.2.

Table 4.2. Cutting Angles for Common Materials

Material	*Cutting Angle*
Aluminium	30°
Brass	50°
Copper	45°
Cast Iron	60°
Cast Steel	65°
Mild Steel	55°

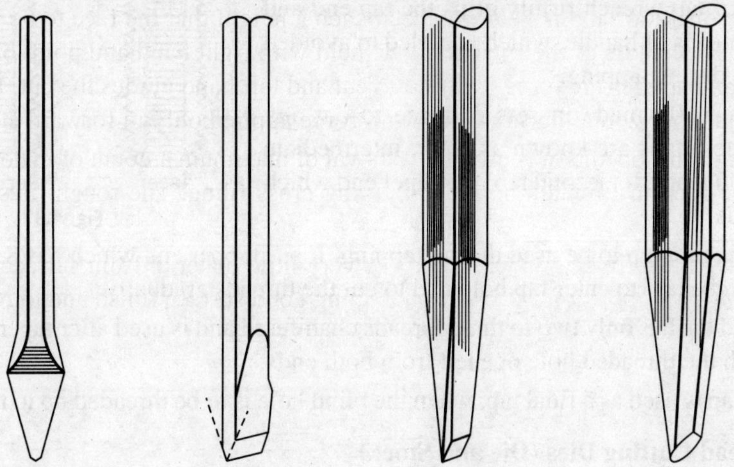

Fig. 4.36 Various types of chisels.

The four most common type of chisels and flat, cross cut, half round and dimond chisel.

(*i*) **Flat Chisel:** It is general purpose chisel and most widely used in cutting work, cutting metal sheets and plate materials. It is tapered and flattened for about 1/3 of its length. The cutting edge is approximately about 2 to 3 mm thick for large chisel and less in small chisels.

(*ii*) **Cross Cut Chisel:** As the name suggest cross cut chisel is used to cut grooves on large surface, before chipping by means of flat chisel. These are also mainly used for making key ways, slots etc. Cross cut chisel is comparatively narrow chisel and having it cutting at the end slightly broader.

This is to allow the body of chisel to clear when the groove is being cut.

(*iii*) **Diamond Point Chisel:** This chisel is used for cutting holes in plates such as boiler plate and for cutting *V*-grooves, chipping sharp corners etc.

(*iv*) **Round Nose Chisel:** It is used for cutting grooves such as oil grooves in bearing, for drawing the eccentric hole back to centre of the hole.

4.8 DRILLING TOOLS

Drilling and hole making tools are discussed in detail in Chapter 12 in Art. 12.12.3 and 12.18.

4.9 THREADING TOOLS

1. Taps

Taps are used for cutting internal thread in a circular hole. Tap may be assumed as a bolt with perfect thread cut on it. Tap cutting edges (threads) are screwed into a hole to cut internal threads. The cutting edge of tap are hardened to avoid wear and tear. It cuts internal thread in the hole which will fit an external thread and the same size. The shank of the tap is circular and edge is squared to accomodate tap in wrench.

The jaws of tap wrench firmly grips the tap end and tap is rotated by means of handle, which is curled to avoided slipping of hands during tapping.

Taps are usually made in sets of three to cut any threaded hole. These taps are known as taper, intermediate and bottoming (plug tap). It (second tap) has taper end which has 8 to 10 threads.

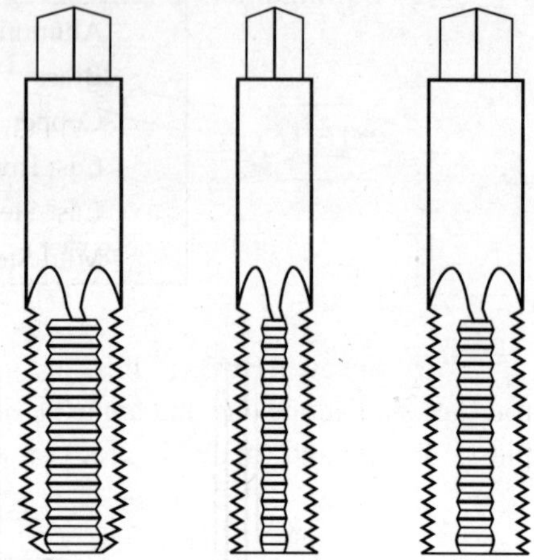

Fig. 4.37 Taps set.

Taper tap is first tap to be used during tapping. It has tapper end which has 8 to 10 threads. The taper end permits the tap to enter tap hole and to cut the threads gradually.

The second tap has only two to three threads chamfered and is used after taper tap. Second tap is also used to finish the threaded hole opened from both ends.

The plug tap is used as a final tap, when the blind hole is to be threaded up to the bottom of hole.

2. External Thread Cutting Dies (Die and Stock)

These tools are used for cutting external threads on the outer round surface of bars or lubes. It consist of nut having internal threads (cutting edges).

Sharpening of Cutting Edges: The cutting edges of the nut are hardened and sharpened. This nut is crewed to the bar on which external threads are to be cut. This die cut is carried and held in die stock. So that it could be easily screwed to the bar. The end of the rod or the bar should be slightly clamfered to allow the nut die to side properly over the bar.

One or two of the die threads may also be chamfered to facilitate the die nut to start the threading operation. The dies have various form. Most common type of die nuts are solid die nut and loose die nut.

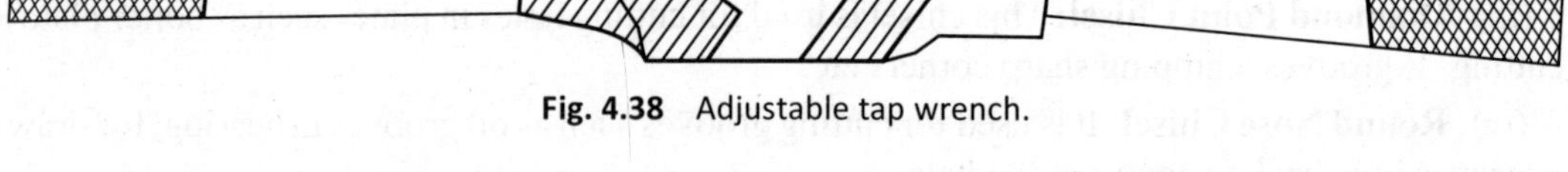

Fig. 4.38 Adjustable tap wrench.

4.10 MISLANIONS TOOLS

(*a*) **Plier:** Pliers are used to hold small jobs. Plier consist of two steel arms which are hinged as shown in figure 4.39. Each arm is having Jaw at one end. These are also used for cutting or bending thin wire.

The commonly used plier are:

(1) Round nose and slip nose pliers

(2) Culting pliers

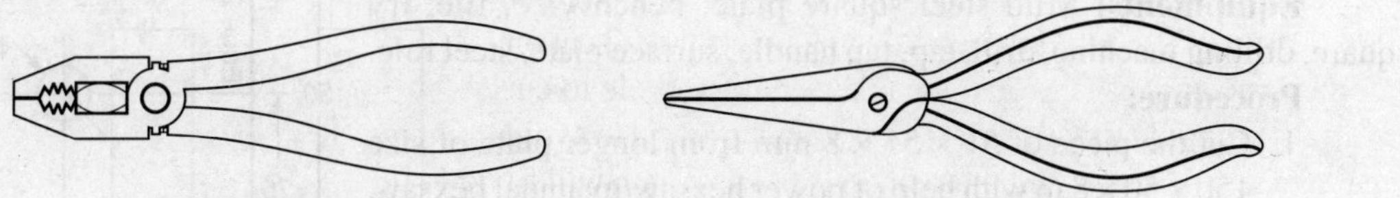

(a) Round nose plier (b) Slip nose plier

Fig. 4.39 Various types of nose pliers.

(*b*) **Screw drivers:** These are used to tighten different types of screws. It consists of a hard steel rod flattened at one end and a plastic or wooden handle at the other end. The screw driver are held with plastic handle and rotated in clockwise direction to tighten the screws.

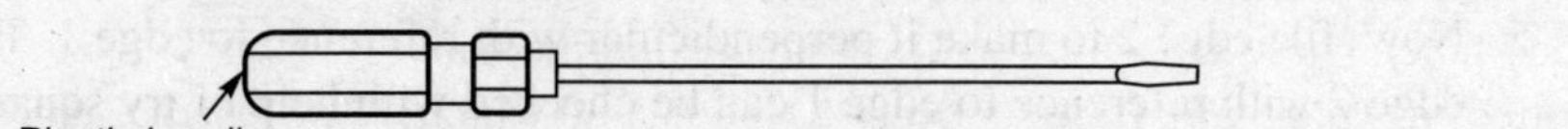

Fig. 4.40 Screw driver.

(*c*) **Spanners:** Spanner is also called wrench. These are used for tightening or loosening the nuts, bolts, screws etc. various types of spanner are shown in figure 4.41. All the spanners are non adjustable.

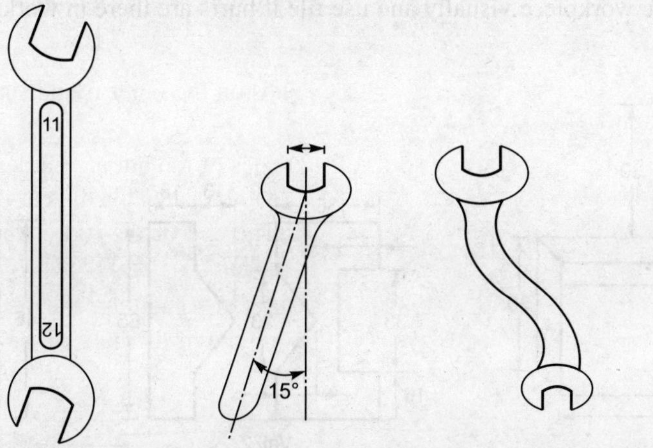

Fig. 4.41 Various types of spanners.

(*d*) **Allen keys:** These are used to tighten or loosen the screws or bolts with Allen heads. (*i.e.* hexagonal socket). These are L shaped tools and have hexagonal cross section. Allen keys are made of hard steel.

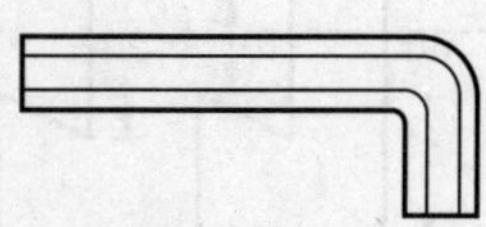

Fig. 4.42 Allen key.

4.11 FITTING SHOP EXERCISE

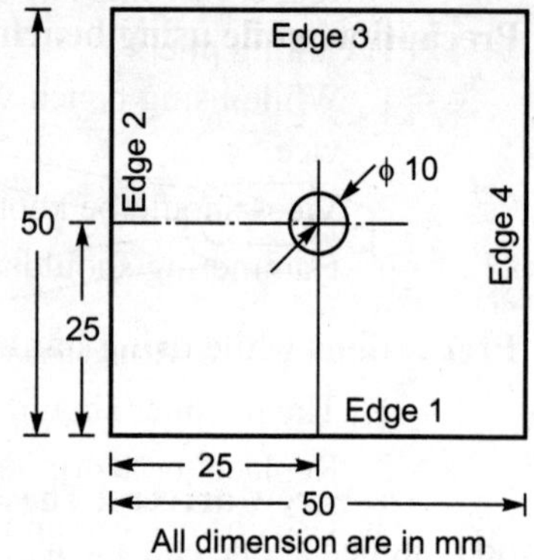

Fig. 4.43

Aim: To fabricate the Mild steel workpiece of 50 × 50 × 8 mm with a threaded hole as shown in Fig. 4.43.

Equipments: Mild steel square plate, bench vice, file, try square, drilling machine, drill, tap, tap handle, surface plate, steel rule.

Procedure:

1. Cut the piece of 51 × 51 × 8 mm from longer plate of size 150 × 50 × 8 m with help of power hexsaw/mannual hexsaw.
2. Punch 1, 2, 3, 4 on 4 edges of M.S. piece (51 × 51 × 8 mm). Also punch your roll number.
3. Hold the work piece in benchvice to file and make edge marked 1 flat.
4. The flatness of edge 1 after filing can be checked with help of try square.
5. Now file edge 2 to make it perpendicular with reference to edge 1. The perpendicularity of edge 2 with reference to edge 1 can be checked with help of try square.
6. Now mark the lines on Edge 3 and Edge 4 with help of scriber to maintain dimensions 50 × 50 × 8 mm.
7. File edge 3 and edge 4 to make the workpiece square.
8. Drill the hole of 10 mm in the MS plate by using bench drilling machine.
9. Perform the tapping operation by rotating the top manually in the hole. The tap should be held in tap handle for easy rotation of tap.
10. Check the workpiece visually and use file if burrs are there in workpiece.

Exercise Problems

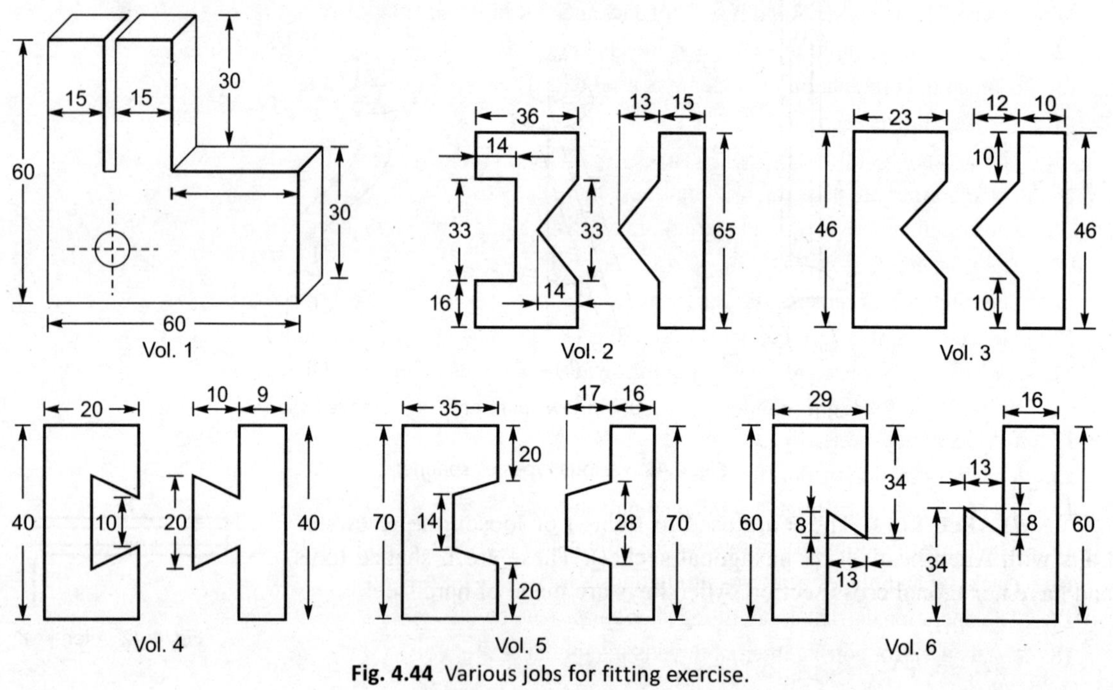

Fig. 4.44 Various jobs for fitting exercise.

4.12 SAFETY PRECAUTIONS WHILE WORKING IN FITTING SHOP

Precaution while using bench vice:

1. While using bench vice it should be ensured that working is properly tightened in Jaws of vice.
2. Vices should be kept clean and occassionally oiled.
3. Hammering should not be done when workpiece is clamped in vice.

Precautions while using hammer:

1. The hammer should be tightly fitted in its handle.
2. Replace the worn out handles immediately to avoid accidents.
3. Use correct size and type of hammer for a given job.

Precautions while using file:

1. Files should be tightly fitted in its handle. Files without handles should not be used.
2. File should be coated with machine oil to prevent rusting. This oil should be removed before using the file.
3. The files should be cleaned using file card or hard wire brush.
4. Files should not be used as hammer.
5. Filing should be done by holding the file properly to obtain a accurate flat surface.
6. Make sure that burrs are removed after filing.

FILL IN THE BLANKS

1. The bench vice is specified by ...... of Jaw and maximum opening between Jaws.
2. The vice used to hold pipe, circular jobs is known as ....
3. V block in combination of U clamp is used to ....
4. Surface plate is used for ....
5. The squareness of try square is tested by ....
6. The marking tool used for scribing lines at any angle is ....
7. The combination set is combination of five different instruments namely ....
8. The least count of vernier caliper = ...
9. The Pitch of micrometer is ....
10. The least count of micrometer is ....
11. The instruments used for measuring inside and outside diameter of circular part and for measuring length of bar are known as ....
12. Dividers are used for ....
13. The cross section of flat file is ....
14. The files used for filing very thin and delicate works are ....
15. The filing operation of removing very little material from relatively narrow space is known as ....
16. Hacksaw and blades perform cutting during ....
17. The operation carried out in fitting shop for producing different types and sizes of holes is known as ....
18. ...... is used for cutting internal threads in circular hole.

19. ...... are used for cutting external threads on bars or tubes.
20. The tool which is used to finish the hole to a size is known as ....
21. ...... is the strip of metal usually steel with one edge straight.
22. The punch is primarily a ...... tool in bench work and fitting.

ANSWERS

1. 'Width' 2. pipe vice
3. support, marking out the shaft and circular jobs.
4. listing the trueness and flatness of finished surface.
5. surface plate and angle plate.
6. Bevel protractor.
7. steel rule, centre head, level protractor, try square and bubble type spirit level.
8. $1 - \dfrac{\text{main scale division}}{\text{vernier scale division}}$.
9. The axial adavance of spindle after one complete revolution of thimble.
10. Pitch/no. of division on circular scale.
11. Calipers
12. transferring dimensions, measuring the distance between two points or lines, drawing circles or arc etc.
13. rectangular
14. needle files 15. draw filing 16. forward stroke
17. drilling operation 18. Tap 19. Thread cutting dies
20. reamer 21. Straight edge 22. marking.

VIVA VOCE QUESTIONS

1. What is function of bench vice?
2. Explain the use of *V* block.
3. What is the purpose of using surface plate and angle plate?
4. Describe the use of try square.
5. How level protector is used for scribing lines at any angle?
6. What is combination set?
7. What is meant by least count of measuring instrument?
8. Explain construction details of vernier caliper.
9. How the least count of vernier caliper is find out.
10. What is the use of vernier height gauge?
11. Describe constructional details of a micrometer.
12. What is meant by pitch of micrometer? How the least count of micrometer is find out?
13. Explain the method of reading a vernier caliper and micrometer.
14. What is the use of inside and outside caliper?
15. Explain the working of dial indicator.
16. What are the various types of files used in fitting shop?
17. Explain the difference between single cut and double cut files.

18. What are the various types of chisels?

19. What is meant by tapping operation?

20. Which instrument is used for cutting external threads on the cylindrical surface.

21. Which instrument is used for cutting internal threads on the internal surface of a cylindrical hole.

22. What is the difference between tap and die.

23. Draw the neat sketch of tap and name its various parts.

24. Explain briefly following tools with neat sketches:

(*i*) *V* block

(*ii*) Bench vice

(*iii*) Calipers

(*iv*) Universal surface gauge

(*v*) Types of files

(*vi*) Various types of chisels

(*vii*) Vernier height gauge.

> *"The greatest mistake you can make in this life is to continually fearing you will make one".*

CHAPTER 5

HOT WORKING PROCESSES (SMITHY AND FORGING SHOP)

INSIDE THIS CHAPTER

5.1 FORGING

Forging is the oldest metal working process known to mankind. In this process material is squeezed between two or more dies to deform its shape and size in such a way that required final shape is obtained. The force applied may be continuous squeezing force or impact blow or combination of both. Dies may be open or closed depending upon the situation. Forging is always considered as hot working unless specified. Raw material used for forging is billet or bar stock which have been previously rolled. When forgings are sufficiently large ingots may be used without rolling.

Grain in Flow in Forging Operations

In forging and most other mechanical working processes metal flows similar to the flow of putty while being squeezed. In forging operation it is possible to closely control the grain flow in specific direction. Strength and toughness of the metal is maximum in the direction of grain flow. The direction of grain flow plays an important role when the parts are subjected to impact or fatigue loading.

Fig. 5.1 (*a*) and Fig. 5.1 (*b*) present desirable grain structure in gear blanks and crane hook respectively. The gear blank in Fig. 5.1 (*a*) is obtained by upsetting operation. The grain flow lines are in radial direction, which is essential for good strength in gear for severe applications. Fig. 5.1 (*a*)

"Success often comes to those who dare and act; it seldom goes to timid who are ever afraid of the consequences."

shows incorrect method for forging the piece to produce a gear blank. As the grain flow lines are not in desired direction so the gear blank will be too weak.

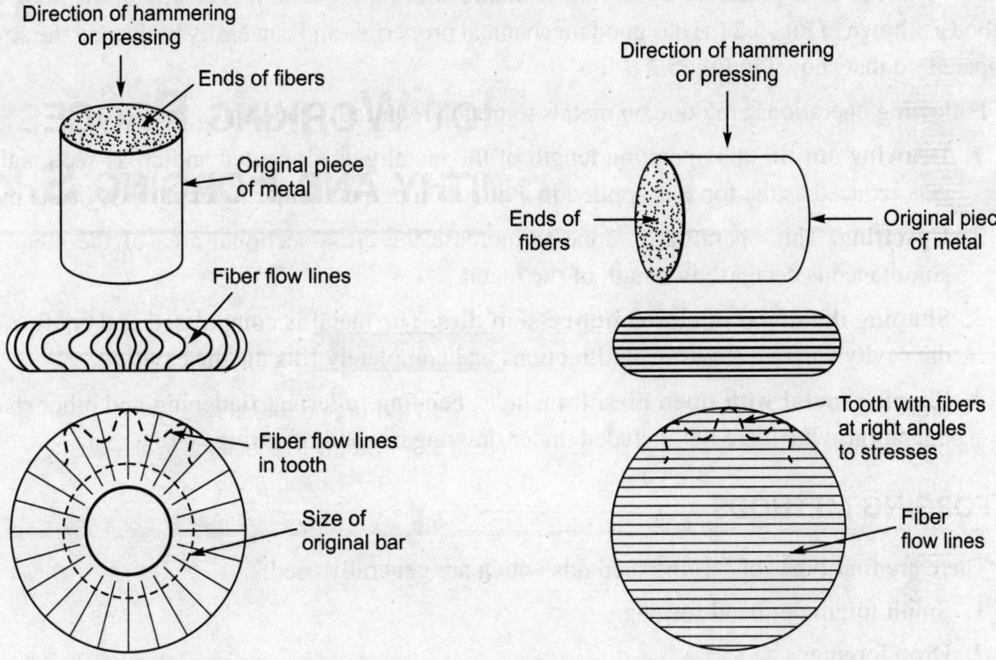

(a) Forging a gear blank by upsetting operation

(b) Incorrect forging of gear blank

Fig. 5.1 Upsetting operation.

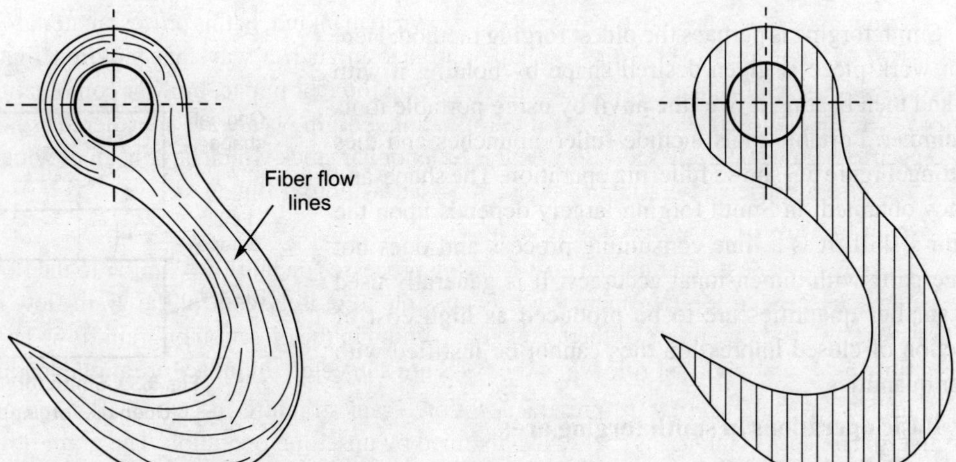

(a) Crane hook produced by bending and drawing out operation. Fiber flow line are bent along the hook to provide better strength to crane hook

(b) Crane hook produced by other manufacturing processes

Fig. 5.2 Drawing out operation.

The crane hook shown in Fig. 5.2 (*a*) is obtained by the bending and drawing out. As a result grain flow is also bent along the hook and thus provide the necessary strength for lifting the loads. In Fig. 5.2 (*b*) crane hook is produced by any other manufacturing process. It is clear that the hook made by method as shown in Fig. 5.2 (*a*) has good mechanical properties and can easily withstand the stresses as compared to that shown in Fig. 5.2 (*b*).

Following operations are done on metals to obtain required shape and size.

1. **Drawing out.** In this operation length of the metal gets elongated and cross-sectional area gets reduced as the force is applied in a direction perpendicular to length axis of the metal.
2. **Upsetting.** This operation is done to increase the cross-sectional area of the metal with simultaneous decrease in length of the metal.
3. **Shaping the metal in closed impression dies.** The metal is compressed and confined in a die cavity where it flows in all directions and completely fills the die cavity.
4. **Shaping metal with open dies.** It includes bending fullering, flattening and other shaping operations which are not included under drawing out and upsetting.

5.2 FORGING METHODS

There are four types of forging methods which are generally used:

1. Smith forging or hand forging
2. Drop forging
3. Press forging
4. Machine forging or upset forging.

5.2.1 Smith Forging

Smith forging is perhaps the oldest forging method. Here the hot work piece is given desired shape by holding it with tongs and then beating it over the anvil by using portable tools and hammer. Portable tools include fullers, punches and dies for cutting. Figure 5.3 shows fullering operation. The shape and accuracy obtained in Smith forging largely depends upon the operator's skill. It is a time consuming process and does not produce parts with dimensional accuracy. It is generally used when smaller quantities are to be produced as high cost of production of closed impression dies cannot be justified with smaller quantities.

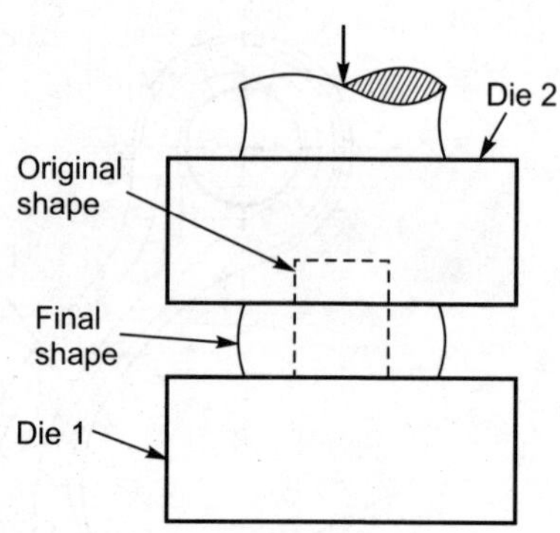

Fig. 5.3 Smith forging (Open die forging).

Basic operations in smith forging are:

(*a*) **Fullering.** In fullering operation, material cross-section is decreased and length is increased. The fullering operation is carried out with a help of a pair of fullers and anvil. Bottom fuller is kept in the anvil hole and heated work piece is kept in between bottom and top fuller. The fullers locate and concentrate the hammer flow, causing the metal to be drawn out more rapidly. Fullering operation is shown in Fig. 5.4.

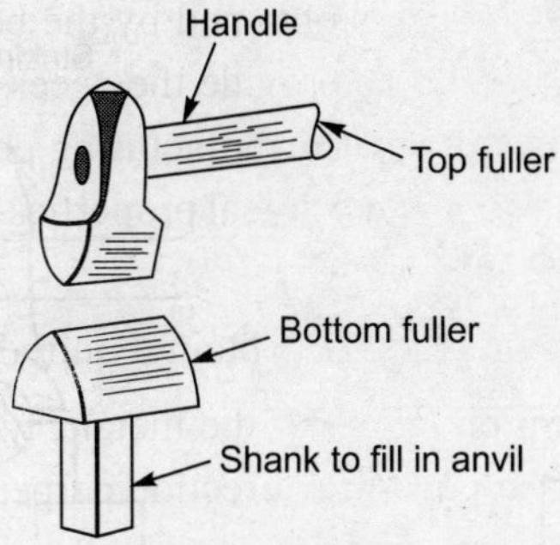

(a) Pair of fullers

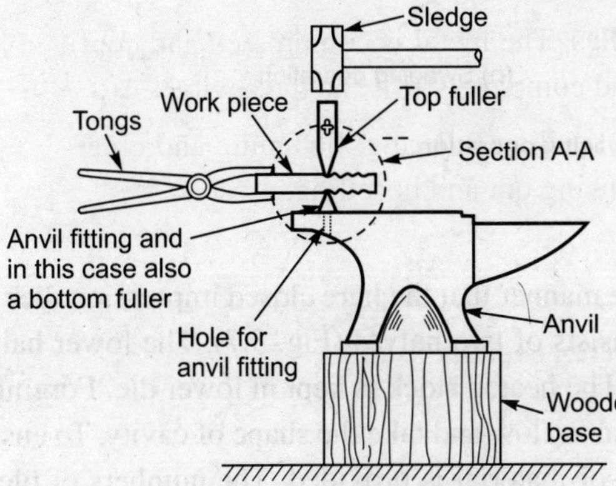

(b) Fullering operation

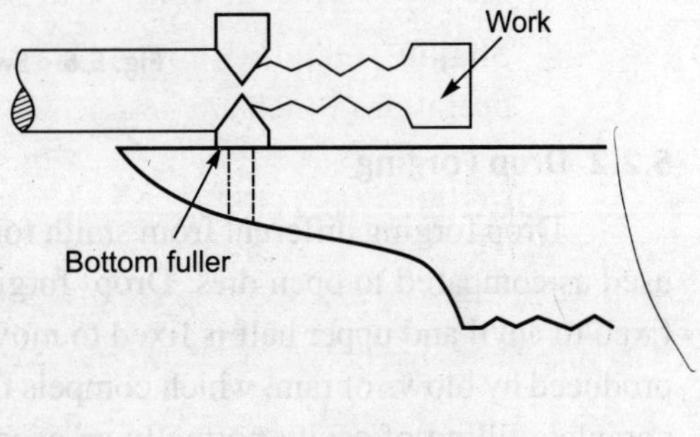

(c) Enlarged view A-A

Fig. 5.4 Fullers and fullering operation.

(*b*) **Flattening.** The flattening operation is carried out after drawing out, fullering or any other operation to remove the fullering marks for levelling and finishing flat surface. Figure 5.5 shows a flattening tool and flattening operation.

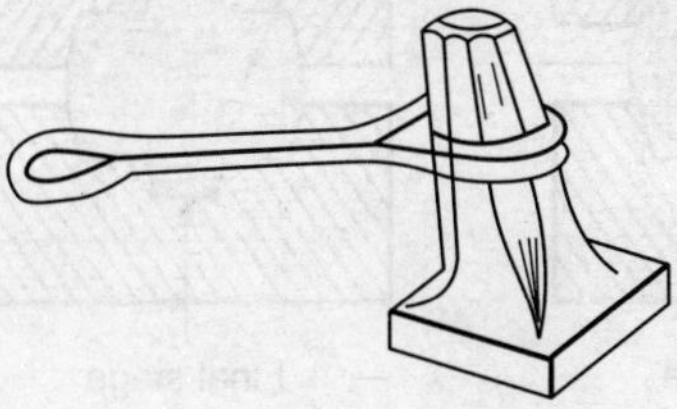

(a) Flatter

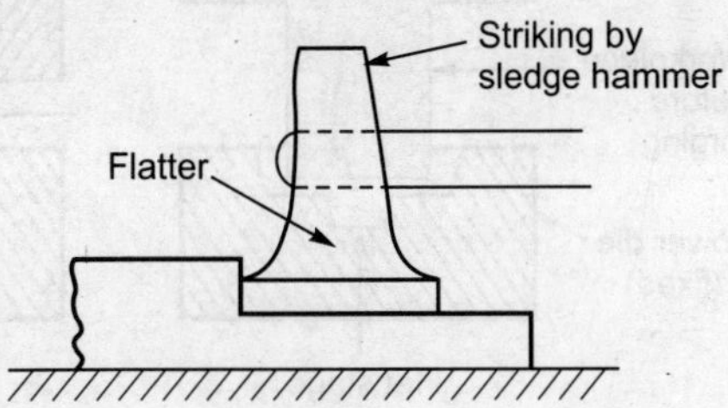

(b) Flattening

Fig. 5.5 Flatter and flattening operation.

(*c*) **Swaging.** Swaging is the operation of changing the shape of cross-sections of jobs to some specific shapes such as round, square, hexagonal etc. (Fig. 5.6). The swaging operation may be carried out with a pair of swages or by swage block.

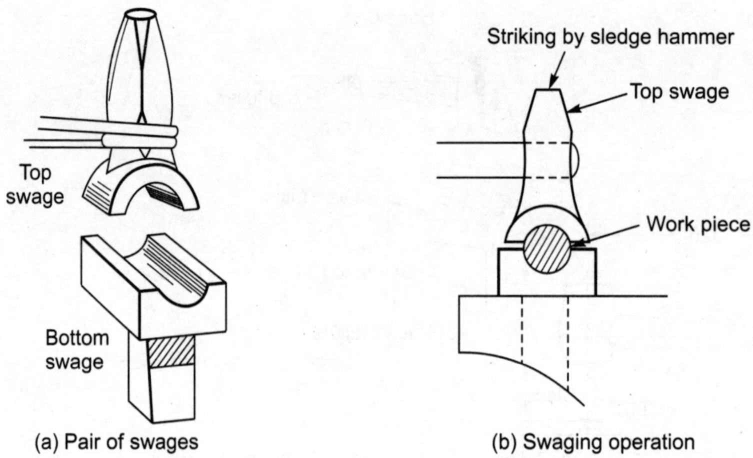

Fig. 5.6 Swages and swaging operation.

5.2.2 Drop Forging

Drop forging different from smith forging in the manner that the here closed impression dies are used as compared to open dies. Drop forging die consists of two halves (Fig. 5.7). The lower half is fixed to anvil and upper half is fixed to movable ram. The heated stock is kept in lower die. Forging is produced by blows of ram, which compels the hot metal to flow and take the shape of cavity. To ensure complete filling of cavity normally an excess amount of material is provided. The numbers of blows required varies according to the shape and size of part, forging qualities of metal and tolerances required. When the product geometry is complicated, a set of dies may be necessary to obtain the final shape.

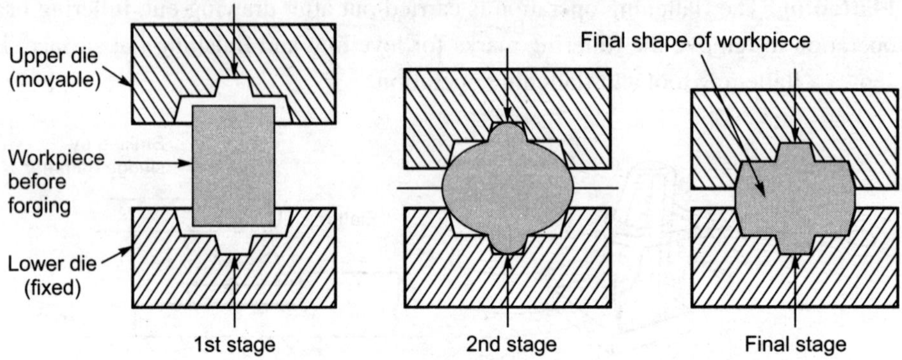

Fig. 5.7 Closed impression dies.

The typical application of drop forging includes crank, crank shaft, connecting rod, wrench, crane hook etc. Fig. 5.8 illustrates the multiple forging dies and various forging sequences like fullering, edging, bending, blocking and finishing to produce the connecting rod.

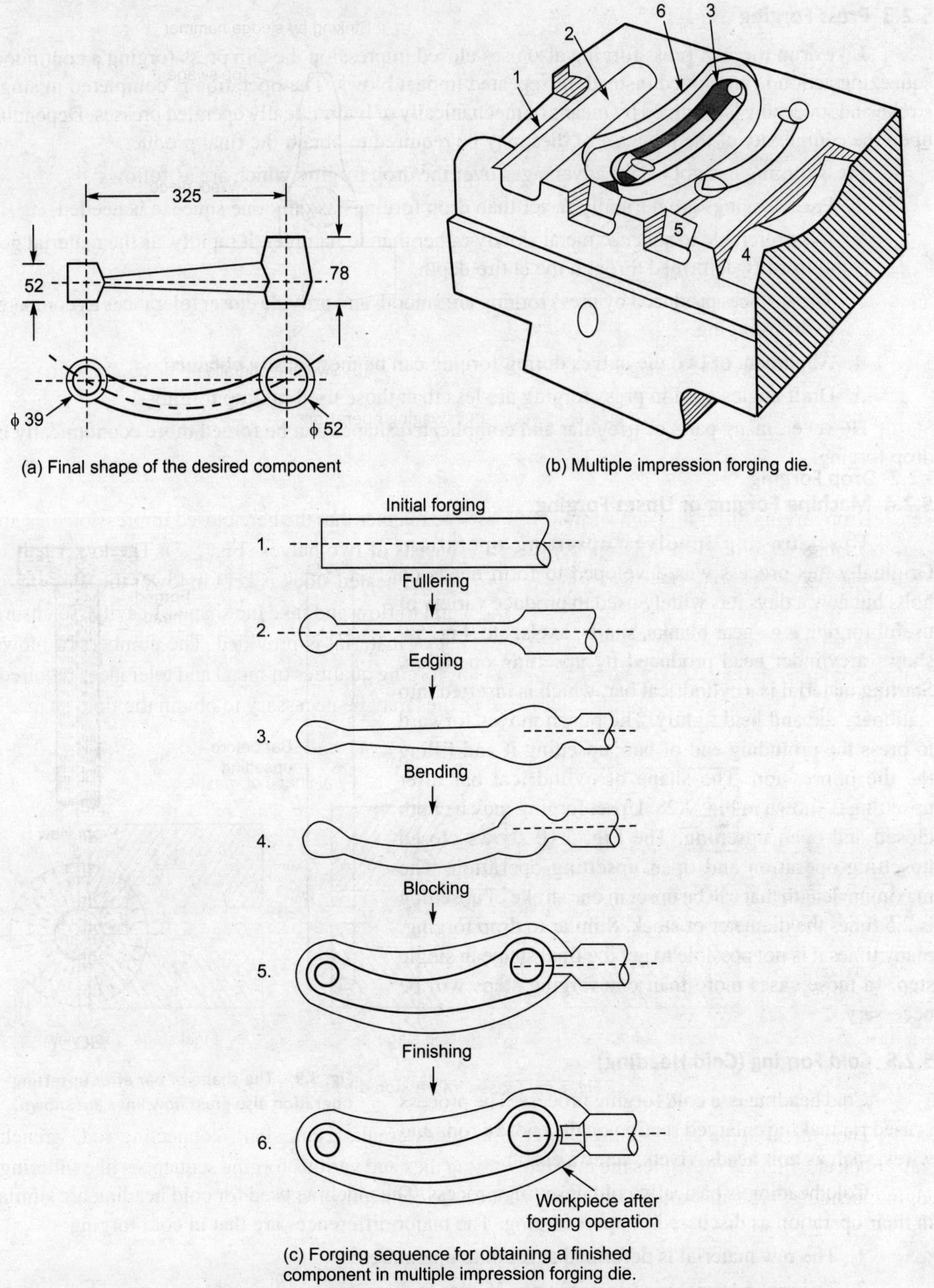

(a) Final shape of the desired component

(b) Multiple impression forging die.

(c) Forging sequence for obtaining a finished component in multiple impression forging die.

Fig. 5.8 Drop forging of a connecting rod in a closed impression die.

5.2.3 Press Forging

Like drop forging, press forging also uses closed impression dies. In press forging a continuous squeezing action is employed instead of repeated impact blows. The operation is completed in single stroke and squeezing is obtained by means of mechanically or hydraulically operated presses. Depending upon the complexity of the job a set of dies may be required to obtain the final product.

Press forging has got cetain advantages over the drop forging which are as follows:

1. Press forgings are normally faster than drop forgings as only one squeeze is needed.
2. It is preferable to squeeze metal slowly rather than to hammer it rapidly, as the material gets uniformly deformed through the entire depth.
3. The surfaces produced by press forging are smooth and provide closer tolerances as compared to drop forging.
4. Alignment of two die halves during forging can be more easily obtained.
5. Draft angles used in press forging are less than those used in drop forging.

However, many parts or irregular and complicated shapes can be forged more economically by drop forging.

5.2.4 Machine Forging or Upset Forging

Upset forging involves upsetting operation. Originally this process was developed to form heads on bolts but now a days it is widely used to produce variety of useful forging *e.g.*, gear blanks, shafts, axels, etc. Fig. 5.9 shows a cylinder head produced by upsetting operation. Starting material is a cylindrical bar, which is inserted into stationary die and held tightly. The punch moves forward to press the protuding end of bar upsetting it and filling the die impression. The shape of cylindrical bar after upsetting is shown in Fig. 7.28. Upset forging may be both closed and open upsetting. The Fig. 5.10 shows closed upsetting operation and open upsetting operation. The maximum length that can be upset in one stroke of upsetting is 2.5 times the diameter of stock. Similar to drop forging, many times it is not possible to get the final shape in single step. In those cases more than one forging steps will be necessary.

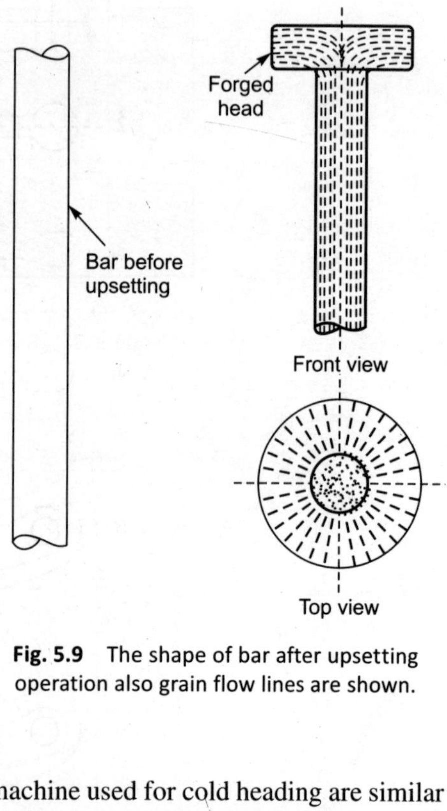

Fig. 5.9 The shape of bar after upsetting operation also grain flow lines are shown.

5.2.5 Cold Forging (Cold Heading)

Cold heading is a cold forging process. The process is used for making enlarged section on the ends or rods and wires, such as bolt heads, rivets, screws etc.

Cold heading is basically cold upsetting process. The machine used for cold heading are similar in their operation as discussed in upset forging. The major differences are that in cold forging:

1. The raw material is deformed at room temperature.
2. The raw material used in wire upto 25 mm in diameter.

3. It is performed on automatic high speed machines. Where number of operations like cutting stock to size, shank entrusion, heading, trimming and threading are performed simultaneously to produce finished component. Due to relatively high tooling cost process is usually limited to mass production.

The advantages of cold headed parts are:

1. Cold headed metals have excellent mechanical properties.
2. Surface finish produced is excellent.
3. There is no wastage of metal because no metal is discarded in the form of chips.

Limitation of cold headed parts:

1. The shape produced are limited.
2. The amount of metal which can be upset is limited.

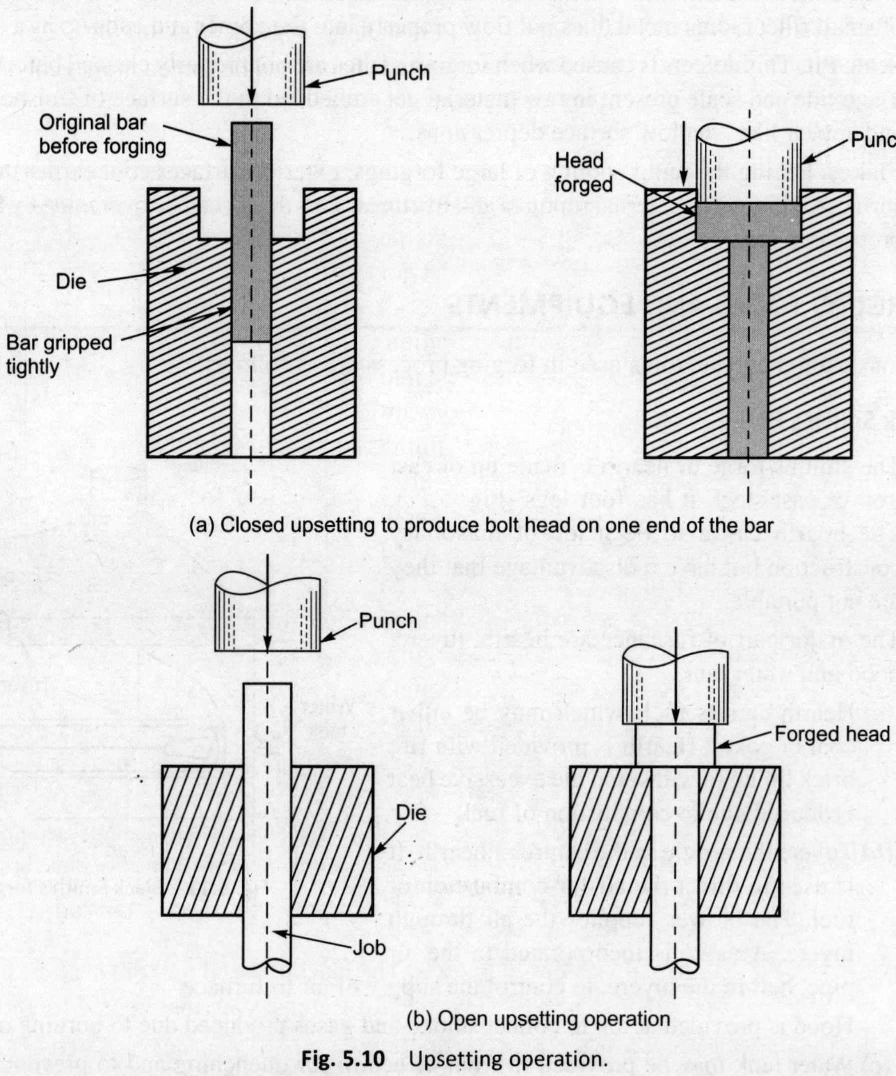

Fig. 5.10 Upsetting operation.

5.3 FORGING DEFECTS

Defects in forging arise due to fault in metal, incorrect die design, improper forging operation or improper heating. To reduce the defects at early stage samples of incoming metal should be tested and inspected for any metallurgical defects before accepting the lot. Following are some of the forging defects:

1. **Unfilled Section.** When some of the section of closed impression dies are not filled with flowing metal then this defect is caused. The causes of this defects may be incorrect die design, improper forging design, lack of heating or misplacing the metal in die.
2. **Mismatched Die Section.** When upper and lower die halves are not aligned then resulting forging sections will have misalignment. This defect can be rectified by modifying die design or by alignment of two die halves.
3. **Cold Shut.** Cold shut appears usually at corners and right angles to the surface. They appear like cracks and are caused when corner and fillet radii are small (improper die design). Because of small fillet radius metal does not flow properly into the corner and ends up as a cold shut.
4. **Scale Pit.** This defects is caused when forging or dies are not properly cleaned before forging. The oxide and scale present in raw material get embedded in the surface of finished forging and appear like shallow surface depressions.
5. **Flakes.** During the rapid cooling of large forgings, exterior surfaces cool earlier than initial surfaces. This causes internal ruptures and fractures. This defect can be overcome by following proper cooling practice.

5.4 FORGING TOOLS AND EQUIPMENTS

Various equipment and tools used in forging process are as follows:

5.4.1 Black Smith's Forge

- The smith's forge or hearth is made up of cast iron or cast steel. It has four legs (Fig. 5.11). The hearth can also be made of masonary construction but have a disadvantage that they are not portable.
- The major part of furnance are hearth, tuyere, hood and water tank.

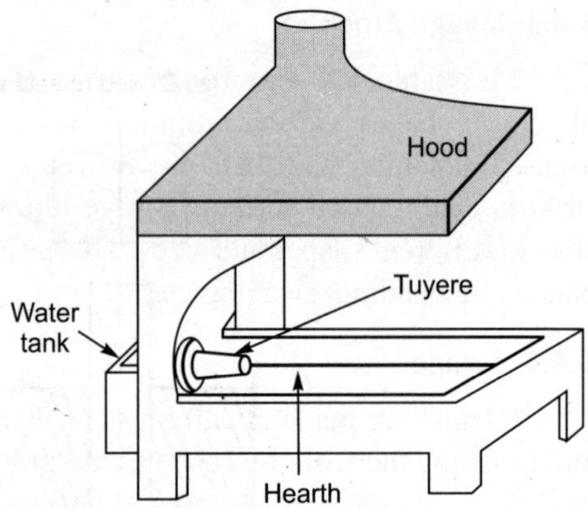

Fig. 5.11 Black Smith's forge.

(*a*) Hearth carries fuel (which may be either coal or coke). Hearth is provided with fire brick lining to withstand the excessive heat produced due to combustion of fuel.

(*b*) Tuyere the nozzle in the centre of hearth. It is used to direct the air for combustion of fuel. The blower supplies the air through tuyere. A valve is incorporated in the air pipe, before the tuyere, to control the supply of air to furnace.

Hood is provided at top to collect smoke and gases produced due to burning of fuel.

(*c*) Water tank may be provided in front of hearth for quenching and to prevent fire from spreading.

5.4.2 Anvil

It is a block made of cast iron or steel on which forging work is carried out. It supports the work piece and is capable of withstanding heavy blows of hammer. Apart from supporting workpiece, its shape helps in doing other forging operations. The thickness of Anvil is 20 to 25 mm and weights approximately 150 kg.

The main parts of anvil are:

Horn or Beak. It is used in bending the metal or forming curve shape.

Chipping Block. It is the flat step provided between horn and anvil face. It may be used for supporting metal when cutting through chisel.

Face. It is the flat surface provided on the top of horn (or beak).

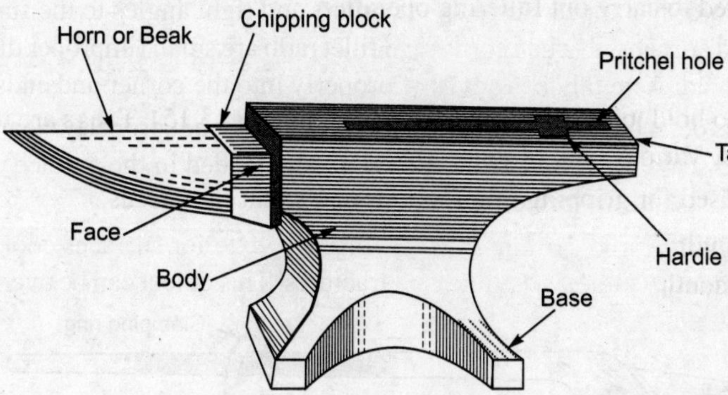

Fig. 5.12 Anvil.

Tail. The projected flat surface at the end of anvil is known as tail. It carries a square hole and round hole to accommodate various jobs and hand tools like fullers, swages etc.

5.4.3 Swage Block

It is a block of cast iron or cast steel carrying number of holes of various shapes and size from top to bottom face. These holes are used in punching and their use prevents punch from spoiling by striking against hard surface. Swage block also carries number of slots of different shapes and sizes. Those slots are useful for holding bars while bending, swaging etc.

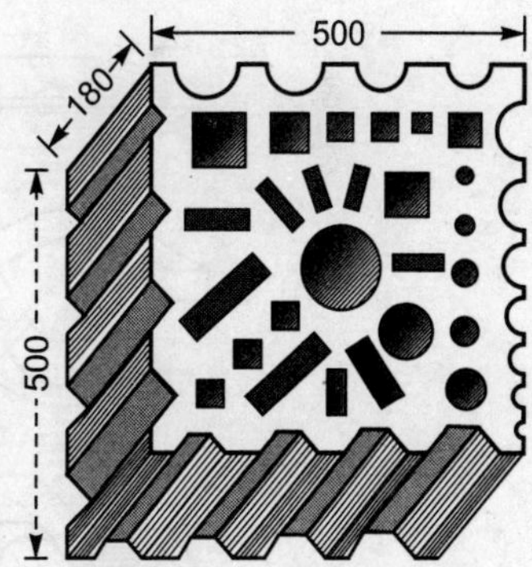

Fig. 5.13 Swage block.

5.4.4 Swages

These are made of carbon steel. These are used for reducing and finishing the work pieces to round or hexagonal form. They are made in two halves called top and bottom swages. The two halves may be separate or connected by a strip of spring steel (Fig. 5.16). Connected swages can be hold by single handed.

5.4.5 Flatters

They are used for finishing flat surfaces. (Fig. 5.15) The flatters may also be used after fulling or drawing to remove fullering marks. These are made with perfectly flat face of about 75 mm square (or round).

5.4.6 Set Hammer

Set hammer works like a flatter but it is a smaller tool, (Fig. 5.14). It does not have enlarged bottom face (like flatter) and is used for finishing in corners and confined spaces. The job may be supported on the anvil.

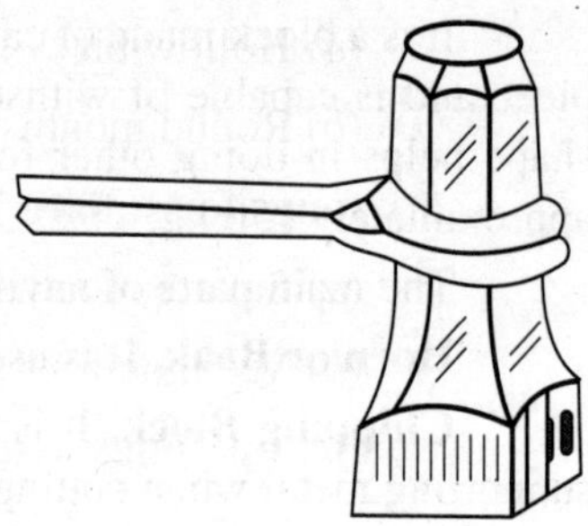

Fig. 5.14 Set hammer.

5.4.7 Fullers

Fullers are used in necking down a work piece. They are made in two halves namely bottom and top fuller. The bottom fuller is held in anvil whereas top fuller is held by means of handles or by tongs (Fig. 7.23). The workpiece is held between top and bottom fuller and then hammered to carry out fullering operation.

5.4.8 Tongs

Tongs are used to hold job during forging operations (Fig. 5.15). Tongs are usually named after the inside shape of jaws. Various type of tongs are:

(*i*) **Flat Tong.** Used for gripping thin sections and small flat pieces.

(*a*) Open mouth

(*b*) Closed mouth.

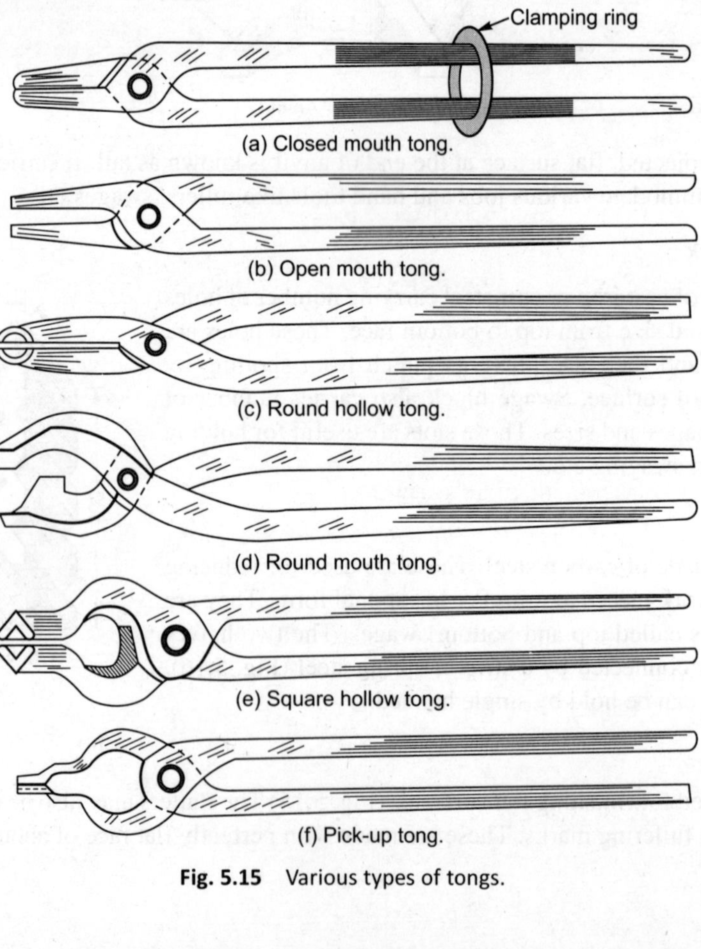

(a) Closed mouth tong.

(b) Open mouth tong.

(c) Round hollow tong.

(d) Round mouth tong.

(e) Square hollow tong.

(f) Pick-up tong.

Fig. 5.15 Various types of tongs.

(*ii*) **Tongs for round jobs.** Used for holding round work pieces.

(*a*) Hollow bit

(*b*) Round mouth

(*iii*) **Tongs for square jobs.** Used for holding square or hexagonal job.

(*a*) Square mouth

(*b*) Vee mouth

(*iv*) **Pickup Tongs.** Used for picking up jobs. Jaws are designed to pickup even small section.

5.4.9 Chisel

These are used for cutting metals and for making a notch before breaking. Chisels may be classified as hot chisel and cold chisels.

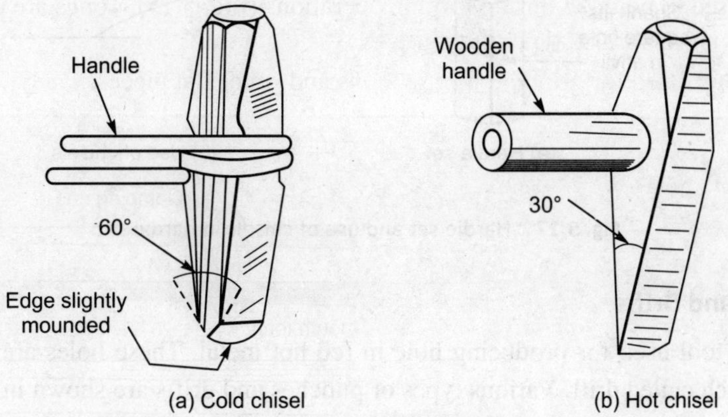

Fig. 5.16 Cold and hot chisel.

Hot Chisel. As name suggest, these are used for cutting hot metals. Included angle for hot chisel is 60°. The edge of chisel are slightly rounded for better cutting action. Hardening of these chisel is not necessary because the hot metal would re-soften it.

Cold Chisel. These are made up of carbon steel. The lip angle of cold chisel is 60° and it is edge hardened and tampered to provide more strength.

5.4.10 Hardie Set

When chisel are used in pair these are called hardie set (Fig. 5.17). Pair consists of top tool which is a chisel and bottom tool, which is known as hardie. The hardie has a square shank which fits in the square hole of anvil face. The top tool (chisel) is held by black smith. Chisel may be fitted with a handle or metal wire to hold.

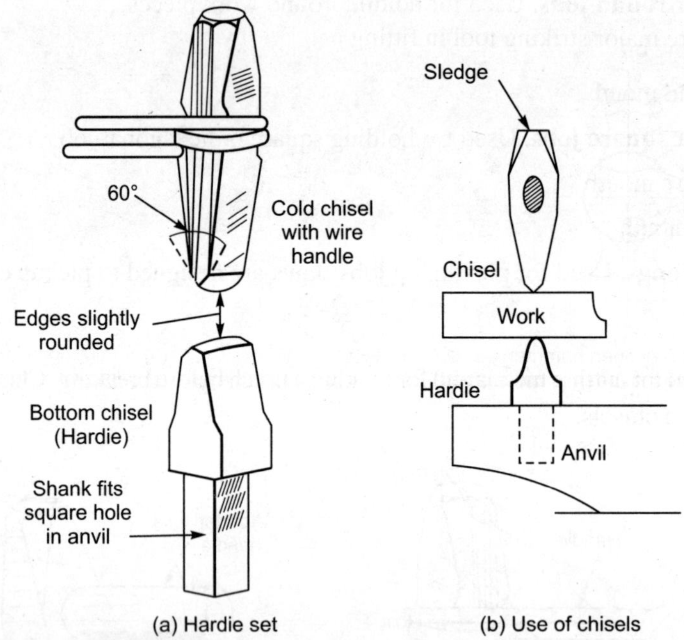

Fig. 5.17 Hardie set and use of chisels in hardie set.

5.4.11 Punches and Drifts

Punch is a tool used for producing hole in red hot metal. These holes are enlarged through a larger tapered punch called drift. Various types of punches and drifts are shown in Fig. 5.18.

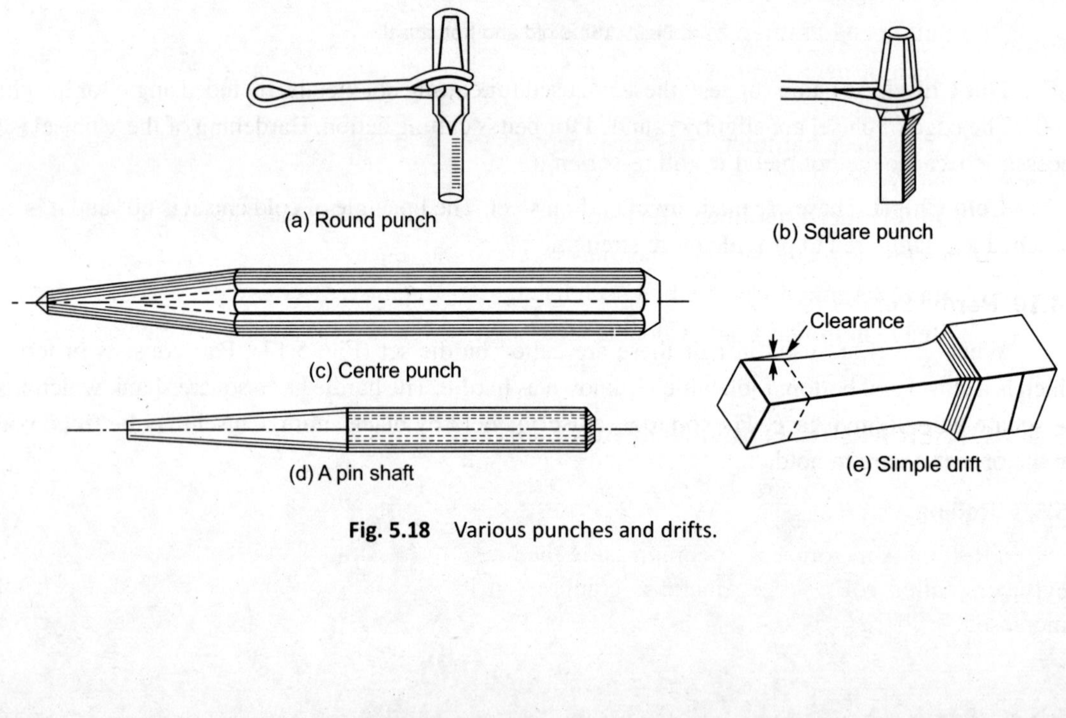

Fig. 5.18 Various punches and drifts.

5.4.12 Hammers

Hammers are major striking tool in fitting and smithy shop (Fig. 5.19).

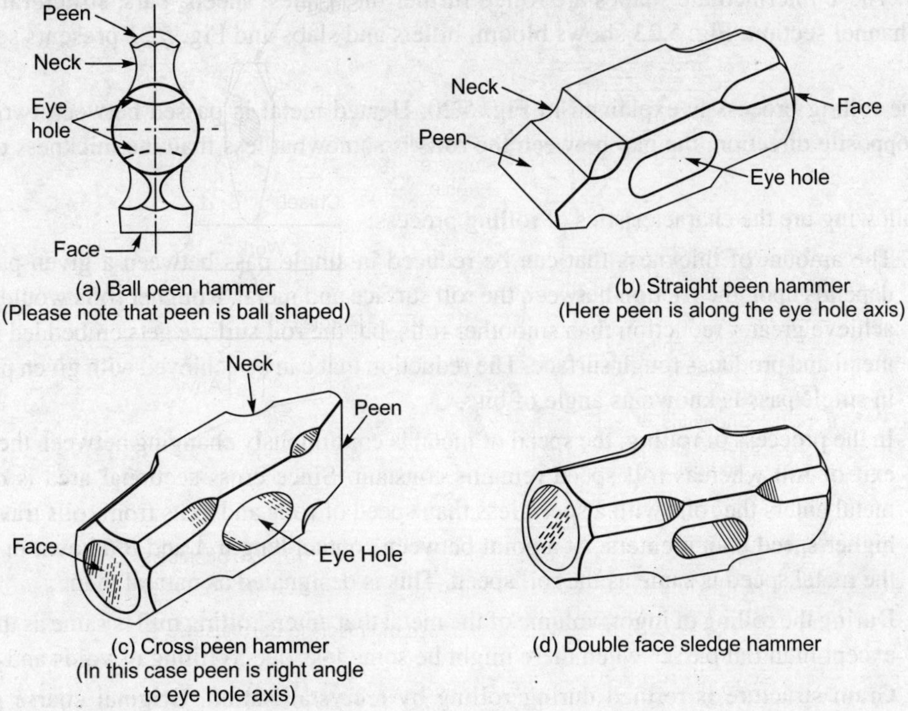

Fig. 5.19 Various types of hammer.

Hammer consists of a heavy iron body and wooden handle. The parts of hammer are peen, eye, neck and face. Depending upon the shape of peen hammers may be classified as

(*i*) **Ball peen hammer.** Shape of peen is hemi-spherical ball.

(*ii*) **Straight peen hammer.** The peen is parallel to the axis of hammer (which passes through the eye hole).

(*iii*) **Cross-peen hammer.** The peen is perpendicular to the axis of hammer (which passes through the eye hole).

(*iv*) **Double faced hammer.** The hammer has no peen formation and has flat faces at both ends.

Depending upon the size hammer may be classified as hand:

(*i*) **Hand hammer.** Usually ball peen hammer of weight 1.0 to 2 kg.

(*ii*) **Sledge hammer.** Usually double face hammer weight 3 to 8 kg. It is used when heavy blows are needed.

5.5 OTHER METAL FORMING PROCESSES

5.5.1 Rolling

Rolling is the process of compressing the metal by passing it between two or more revolving cylinders called **rolls.** As the metal is compressed its cross-section area is reduced and length is increased.

Rolling is normally a hot working process unless specifically mentioned as cold working. The starting material is the moulded ingot which are rolled into intermediate shapes like blooms, billets and slabs. These intermediate shapes are rolled further into plates, sheets, bars, structural shapes, I, L, T, or channel section. Fig. 5.23 shows bloom, billets and slabs and Fig. 5.22 presents some rolled shapes.

The rolling process is explained in Fig. 5.20. Heated metal is passed between two rolls that rotate in opposite direction, the gap between the rolls is somewhat less than the thickness of metal at entrance.

Following are the characteristics of rolling process:

1. The amount of thickness that can be reduced in single pass between a given pair of rolls depends upon the friction between the roll surface and metal. Rougher rolls would be able to achieve greater reduction than smoother rolls, but the roll surface gets embedded into rolled metal and produces rough surface. The reduction that can be achieved with given pair of rolls in single pass is known as angle of bite.
2. In the proccess of rolling, the speed of metal is continuously changing between the entry and exit of roll whereas roll speed remains constant. Since cross-sectional area is decreasing metal enters the rolls with a speed less than speed of rolls and exits from rolls travelling at a higher speed than it enters. At a point between contact length *A* and *B* (shown in Fig. 5.20) the metal speed is same as the roll speed. This is designated as natural plane.
3. During the rolling of ingot, volume of the metal that enters rolling mill is same as that leaving except in initial passes when there might be some loss due to filling of voids and cavities.
4. Grain structure is refined during rolling by recrystallisation. Original coarse grains are elongated by rolling action. Because of high temperature recrystallisation starts and small grains begin to form. (Fig. 5.21)

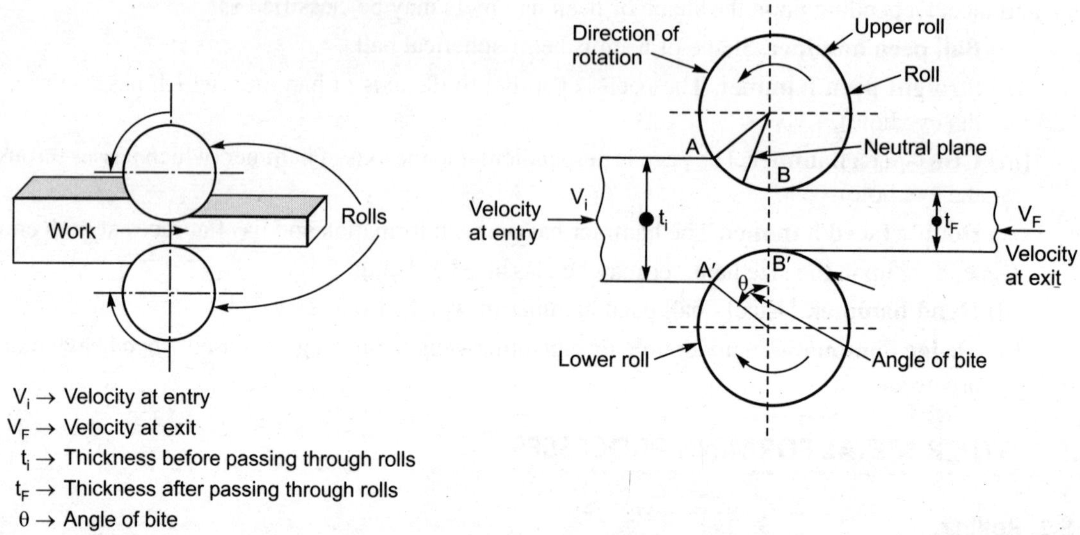

Fig. 5.20 Rolling process.

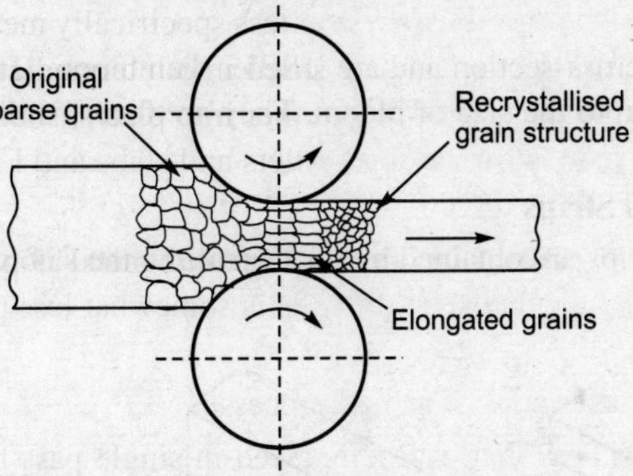

Fig. 5.21 Hot rolling process showing deformation and recrystallisation of metal being rolled.

5.5.1.1 Terminology Associated with Rolled Products

(1) Ingot

This is the raw material used in rolling. This is obtained by pouring molten metal into ingot moulds having standard dimension. The ingot is rolled in intermediate shapes like slabs, billets and blooms.

(2) Slab

The standard dimensions of slab are 50 to 150 mm thickness and 600 to 1500 mm width.

(3) Bloom

Blooms have square cross-section with minimum 150 × 150 mm. Blooms are further rolled in structural shapes such as I-beam, rail road rails etc.

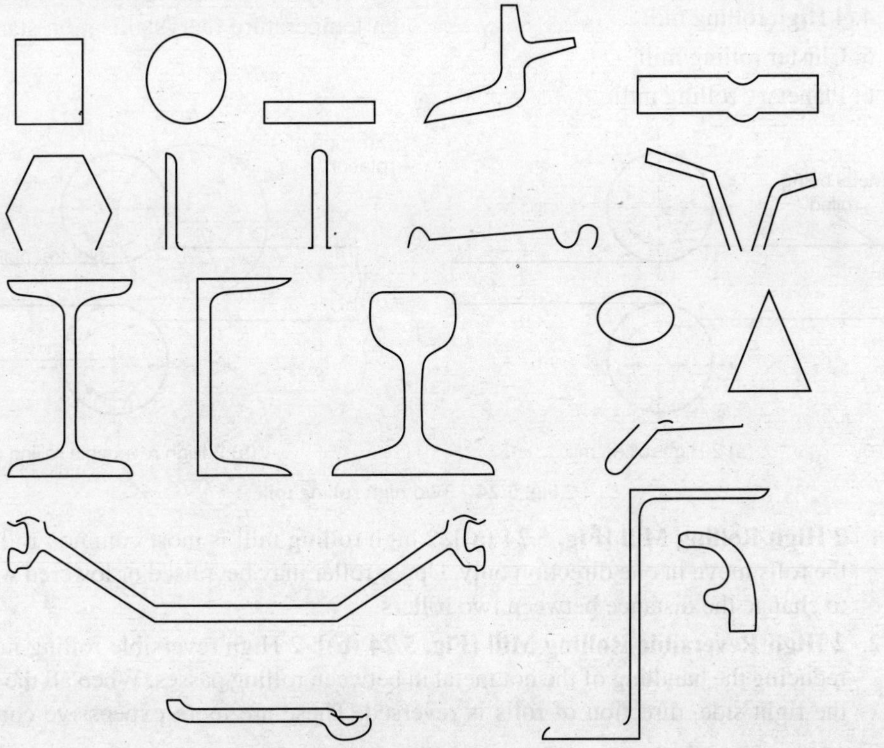

Fig. 5.22 Various rolled products.

(4) Billet

Billets have square cross-section and are smaller than bloom. They can have any cross-section of from (38.1 × 38.1 mm) upto the size of bloom. They are further rolled into shapes such as rounds, rods or bars.

(5) Plate Sheets and Strips

Plates, sheets and strips are obtained by further rolling the slab with flat rolls.

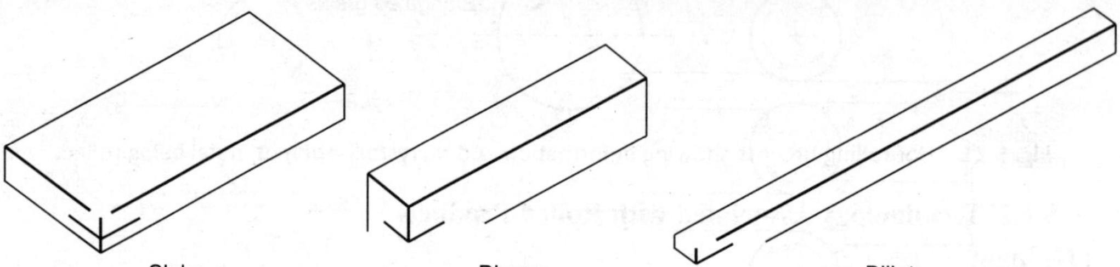

Fig. 5.23 Various starting material for obtained different cross-sections and structural shapes.

5.5.1.2 Rolling Stand Arrangements

The arrangement and number of rolls in a rolling mill depends upon the applications. Various rolling stand arrangements are shown in Fig. 5.24. The various rolling stand arrangements are:

1. 2 High rolling mill
2. 2 High reversible rolling mill
3. 3 High rolling mill
4. 4 High rolling mill
5. Cluster rolling mill
6. Planetary rolling mill.

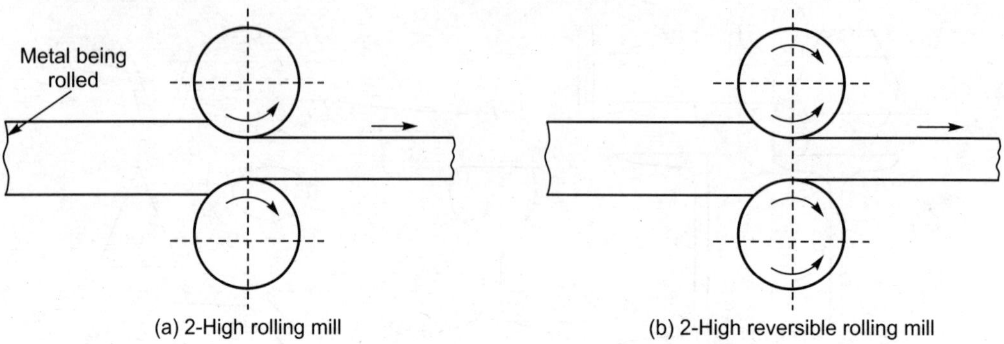

Fig. 5.24 Two high rolling mill.

1. **2 High Rolling Mill [Fig. 5.24 (a)].** 2 high rolling mill is most common rolling mill. Here the rolls move in one direction only. Upper roller may be raised or lowered when necessary to change the distance between two rollers.
2. **2 High Reversible Rolling Mill [Fig. 5.24 (b)].** 2 High reversible rolling mill is useful in reducing the handling of the hot metal in between rolling passes. When all the metal reach to the right side, direction of rolls is reversed. These are more expenssive compared to non

reversible type as reversible drive is needed. Two high reversible rolling mills are often used for first rolling of ingot.

3. **3 High Rolling Mill [Fig. 5.24].** 3 High rolling eliminates the need for reversing the rolls. After the metal has passed through bottom rolls end of metal is entered into upper roll set for next pass. If a job is sufficiently long and flexible a three high rolling mill can be provided with some arrangement for feeding the second pass even before first pass is completed.

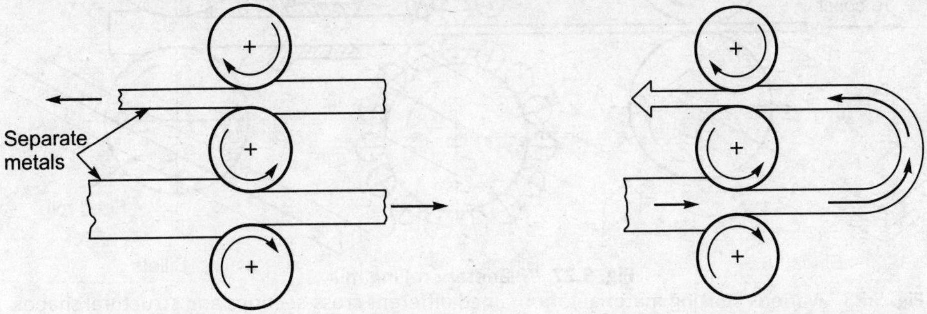

(a) Three high roll mill with separate metals (b) Three high rolling mill with same metal

Fig. 5.25 Three high rolling mill.

4. **4 High Rolling Mill [Fig. 5.26 (a)].** A 4 high rolling mill is essentially a 2 high rolling mill but with small sized rolls. The other two rolls are back rolls which provide the necessary rigidity to small roll and reduce the roll deflection.

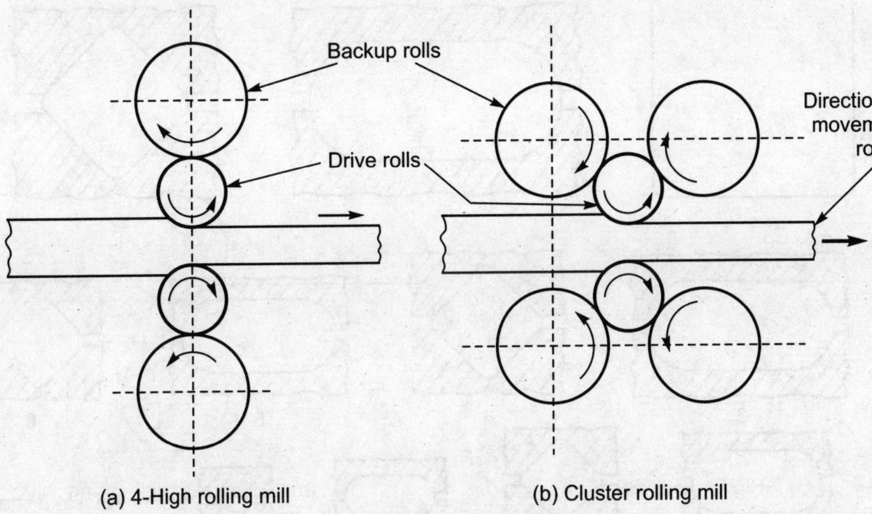

(a) 4-High rolling mill (b) Cluster rolling mill

Fig. 5.26 Use of backup rolls to reduce roll deflection.

5. **Cluster Rolling Mill [Fig. 5.26 (b)].** A better backup can be provided with 4 rolls. These are known as cluster rolling mill.
6. **Planetary Rolling Mill [Fig. 5.27].** In a rolling arrangement when large reduction is required, number of free rotating wheels (instead of single small roll) are fixed to large back up roll. Figure 5.27 represents Planetary rolling mill.

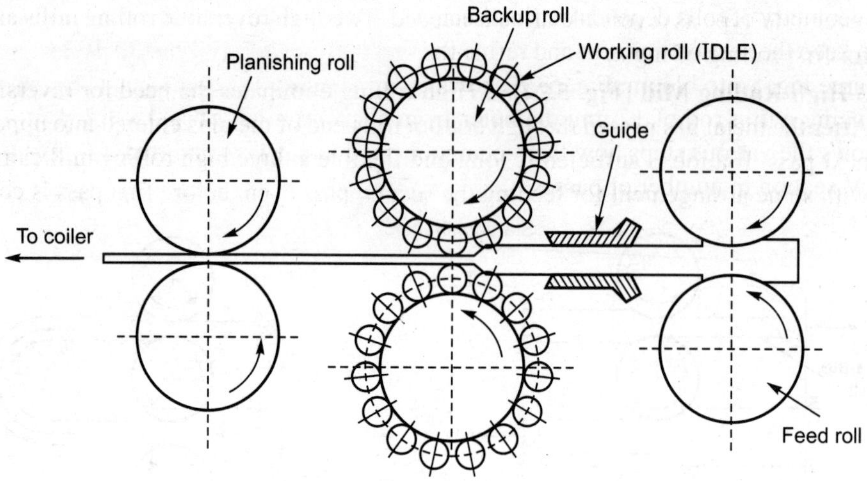

Fig. 5.27 Planetary rolling mill.

5.5.2 Rollers for Obtaining Different Cross-section

The input stock to a rolling mill is normally of rectangular cross-section (ingot, billet, bloom, or slab).

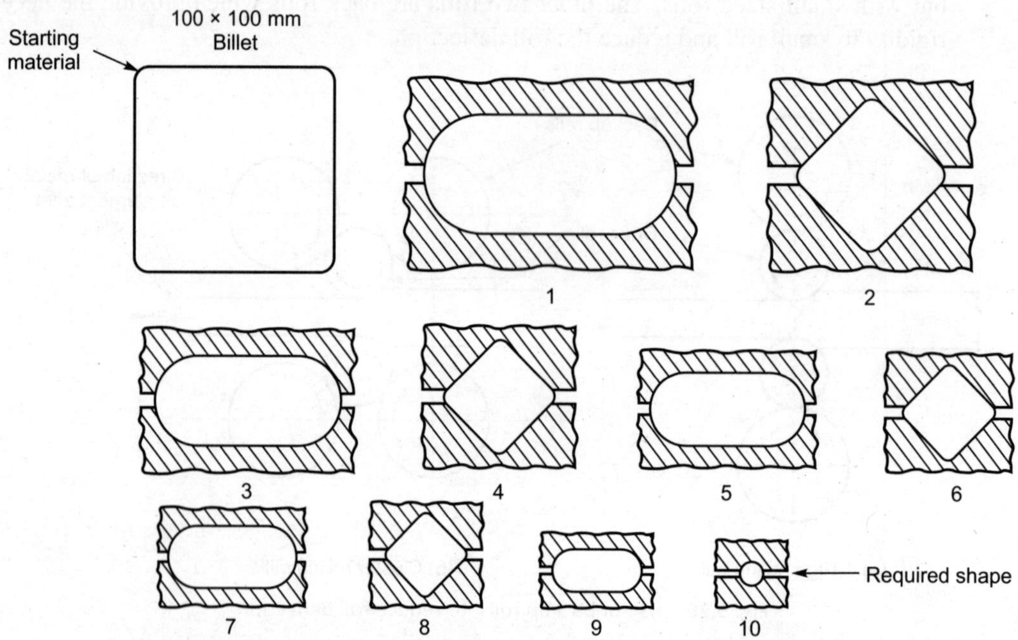

Fig. 5.28 Rolling process for obtaining different cross-sections.

Figure 5.28 illustrate the conversion of square billet 100 × 100 mm into circular rod of 10 mm diameter in 10 roll passes. The gradual changing geometry of rolls are shown by cross lines.

The final rolled components such as plates, flats, sheets, rounds and sections are obtained in number of passes. For rolling flat products, plain cylindrical rolls are used but to obtain different cross-section (like I-section, T-section or channel section) rolls with gradually changing geometry are used.

The type of geometry of rolls depends on final section desired. At frequent intervals metal is turned 90° on its side to keep the section uniform and refine the metal throughout. About 30 passes are required to reduce the large ingot into bloom. Fig. 5.28 shows how the geometry of the gap between rolls change while producing a thin rod of 12 mm diameter from a square billet of 100 × 100 mm in 10 passes. Fig. 5.29 shows the various steps how the rectangular section can be converted into I-section, channel section and V-section in number of passes.

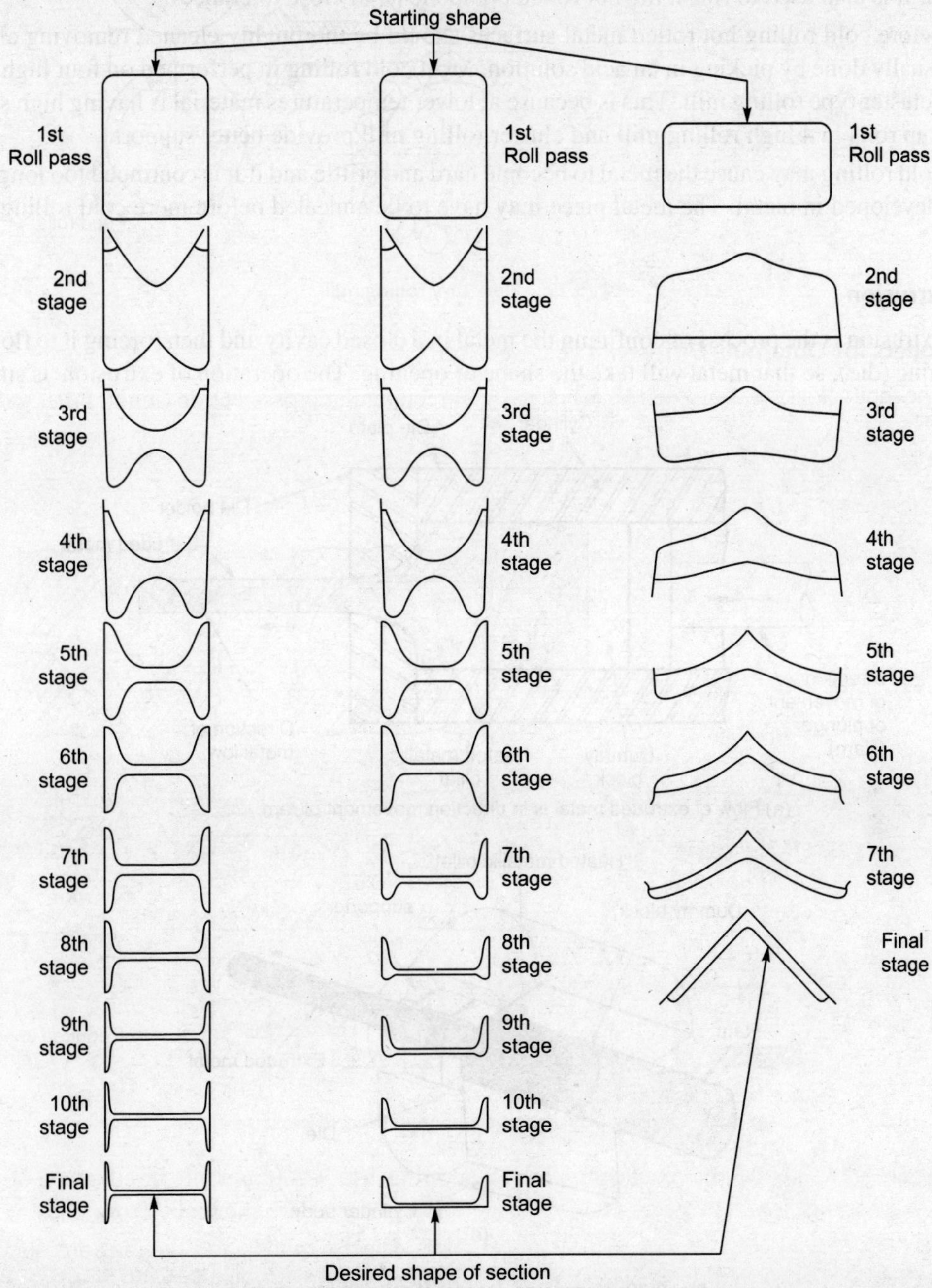

Fig. 5.29 Billets, blooms and slabs can be converted in various structural shapes in various roll passes.

5.5.2.1 Cold Rolling

The rolling of metal below its recrystallisation temperature is called cold rolling.

1. Cold rolling is generally done to provide a smooth and bright surface finish to previously hot rolled steel.
2. Cold rolling is done to improve the hardness and strength of previously hot worked metals.
3. It is also used to finish the hot rolled components to close tolerances.

Before cold rolling hot rolled metal surfaces should be thoroughly cleaned removing all scale. This is usually done by picking in an acid solution. Most cold rolling in performed on four high rolling mill and cluster type rolling mill. This is because at lower temperatures material is having high strength and backup rolls in 4 high rolling mill and cluster rolling mill provide better support.

Cold rolling may cause the metal to become hard and brittle and if it is continued too long cracks may be developed in metal. The metal piece may have to be annealed before more cold rolling can be done.

5.5.3 Extrusion

Extrusion is the process of confining the metal in a closed cavity and then forcing it to flow from one opening (die), so that metal will take the shape of opening. The operation of extrusion is similar to

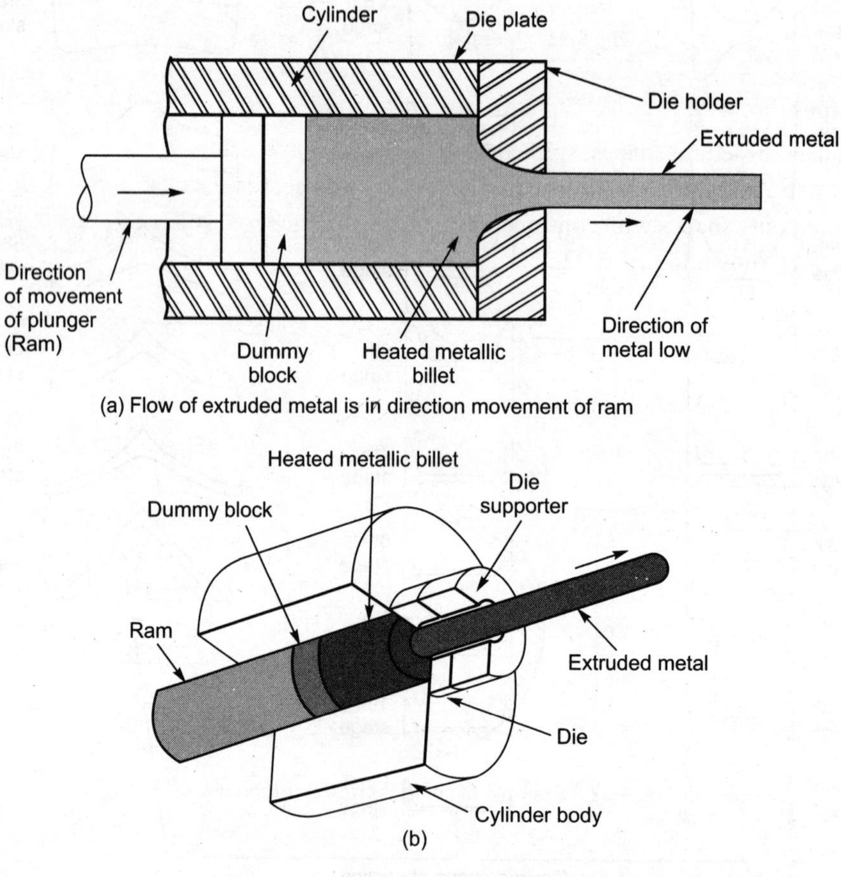

Fig. 5.30 Extrusion process (Forward extrusion).

the squeezing of tooth paste out of toothpaste tube. The paste inside the toothpaste has no shape, when the toothpaste tube is squeezed the paste flows out of the circular opening taking the shape of the opening.

Fig. 5.30 represents extrusion process. The equipment consists of a cylinder or container into which dummy block and heated metal billet are placed into position. On one end of the cyliner the die plate with necessary opening is fixed. The shape of the opening resembles the shape of the cross-section of extruded component. From the other end plunger compresses the hot billet against the container walls and die plate thus forcing it to flow through the die-opening acquiring the shape of opening. Dummy block of steel approximately 40 mm thick is kept between ram and billet to protect ram from heat and pressure. The flow of metal during the extrusion process is shown in Fig. 5.31.

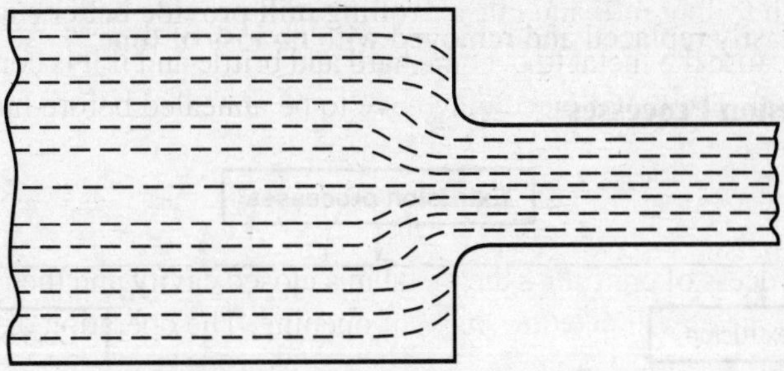

Fig. 5.31 Metal flow in direct extrusion.

Extruded Shapes

Rods, tubes, structural shapes, splined shafts, and bars of different cross-sections are the typical applications of extrusion process. Almost any length of cross-section can be produced with extrusion process. However, only shapes with constant corss-section can be produced in this process. Some of the extruded shapes are shown in Fig. 5.32.

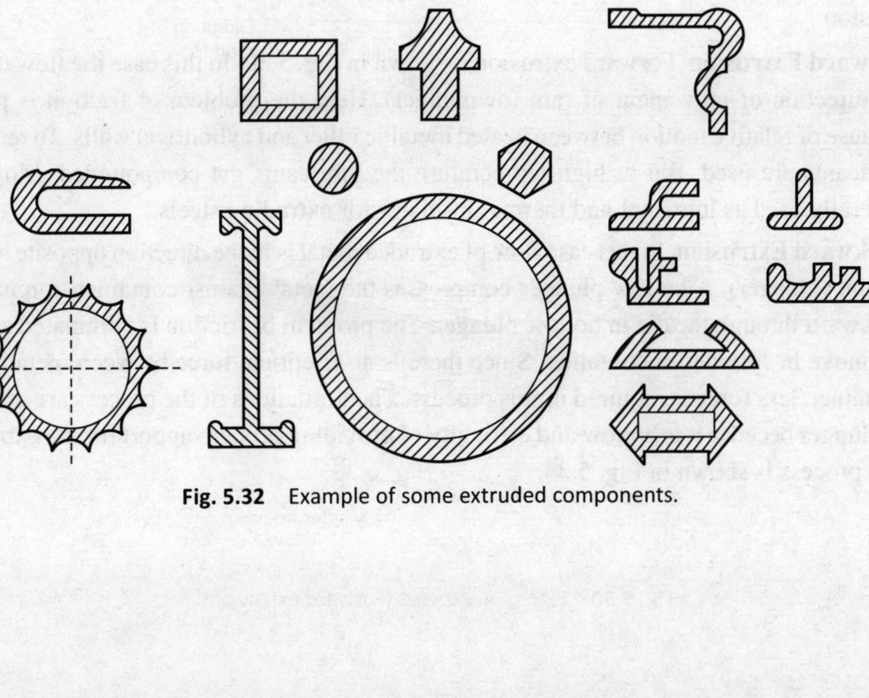

Fig. 5.32 Example of some extruded components.

Advantages of the extrusion process are as follows:

1. Extrusion is a single pass process unlike rolling.
2. Dies are easy to manufacture.
3. Variety of shapes of high strength, good accuracy and surface finish can be obtained.
4. High production rate with relatively low die cost.
5. Complicated cross-sectional shapes which are not possible to achieve by rolling can be achieved by extrusion.
6. Larger deformation can be achieved by this process than other processes (except casting).
7. Dies can be easily replaced and removed with no loss of time.

Classification of Extrusion Processes

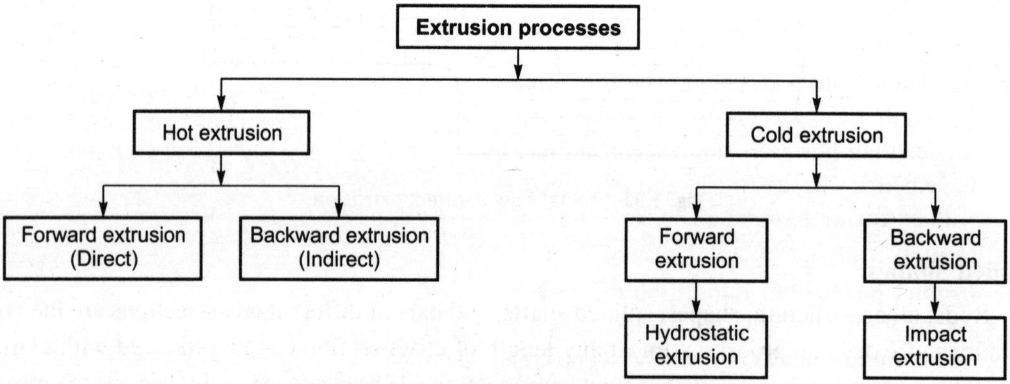

Fig. 5.33 Classification of extrusion process.

Hot Extrusion

(*a*) **Forward Extrusion.** Forward extrusion is shown in Fig. 5.30. In this case the flow of metal is in the direction of movement of ram (or plunger). Here the problem of friction is predominant because of relative motion between heated metallic billet and cylindrical walls. To reduce friction lubricants are used. But at high temperature the lubricants get compounded. Molten glass is generally used as lubricant and thermal insulator for extruding steels.

(*b*) **Backward Extrusion.** In this case flow of extruded metal is in the direction opposite to movement of ram (plunger). A hollow plunger compresses the metal against container, forcing it to flow backward through the die in hollow plunger. The problem of friction is eliminated as billet does not move in cylinder or container. Since there is no frictional force between metallic billet and container, less force is required in this process. The limitations of the process are the weakening of plunger because it is hollow and difficulty of providing proper support for the extruded metal. The process is shown in Fig. 5.34.

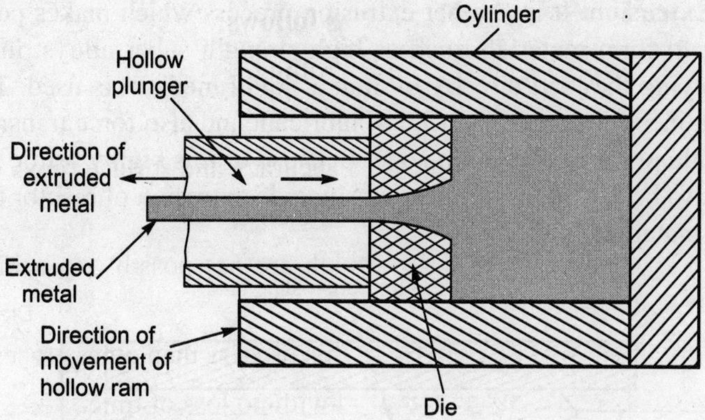

Fig. 5.34 Backward hot extrusion process.
(The direction of extruded metal is opposite to the movement of ram)

Cold Extrusion

(*a*) **Forward Cold Extrusion.** The forward cold extrusion is similar to forward hot extrusion (as shown in Fig. 7.12) except that temperature is comparatively lower and extrusion pressures are higher than hot extrusion. It is usually used for simple shapes. The cold extruded products have better surface finish and improved mechanical properties. The common applications of cold extrusion has also been used for forming mild steel parts often in combination with cold forging.

(*b*) **Impact Extrusion.** Impact extrusion is used for manufacturing collapsible tubes for shaving cream, toothpaste and thin walled cans. The process is limited to soft and ductile materials such as aluminium and its alloys. The set up consists of die and punch as shown in Fig. 5.35. A slug of metal of suitable thickness is kept on the die. The outside dia of tube is same as that of diameter of hole in die. Punch strikes blow causing the metal of flow through the annular space between the punch and die. The metal is extruded in the direction opposite to punch movement because of impact force. The process is conducted at room temperature. In recent years the process has been developed and now also applied to production of components for aircraft, cars and domestic applicances of Al, copper alloy and steels.

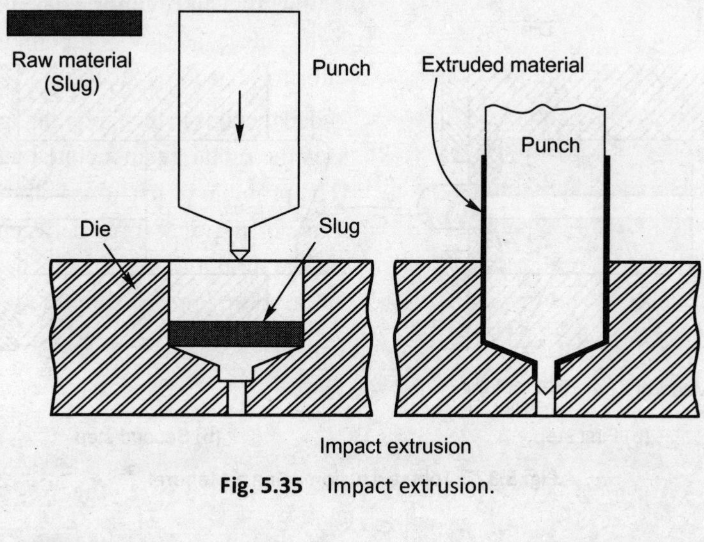

Fig. 5.35 Impact extrusion.

(*c*) **Hydrostatic Extrusion.** It is another extrusion process which makes possible to cold extrude many difficult to form materials such as high strength super alloys, mollybdenum etc. Here instead of applying the load directly by ram a liquid medium is used. The presence of liquid inside the container eliminates the need of lubricant and also force transmitted is uniform from all sides throughout the deformation zone. Because of this highly brittle materials such as grey cast iron can also be extruded. Applications include extrusion of reactor fuel rods, making wire of less ductile materials.

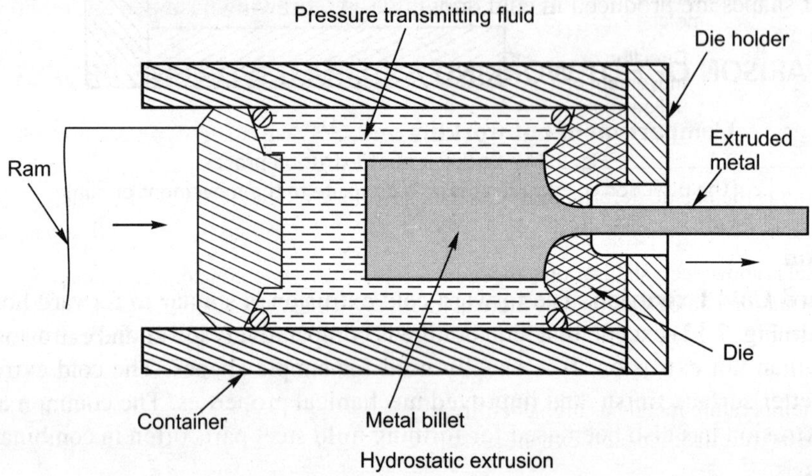

Fig. 5.36 Hydrostatic extrusion.

5.5.4 Tube Extrusion

Hollow objects such as tubes and other shapes can be produced by forward extrusion. For tubular sections mandrel is used alongwith the ram to shape the inside of the tube as shown in Fig. 5.37. After the billet is placed inside the extrusion container, the ram with mandrel advances and extrudes the metal through the die and around the mandrel. The entire operation must be rapid. Speed upto 3 m/second is used in making steel tubes. Low carbon steel can be extruded cold but for most of alloy billet is heated to around 1250°C.

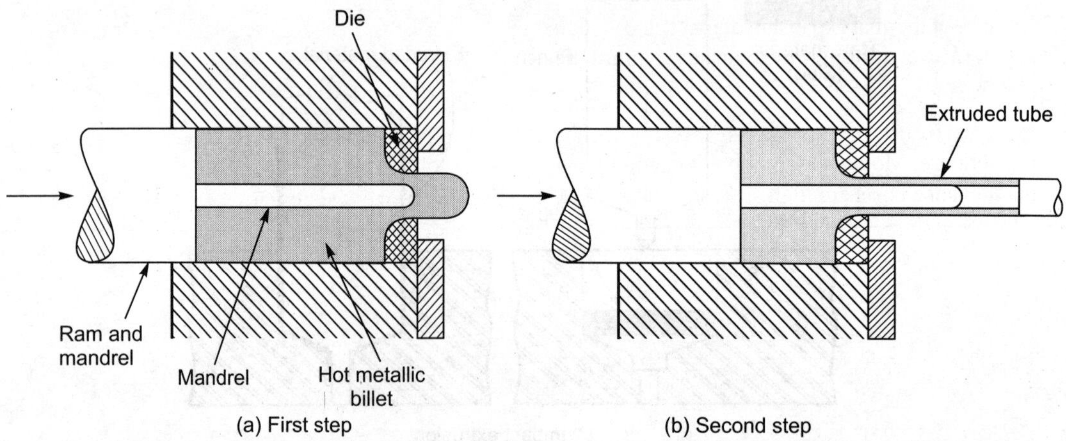

Fig. 5.37 Tube extrusion using a Mandrel.

5.5.5 Drawing

Drawing is the plastic deformation process in which a flat sheet or plate is formed into a three-dimensional part with a depth more than thickness of metal. Drawing process may be hot drawing or cold drawing. In hot drawing relatively thick walled parts of simple geometries, usually cylindrical are formed. Because the material is hot there is often considerable thinning as it passes through the dies. In case of cold drawing relatively thin metals are used and thickness changes very little or not at all. A wide variety of shapes are produced in cold drawing. (cold drawing has been discussed in 8.8.17)

5.6 COMPARISON OF HOT WORKING AND COLD WORKING PROCESSES

Comparison of hot working and cold working processes

1. Hot working is the plastic deformation of metals above recrystallisation temperature.	1. Cold working is the plastic deformation of metals below recrystallisation temperature.
2. As the metal is above recrystallisation temperature large deformation can be obtained as there is no strain hardening.	2. At lower temperature the amount of deformation that can be achieved is less. Also materials gets strain hardened. Any further deformation can be achieved by successive annealing process.
3. At higher temperature metal has high ductility so less force is required to achieve necessary deformation.	3. At lower temperature material has lower ductility, so more force is required to achieve necessary deformation.
4. Because of grain refinement ductility and toughness are improved, strength is increased and greater homogeneity is developed in metals.	4. Cold working increases hardness and strength at the expense of ductility which is useful in some applications.
5. The problem of brittleness and development of internal stresses is not there in hot working processes.	5. As the metal gets strain hardened and internal stresses are developed during cold working. The metal becomes brittle if it is over worked.
6. There is rapid oxidation or sealing of surfaces due to high temperature of metal which results in poor surface finish.	6. Scaling or oxidation problem is not there in cold working processes because of lower working temperatures, so surface finish achieved is good.
7. Because of thermal expansion of metal and at high temperature dimension accuracy cannot be obtained.	7. Better dimensional accuracy is achieved.
8. It is difficult to handle metals at higher temperature. Moreover, equipment handling and maintenance costs are high.	8. It is easier to handle cold metals and also economical for some cases like menufacturing small size components.

5.7 FORGING EXERCISE

Aim. To prepare a mild steel ring of diameter 90 mm with forging operation.

Ram material. Mild steel rod (ϕ12 mm thickness).

Equipments. Forging hammer, forging hearth, anvil, swage block, long chiesel.

Procedure:

(*i*) Cut the mild steel rod of length 283 mm. (*i.e.*, $\pi D = 3.1416 \times 90$) and punch roll number on the rod.

(*ii*) Heat the rod in forging hearth.

(*iii*) Take out the rod from forging hearth with help of tong when it is red hot.

(*iv*) Keep the red hot-rod on swage block at a suitable hole by holding it with tong and beat it with hammer.

(*v*) While hamming the hot rod, bend it progressively so that it takes the shape of the ring.

(*vi*) Ensure that complete circular ring is formed.

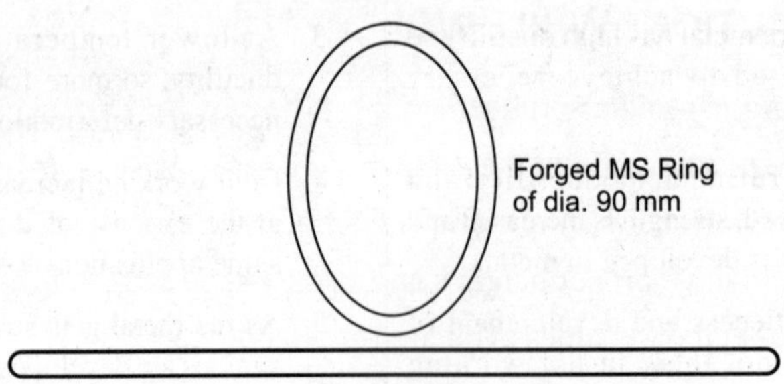

Fig. 5.38 Forging Exercise

FORGING JOBS EXERCISE

Job 1. To Forge Mild steel (MS) round rod of 15 mm into square bar of 15 × 15 mm. Also find out the length of square bar given that length of round bar is 80 mm.

Job 2. To make a round bar with Hexagonal head from given round bar by forging.

Job 3. To make a round bar with square head from given round bar by forging.

Job 4. To make a hook from given rand bar by forging.

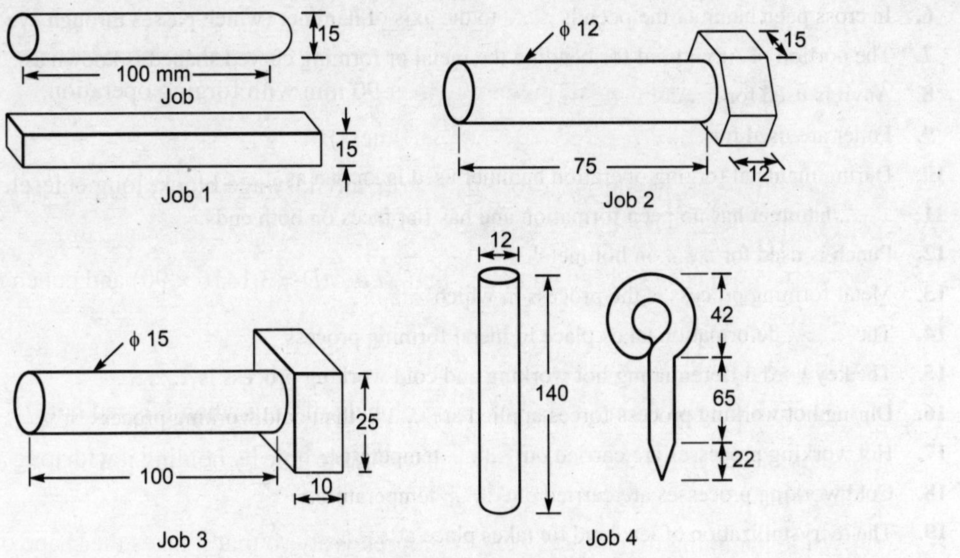

Fig. 5.39 Various jobs for forging exercise.

5.8 SAFETY PRECAUTIONS WHILE WORKING IN FORGING SHOP

(*i*) Wear proper clothes and use rubber sole closed shoes.

(*ii*) Hold the work piece tightly in the tong.

(*iii*) Always use proper tong according to type of work.

(*iv*) Hammer should have proper handles and while hammering handle should be held tightly.

(*v*) Do not touch metal parts unless you are sure that it is not hot.

(*vi*) The fire in forging hearth should be lightened very carefully.

(*vii*) All the exhaust gases and smoke from forging hearth should go to chimney.

(*viii*) Always put off fire in forging hearth properly before leaving the forging shop.

(*ix*) Keep the working space clean and ensure that forging tool and equipments should be free from greece, dust and moisture while in use.

(*x*) While working with machine forging, strictly follow the safety norms prescribed for each forging machine. One must know about the operating parameters (like pressure, temperature) of the forging machines.

OBJECTIVE TYPE QUESTIONS

1. The process used for making crane hook is ....
2. The forging process used for producing gear blank is ....
3. The tools used to hold the job during forging operations are termed as ....
4. In ball peen hammer shape of peen is ....
5. In straight peen hammer the peen is .... to the axis of hammer (which passes through eye hole).

6. In cross peen hammer the peen is .... to the axis of hammer (which passes through eye hole).
7. The portion of Anvil used for bending the metal or forming curved shaped is known as ....
8. Anvil is used for ....
9. Fuller are used for ....
10. During mannual forging operation hammer used is known as ....
11. .... hammer has no peen formation and has flat faces on both ends.
12. Punch is used for .... in hot metal.
13. Metal forming process is the process in which ....
14. The .... deformation takes place in metal forming process.
15. The keyword differentiating hot working and cold working process is ....
16. During hot working process forces applied are .... than cold working processes.
17. Hot working processes are carried out .... temperature.
18. Cold working processes are carried out .... temperature.
19. The recrystallization of lead and tin takes place at ....
20. In cold working processes hardness is increased due to ....
21. In cold working processes hardness and yield strength of material is ....
22. In hot working processes .... deformation can be obtained.
23. Temperature at which entirely new crystal or grains are formed is known as ....
24. Extrusion and rolling are ....
25. The process of confining the metal in closed cavity and thin forcing it to flow from one opening is termed as ....
26. Seamless tube can be made by .... process.
27. The process of compressing the metal by passing it between two or more revolving cylinder (called rolls) is ....
28. In .... rolling mill two rolls are drive rolls and two are back up rolls.
29. In .... rolling mill .... number of free rotating wheels as are fixed to large backup roll.
30. During .... extrusion .... the flow of extruded metal is in the direction of movement of Ram.
31. During .... extrusion the flow of metal is opposite to the movement of Ram (plunger).
32. The operation done in forging processes to increase cross-sectional area and reduce length is ....
33. The operation done in forging process to increase the length and reduce the cross-sectional area is ....
34. During drop forging process .... dies are used.
35. The force applied during press forging is ....
36. The force applied during drop forging is ....
37. The forging process is done above .... temperature.
38. In punching operation sheet with the holes is .... product and removed piece is .... product.
39. In blanking operation the shapes cutout from sheet is .... product and the sheet left with holes is .... product.
40. The operation of removing a small portion of metal along the edge of sheet is ....
41. The process of making cup shaped articles from flat sheet metal is ....
42. The process of obtaining wires from rods of bigger diameter is ....

ANSWERS

1. drawing out	**2.** upsetting	**3.** tongs
4. Hemi-spherical ball	**5.** along	**6.** perpendicular
7. Horn or beak	**8.** supporting the workpiece during forging	
9. forming necks or grooves on the workpieces		**10.** sledge hammer
11. sledge hammer	**12.** making hole.	
13. metal is shaped by apply forces through various dies and holes		
14. plastic	**15.** recrystallization temperature	**16.** comparatively less
17. above recrystallization	**18.** below recrystallization	**19.** room temperature
20. work hardening	**21.** increased	**22.** large
23. recrystallization temperature		**24.** metal forming processes
25. extrusion	**26.** tube extrusion	**27.** Rolling
28. 4 high	**29.** planetary	**30.** forward
31. backward	**32.** upsetting	**33.** drawing out
34. close impression	**35.** continuously squeezing force	**36.** impact blow
37. recrystallization	**38.** required, waste	**39.** required, waste
40. notching	**41.** sheet metal drawing	**42.** wire drawing

MULTIPLE CHOICE QUESTIONS

1. Which of following tool used for supporting the workpiece during forging
(*a*) tong (*b*) swage block
(*c*) hammer (*d*) fuller.

2. Which of the following tool is used for forming neck and groove in workpiece
(*a*) hammer (*b*) fuller
(*c*) mallet (*d*) flatter.

3. Which of the following tool is used for holding the workpiece during forging
(*a*) anvil (*b*) set hammer
(*c*) tong (*d*) chisel.

4. Chisels are used for
(*a*) supporting the workpiece (*b*) cutting the metal and for making notch
(*c*) for making holes in workpiece. (*d*) for holding the workpiece.

5. Open dies are used in
(*a*) smith forging or leand forging (*b*) Drop forging
(*c*) Press forging (*d*) Machine forging or upset forging.

6. The forging operation used for making head of bolt is
(*a*) fullering (*b*) upsetting
(*c*) drawing out (*d*) swaging.

7. Fullers are used for
(*a*) holding the workpieces (*b*) supporting the workpieces
(*c*) decreasing the length of workpieces (*d*) forming neck or groove on the workpieces.

8. The hammer in which the peen is perpendicular to the axis of hammer (which passes through eye hole) is termed as
   (*a*) ball peen hammer (*b*) straight peen hammer
   (*c*) cross peen hammer (*d*) sledge hammer.
9. The hammer in which peen is along the axis of hammer (which passes through eye hole) is termed as
   (*a*) cross peen hammer (*b*) double face hammer
   (*c*) straight peen hammer (*d*) sledge hammer.
10. Swage block and Anvil are used for
   (*a*) holding the job (*b*) supporting the job (work piece)
   (*c*) finishing the job (*d*) holding the forging tools.
11. The various operation like fullering, flattening, swaging, bending, performed on open dies belong to which of following forging methods
   (*a*) smith forging (*b*) drop forging
   (*c*) machine forging or upset forging (*d*) press forging

ANSWERS

1. (*b*)	**2.** (*b*)	**3.** (*c*)	**4.** (*b*)	**5.** (*b*)
6. (*b*)	**7.** (*d*)	**8.** (*c*)	**9.** (*c*)	**10.** (*a*)
11. (*a*).				

VIVA VOCE QUESTIONS

1. What are the metal forging processes?
2. Compare the manufacturing of component using casting, machining and forging. Also compare the properties of component material manufacture with these processes.
3. How would you classify forging processes?
4. What is recrystallisation? Why is it important in metal forming processes?
5. Describe the advantages and disadvantages of hot working processes.
6. What for is the hot working process chiefly used?
7. What is the principle of rolling process? Why is the strength of a rolled part considered usually better than a cast piece?
8. Explain various rolling stand arrangements with neat sketches.
9. How the difference shapes like I-section, channel section are obtained in rolling process?
10. What is the difference between hot rolling and cold rolling.
11. Explain what you understand by terms ingot, slab, bloom and billet.
12. Describe extrusion process with neat sketch and show the flow of material.
13. How will you classify extrusion processes?
14. What is the difference between forward hot extrusion and backward hot extrusion?
15. What are the advantages of using hydrostatic extrusion?
16. Explain forging processes and their types.
17. What is meant by upset forging?
18. What is difference between hot forging and cold forging?

CHAPTER

6

SHEET METAL PROCESSES (SHEET METAL SHOP)

INSIDE THIS CHAPTER

6.1 INTRODUCTION

Sheet metal processes are the one of the cheapest and fastest way of the complete manufacturing of components. Sheet metal processes or press working are an essential part of modern manufacturing system. Sheet metal components are extensively used in industrial as well as house hold applications. The common examples of product made from sheet metal processes are automobiles, almirahs, heating ducts, domestic heating appliances, tubelight fixture, chemical plants applications, containers for ghee and edible oils, kitchenware dairy equipments, tanks, pans, trunks etc.

Sheet metal is considered to be a metallic plate with thickness less than about 6 mm. The starting material for sheet metal processes are plates, bars, tubes, sheets and similar shapes, which are obtained from metal forming processes. These shape are converted in further shape. These materials are processed on the presses to get the sheet metal components. Presses are the machines on which cutting, bending or drawing operations on sheet metal materials are performed using punches and dies. The punch is usually attached to the ram of the press and forced into die cavity.

The die has an opening to receive the punch and two much be in alignment for quality operation. Punches and dies are not interchangeable. They are mated for particular operations. A single press may do a large variety of operations depending on the dies used.

"Success often comes to those who dare and act; it seldom goes to timid who are ever afraid to the consequences."

Sheet metal processes are mostly cold working processes. Magnesium is heated slightly to make it more workable. The stresses caused in the metal during the operations are shearing, tension, compression or bending (lension and compression) depending upon the name of process. Various sheet metal operations and the type of stresses induced in these operations are shown in Table 6.1.

Table 6.1. Stresses in Sheet Metal Operations.

Stresses caused in the metal	Operations
Shearing	Shearing, punching, blanking, shaving, notching trimming, nibbling
Tension	Stretching
Compression	Coining, sizing,
Tension and compression	Drawing, embossing, spinning
Bending (Tension and compression)	Bending, forming

6.2 METALS USED FOR SHEET METAL WORK

There is a large variety of metals used in sheet metal work in form of sheets and plate. As we know that sheet thickness upto 6 mm may be considered in sheet metal work. The sheet metals are specified in term of gauge number, length and width. Gauge number can be known using a standard wire gauge. The thickness of sheet varies inversely proportional to the gauge number *i.e.*, higher the gauge number, the small is the thickness and vice-versa. Following are some of metals used as sheet metals.

1. Galvanised Iron

Galvanised iron is popularly known as G.I. Sheet.These sheets are made using galvanisation process *i.e.*, coating of zinc on the surface of soft iron. Zinc coating provide corrosion resistance surface and also imporves the appearance. It also enables easy soldering. However, welding of zinc offers problems due to generation of toxic fumes due to burning of zinc coating.

G.I. sheets are widely used in both industrial and domestic applications. The application of galvanised iron include heating ducts, furnaces, storage tanks, buckets, trunks, pans etc.

2. Copper

Copper is costlier than iron or steel but, it has certain specific advantages. Copper has primarily three important properties:

- High thermal and electrical conductivity
- Useful strength with high ductivity
- Corrosion resistance.

Because of these properties copper and its alloy are widely used in electrical applications, applications where extensive forming is involved and also for other manufacturing processes including casting welding and machines. Applications of copper include domestic heating appliances, chemical plant applications, electrical appliances etc.

3. Stainless Steel

Stainless steel is alloy of iron, carbon nickel and chromium. It is highly corrosion resistant and has bright surface without any coating. It is widely used for making kitchen ware, food handing equipment, dairy equipment, chemical plant, food processing, plants and house hold items.

4. Aluminium

Aluminium has number of unique and attractive properties. These include workability, light weight, corrosion resistance and good electrical and thermal conductivity. The specific weight of aluminium is one third of the weight of steel for equivalent volume. It is used in the manufacturing of aeroplane bodies, kitchen ware and domestic appliances, tube light fixtures, electrical applicances etc. The serious limitation of aluminium is relatively low modulus of elasticity.

5. Tin

Tin is widely used as coating on iron and steel sheets to improve corrosion resistance. Tin coating sheets provide a bright silver colour appearance. As tin provide resistance to corrosion and rusting it is widely used in *packet food containers*, containers for edible oils and ghee, dairy equipments cans, etc.

6. Lead

It is very weak, soft, malleable and low melting point point metal. It possesses high resistance to corrosion. It finds application in radiation shielding sheet, inner linning for acid tank etc.

6.3 SPRING BACK IN SHEET METAL OPERATIONS

In metal working processes when the metal is deformed, it is first elastically deformed and then plastically. We know that elastic deformation is recoverable and whereas plastic deformation is plastically. We know that elastic deformation is recoverable and whereas plastic deformation is permanent. In many sheet metal operations when the load is removed there is an elastic recovery by the metal and total deformation will get reduced a little. This phenomenon is called as spring back and is of more importance in operations such as bending. The spring back is shown in Fig. 6.1. Here the beneding of sheet metal is

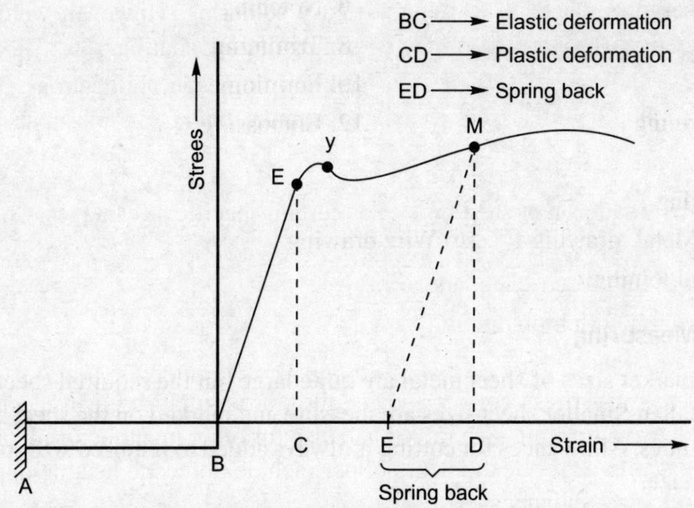

Fig. 6.1 Spring back in sheet metal processes.

shown. When the bending forces are removed the work piece moves back slightly towards its original shape as a result of elastic deformation.

Fig. 6.1 shows a stress strain curve, where a part *AB* is stretched. If part *AB* is stretched upto length *AC* (elastic limit) and load is removed, it will return to its original length *AB*. If *AB* is stretched beyond elastic limit *E* on the stress strain curve to a length *AD* and then load is removed it final length will be *AE*. The spring back in this case is *ED*. Spring back is the property of material which depends on modulus of elasticity and yield strength. Higher the modulus of elasticity lower would be spring back whereas high yield strength of metal means more spring back.

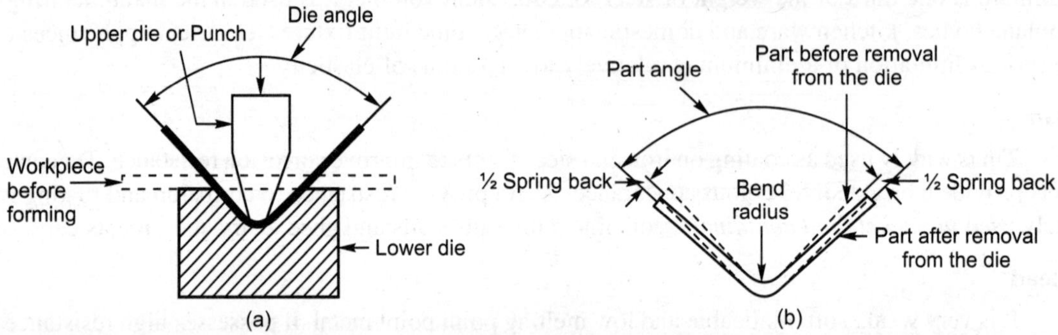

Fig. 6.2 Spring back in bending operation.

Fig. 6.2 (*a*) shows bending operation and Fig. 6.2 (*b*) shows that how workpiece springs back after the removal of bending forces.

6.4 SHEET METAL OPERATIONS

Various sheet metal opeations are:

1. Sheet metal measuring
2. Laying out
3. Punching
4. Blanking
5. Shaving
6. Notching
7. Nibbling
8. Trimming
9. Bending
10. Forming
11. Stretch forming
12. Embossing
13. Coining
14. Cold drawing
    (*a*) Sheet Metal drawing (*b*) Wire drawing
15. Sheet metal joining.

6.4.1 Sheet Metal Measuring

The standard market sizes of sheet metal are quite large but the required sheet size for making a part may be much smaller. Smaller sheet sizes are measure and marked on the sheet and then sheets are cut along the marked lines. Allowances for cutting is always added to required size so that cut pieces are not oversize or undersize.

- Marking of straight lines is done with the help of steel rule, and straight edge scriber.
- Circles and arcs are marked with a divider. Large circle and arcs may be drawn with trammel.

- ➢ Irregular curves may be marked with the help of french curves or by flexible rule made of spring steel.
- ➢ Sheet metal surface may have to be coated with a colouring media such as cellulose lacquer so that scribed lines are clearly visible.
- ➢ In mass production of identical items marking is done with the help of a template and scribe. Fig. 8.5 shows the scriber.

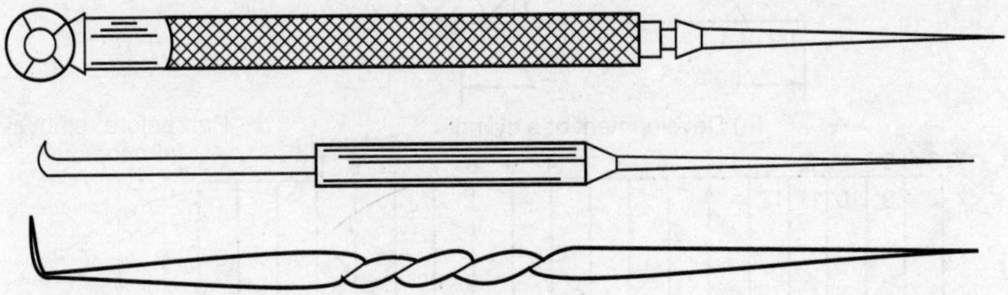

Fig. 6.3 Scriber.

6.4.2 Laying Out

All sheet metal components are fabricated from flat pieces of sheets. Development of surfaces is scribed on the sheet considering allowances for overlapping, bending, cutting etc. Now these sheets are cut from the blank along the marked lines and folded and joined to fabricate the required component.

The laying out means scribing the development of surfaces on the sheet blank. The process of laying out is also known as pattern layout. Let's see what is meant by development of surfaces.

Development of Surfaces (Fig. 6.4)

In development of surfaces the surfaces of an object are opened out and laid on a flat plane. These opened surfaces will take certain shape. The shape so obtained is development of surfaces of that particular object. For example, if we develop the surface of cylinder we will get rectangle. The width of rectangle is equal to the circumference of cylinder ($2\pi r$) and height is the axis height of cylinder.

Development of right cone is a sector of circle having radius equal to the slant height of the cone and an arc length equal to the circumference of its base.

There are three common methods of development (or laying out):

1. **Parallel Line Method (Fig. 6.5).** This method is used for development of cylinders and prismatic components.

2. **Radial Line Development (Fig. 6.6).** This method is used for components which carry radial edges or slant surfaces like cones and pyramids.

3. **Triangulation Method (Fig. 6.7).** This method is combination of parallel line development and radial line development. This method is used for developing the surfaces of transition prices *e.g.* the filing which is used to connect opening of different shapes like circular cross section to square corss section. These methods are covered in details in engineering graphics and drawing books.

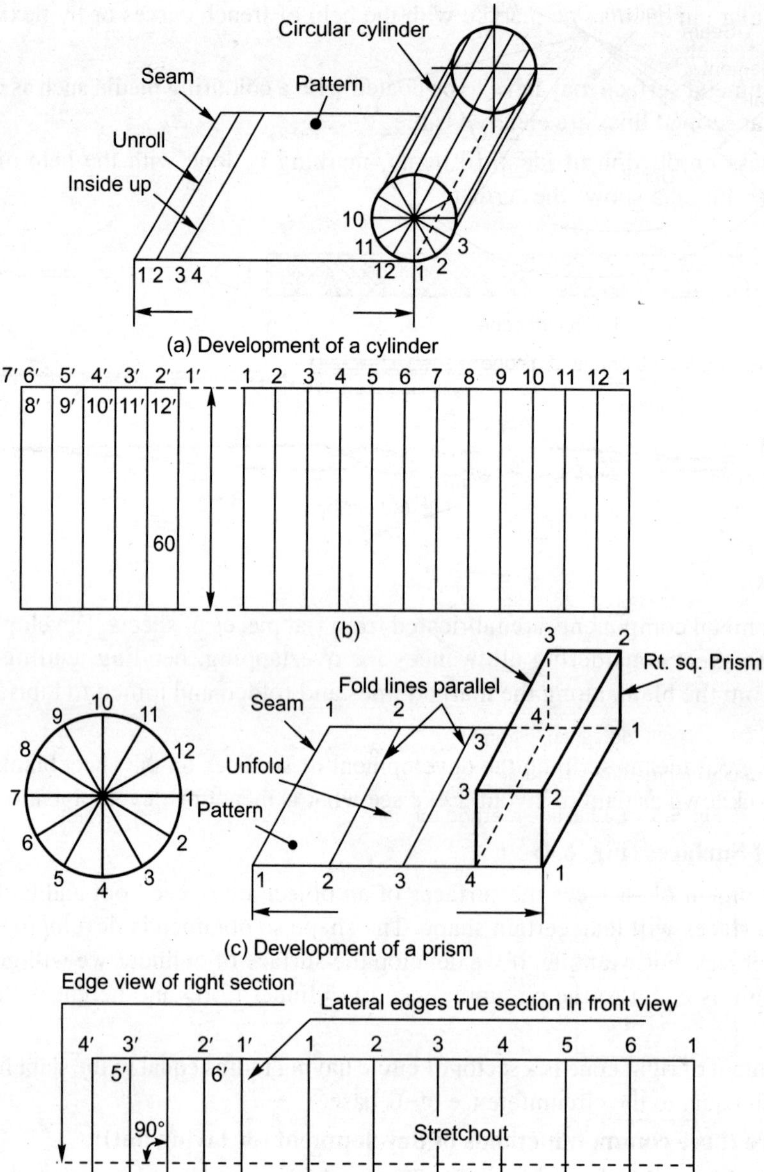

Fig. 6.4 Parallel line method for development of cylinder and prism.

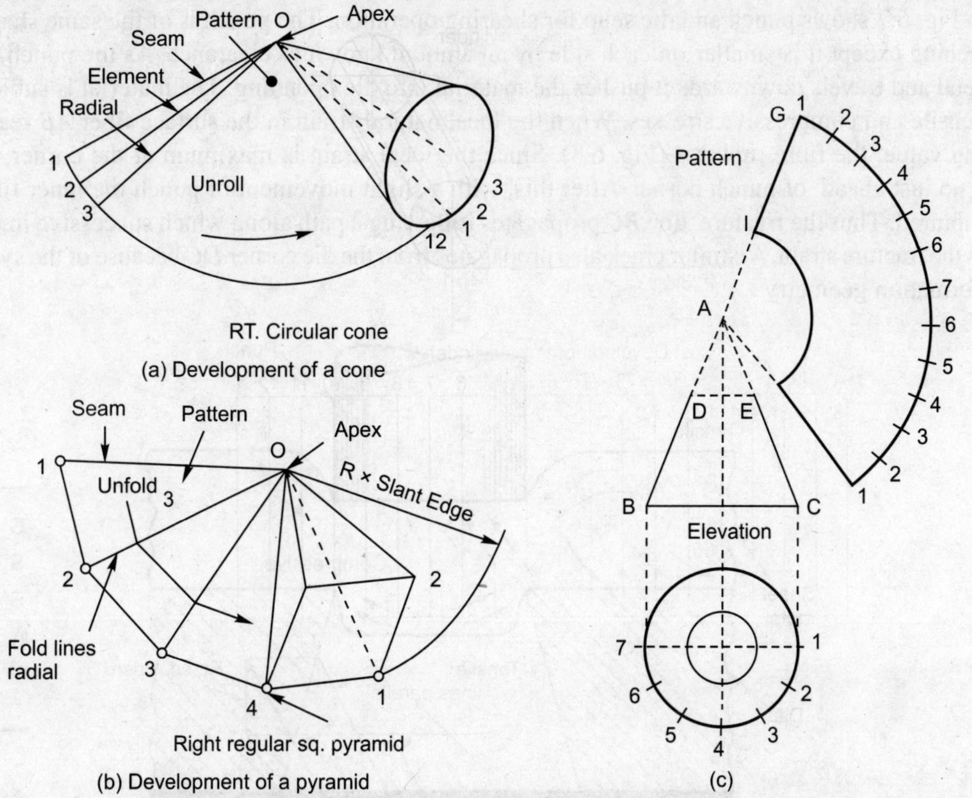

Fig. 6.5 Radial line method for development of cones and pyramids.

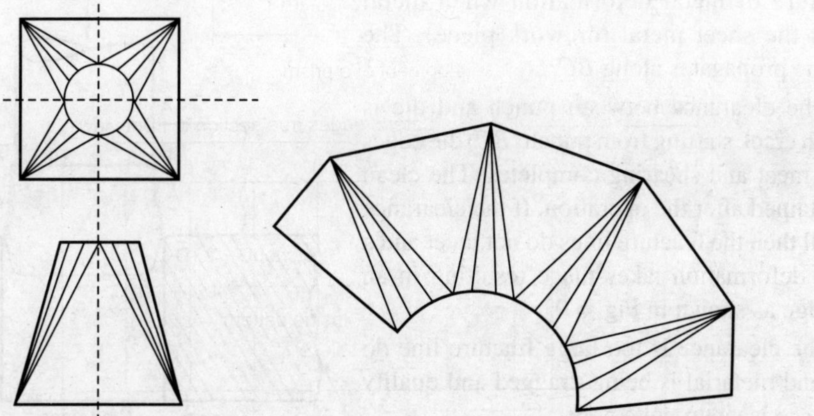

Fig. 6.6 Triangulation method for development of transition piece.

6.4.3 Shearing Process (Cutting Process)

Cutting of sheet metal in sheet metal process is a shearing process. Let's see what exactly happens during the shearing process.

Fig. 6.7 shows punch and die seup for shearing operation. The punch is of the same shape as of die opening except it is smaller on each side by an amount known as clearance. As the punch touches the metal and travels downwards it pushes the material into die openning. The material is subjected to both tensile and compressive stresses. When the local natural strain in the surface fiber *AB* reaches its limiting value, the fibre ruptures (Fig. 6.8). Since the local strain is maximum at the corner, a crack opens up just ahead of punch corner. After this, with a slight movement of punch the inner fibre also gets ruptured. Thus the fracture line *BC* propagates following a path along which successive inner fibre attains the fracture strain. A similar crack also propagates from the die corner *DC* because of the symmetry of deformation geometry.

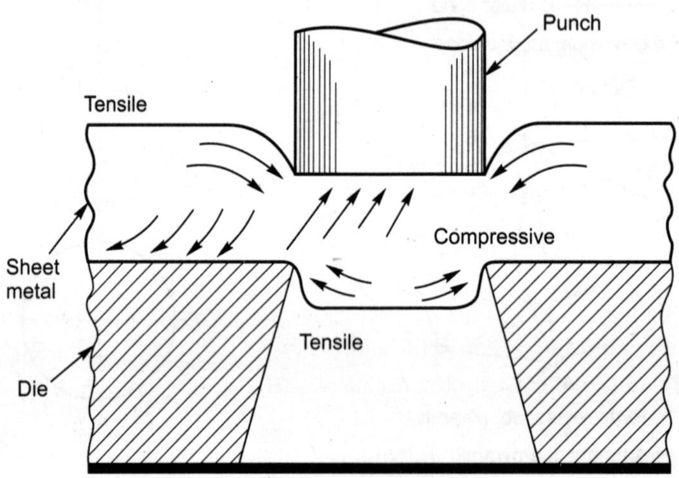

Fig. 6.7 Stresses in cutting operation (shearing).

Nature of metal deformation when metal penetrates the sheet metal (or work piece). The fracture line propagates along *BC*.

If the clearance between punch and die is correct then crack starting from punch (*BC*) die edges (*DC*) will meet and shearing completes. The clean edge is obtained after the operation. If the clearance is too small then the feacture lines do not meet and a secondary deformation takes place resulting in an unclean edge as shown in Fig. 6.9.

It the clearance is too large fracture line do not meet and metarial is being dragged and quality of work piece is again quite poor.

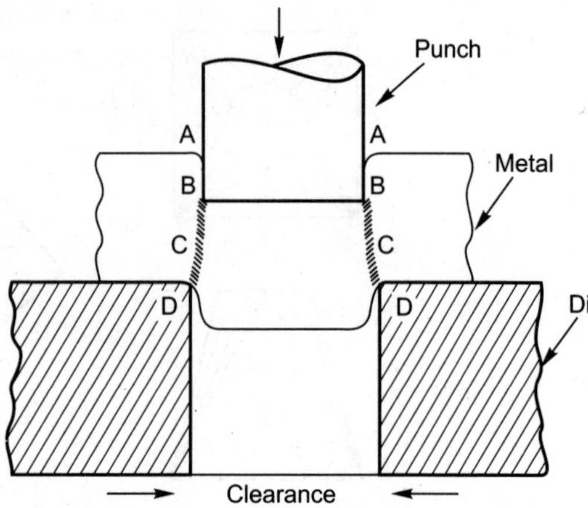

Fig. 6.8 Deformation of workpiece with punch travel.

The designed clearance depends on thickness, hardness and strength of material. Deuctile materials require smaller clearances and longer penetration of punch compared to harder material. The various steps in the fracture of material can be written as:

(*i*) Stressing the material beyond its elastic limit

(*ii*) Plastic deformation

(*iii*) Reduction in area

(*iv*) Fracturing starts in the reduced area and becomes complete.

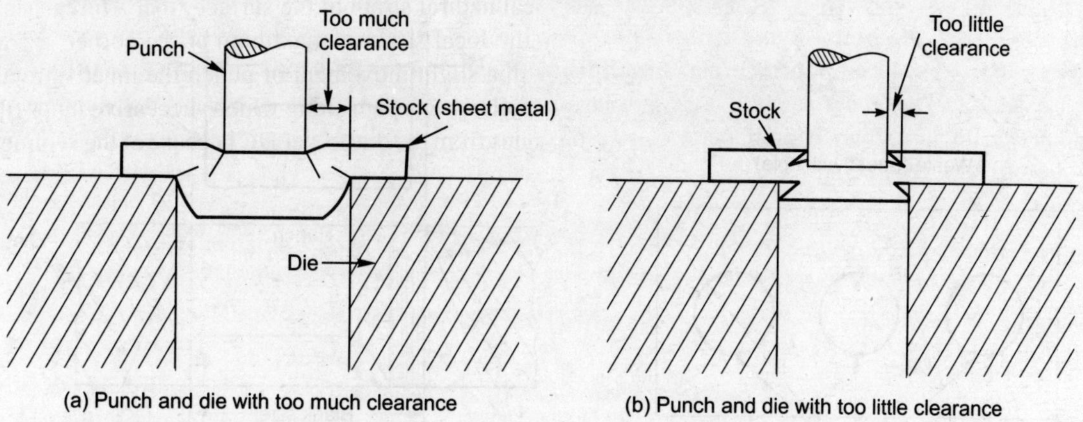

Fig. 6.9 Effect of clearance between punch and die.

6.4.4 Punching

The punching is cutting operation by which holes of various shapes are made in sheet metal strip. In punching operations sheet with the hole is desired product and material punched out is scrap. Normally punching opeartion is performed after blanking.

- The punch and die are made of a suitable hard and strong material, such as hardened high carbon steel, high chromium oil hardening steel or tungsten carbide.
- In punching operation the punch size equals the size of hole and die opening size is obtained by adding the clearance to punch size.
- In punching opeartion shear is applied to punch.

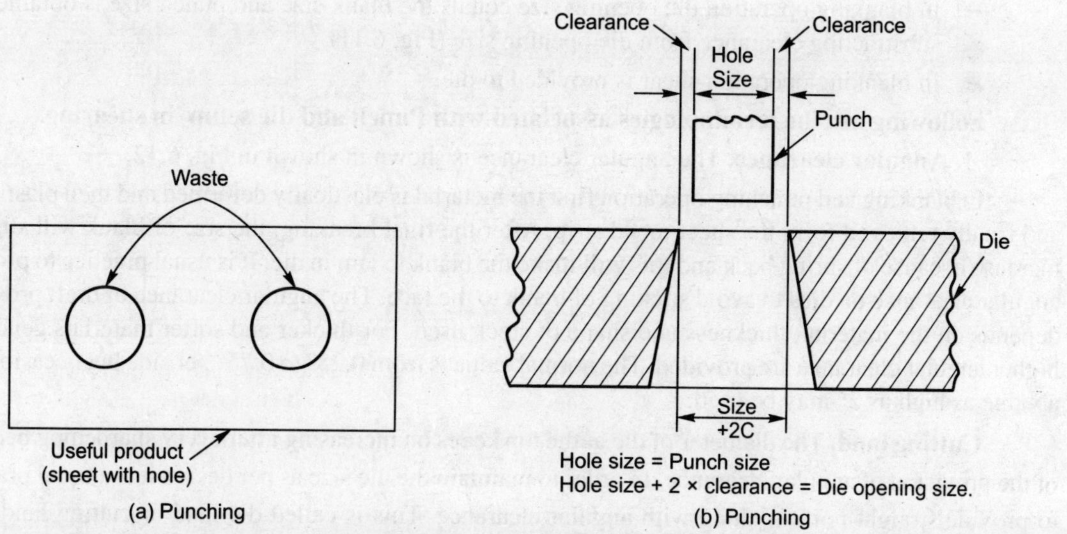

Fig. 6.10 Punching operation.

6.4.5 Blanking

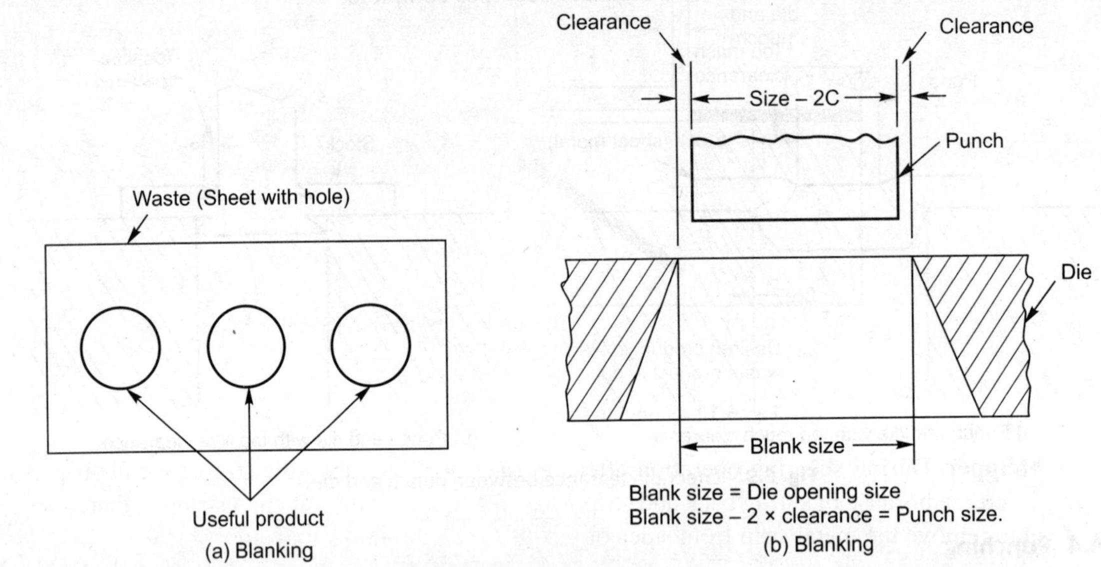

Fig. 6.11 Blanking operation.

Blanking is the operation of cutting out various flat shapes from sheet metal strip. The removed portion is called blank which is required product of the operation and metal with hole left behind is scrap as waste. Blanking is usually first opeartion and blank is further processed to produce desired part. Blanking is identical to the punching exept that in blanking removed portion is desired product where as in punching sheet with the hole is desired product.

- ➢ The punch and die are made of a suitable hard and strong material such as hardened high carbon steel, high chromium oil hardening steel, or tungsten carbide.
- ➢ In blanking operation die opening size equals the blank size and punch size is obtained by substracting clearance from die opening size (Fig. 6.11).
- ➢ In blanking operation shear is provided to die.

Following are the terminologies associated with Punch and die setup in shearing.

1. **Angular clearance.** The angular clearance is shown in shown in Fig. 6.12.

In blanking and punching operation first the metarial is elastically deformed and then plastically and finally removed form the sheet metal strip. After the final breaking, the size of blank will slightly increase because of spring back and this will make the blank to jam in die. It is usual practice to provide angular clearance on dies to avoid sticking of blank to die face. The angular clearance or draft provided depends on the material, thickness and shape of stock used. For thicker and softer materials generally higher angular clearance are provided. The normal value is from 0.25° to 0.75° per side but occasionally a value as high as 2° may be used.

Cutting land. The diameter of die at the top keeps on increasing after every sharpening because of the provision of angular clearance. In order to maintain the die size as per design it is a usual practice to provide straight portion along with angular clearance. This is called die land or cutting land. The length of cutting land is about 3 mm for sheets which are less than 3 mm thick. For greater thickness die land same as the material thickness has been found to be a good practice.

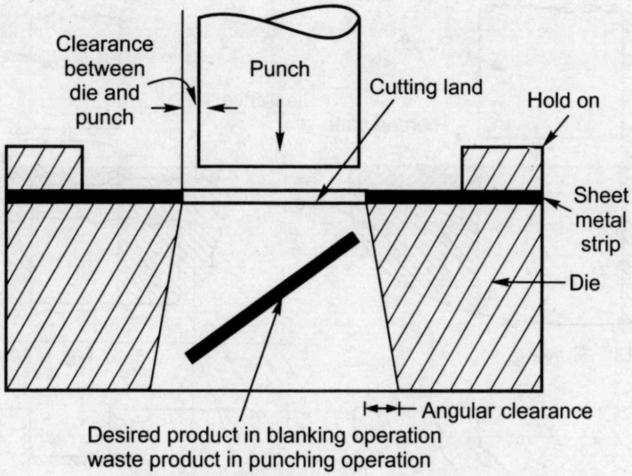

Fig. 6.12 Punching and blanking operations.

Stripper. During shearing operation after the final breaking, the hole size of metal strip will slightly decrease because of spring back and strip will cling tightly to the punch. A stripper, therefore, is provided to remove the metal strip from sides of punch as punch returns upwards.

6.4.6 Shaving

In blanking and piercing, the edge of blank or the hole is not perfectly clean because of burr generated in the shearing process. Shaving is basically a finishing operation in which the small amount of material is sheared away from the edge of an already blanked or pieced part. Its primary use is to obtain greater dimensional accuracy. It is also used to produce part. Its primary use is to obtain greater dimensional accuracy. It is also used to produce a smoother edge. Because of only small amount of metal is removed the punches and dies must be made with very little clearance. Blanked parts such as small gears can be shared to produce dimensional accuracy within 0.025 mm.

6.4.7 Trimming

In operations such as forging, die casting and drawing of sheet metals the small amount of excess metal gets spread out near the parting line. This extra metal is called flash. The flash is to be trimmed before forging or casting is to be used. The trimming may be defined as removal of excess material generated during forging and casting operation. The dies used for this purpose are similar to blanking dies. The main difference being the type of presses used for trimming which should normally have a large table. The Figure shows trimming operation.

6.4.8 Nibbling

In nibbling operation a specific contour is cut by producing a series of overlapping holes or notches as shown in Fig. 6.15. In this manner simple punches can be used to cut a complex shape from sheet metals. Nibbling is used when contour is long and separate punch is impractical and uneconomical. The punches used may be square, round or triangular depending upon the applications.

6.4.9 Notching

Notching is essentially same as the piercing except that it removes small portion of metal along the edge of the sheet.

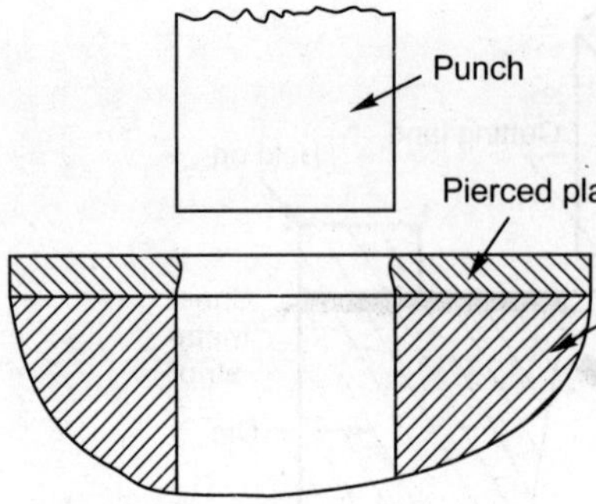

Fig. 6.13 Shaving.

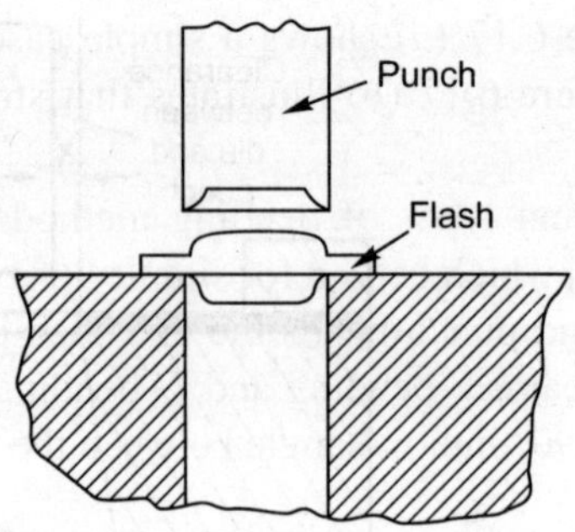

Fig. 6.14 Trimming.

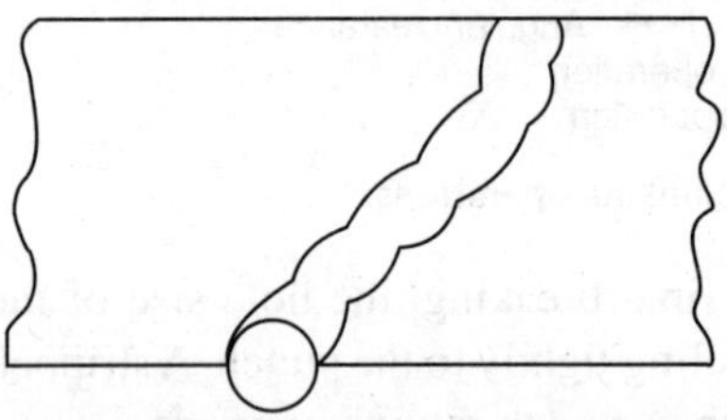

Fig. 6.15 Nibbling.

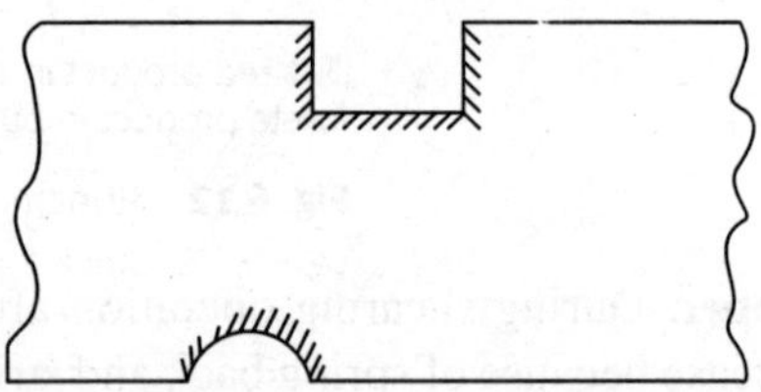

Fig. 6.16 Notching.

6.4.10 Bending

Bending is the plastic deformation of metals about a linear axis with little or no change in surface area. The bending of flat sheet is shown in Fig. 6.17. Here due to applied forces, metal on the outside is stretched while that on the inside is compressed. There is a plane in between which is neither stretched nor compressed. This plane is known as neutral plane. Neutral plane should be at the centre when the material is elastically deformed. When the material is plastically deformed the neutral axis move downward towards compressed layer. Since materials oppose compression much better than tension. The thickness is slightly decreased.

In bending operations there is some elastic recovery after the punch is removed. This is known as spring back (Fig. 6.18). Thus if metal is to be bent at specific angle the metal must be slightly over bent to compensate for the unbending action of spring back. Bending operation is performed while manufacturing many components such as trays, boxes, brackets, clips etc.

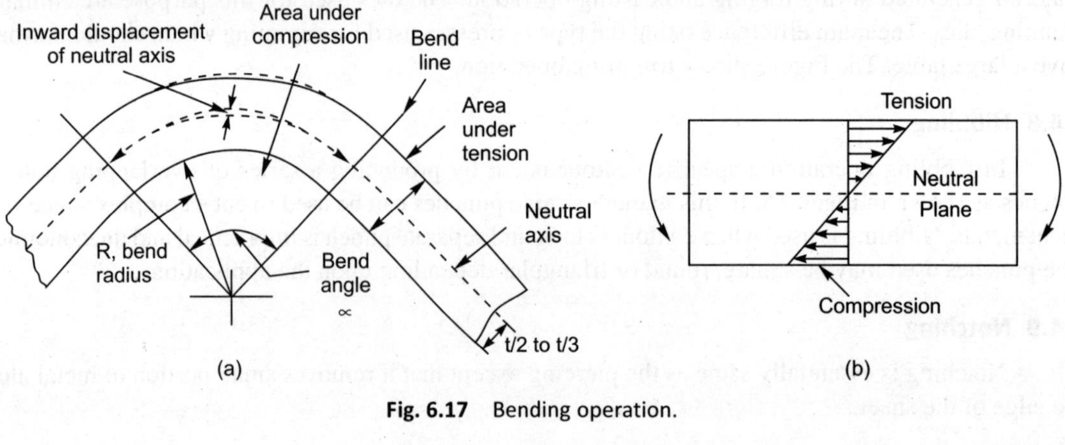

Fig. 6.17 Bending operation.

Figure 6.17 (*a*) shows a simple case of bending and inward displacement of neutral axis in bending. Figure 6.17 (*b*) illustrates that stresses on the outer most sides are maximum and zero at netural plane

Different types of bending methods are shown in Fig. 6.19. The first one is edge bending [Fig. 6.19(*a*)] which is used for simple 90° bend only. Here the workpiece is held firmly to the die with hold down and punch bends the extended portion of metallic blank. Here the work piece behaves a cantilever beam. *V*-bending and *U*-bending is shown in (Fig. 6.17 (*b*), 6.20 (*c*)). The punch and die often form a 90° internal angle between the faces of the metal.

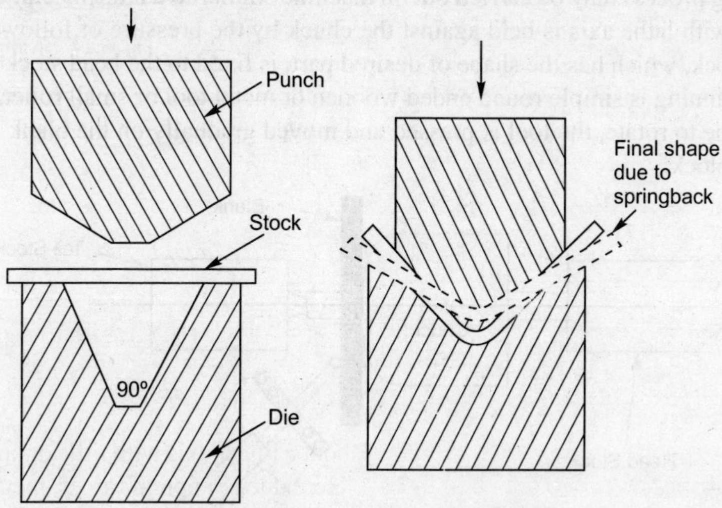

Fig. 6.18 Spring back in bending operation.

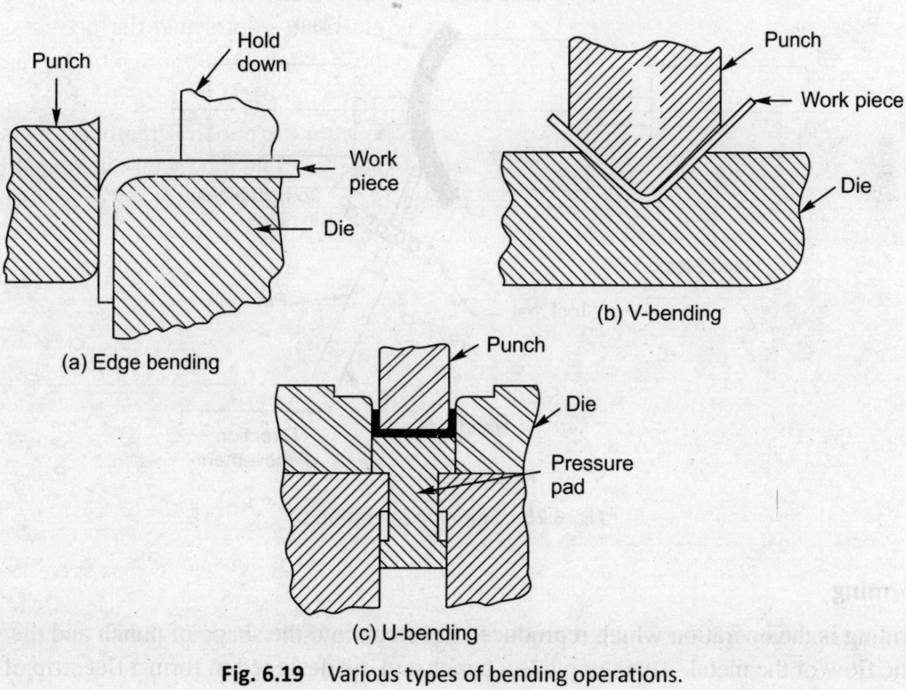

Fig. 6.19 Various types of bending operations.

Bending operation is closely related to forming operation. Bending is some times called forming operation, when multiple bends are made with single die. If the axes of deformation are not linear or are independent, the process becomes drawing and/or stretching.

6.4.11 Spinning

Spinning is the process used for making cup sharped parts having symmetrical shapes. In this process a blank is rotated, fixed against the forms block and then gradually force is applied on blank so that, blank takes the shape of form block (Fig. 6.20).

The spinning process may be carried out on machine similar to a lathe machine. A circular blank properly centered with lathe axis is held against the chuck by the pressure of follower attached to tail stock. The form block, which has the shape of desired part, is fixed to the head stock of lathe machine. The tool used in spinning is simple round ended wooden or metal tool or small roller. As the blank and form block are made to rotate, the tool is pressed and moved gradually on the blank so that blank take the shape of form block.

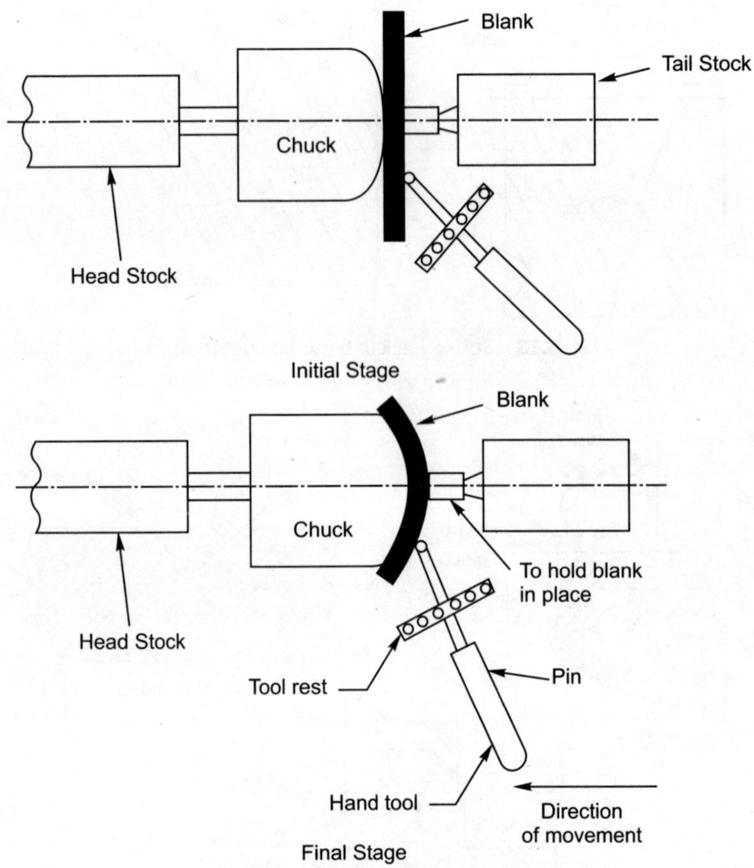

Fig. 6.20 Spinning operation.

6.4.12 Forming

Forming is the operation which reproduces flat stock into the shape of punch and die with little or no plastic flow of the metal. An example is a punch and die designed to form a flat strip of steel into a *U*-shape.

6.4.13 Stretch Forming

Stretch forming is an attractive way of producing the large sheet metal parts in limited or low quantities. In this process the sheet of metal is gripped by two or more sets of jaws that stretch it and warp it around a single form die block. This process of simultaneously stretching and bending is called stretch forming.

- In stretch forming most of the deformation is induced by the tensile stretching, therefore, forces on the die are far less than those normally present in bending or forming because forces are so low that the dies can often be made of wood, low melting poing metal or even plastic.

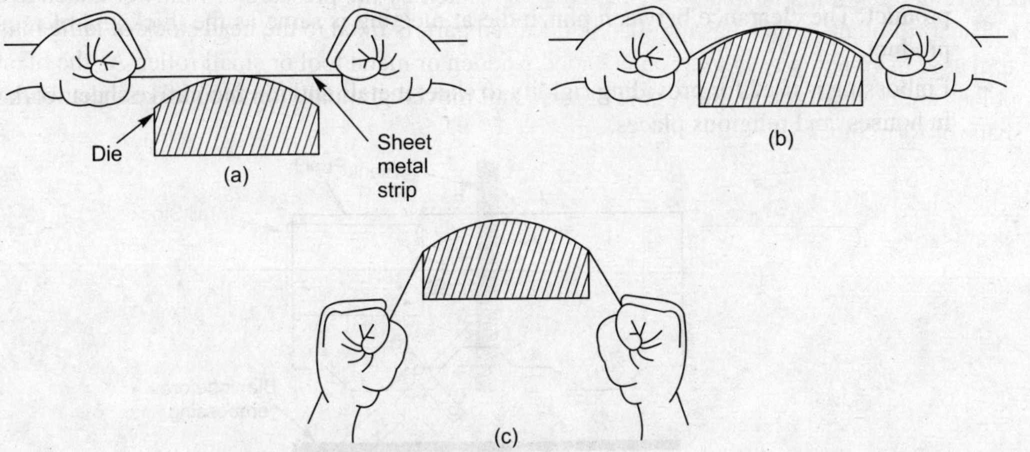

Fig. 6.21 Stretch forming.

Various steps to illustrate the process of stretch forming. It is also shown the how metal is subjected to stretching and bending in stretch forming.

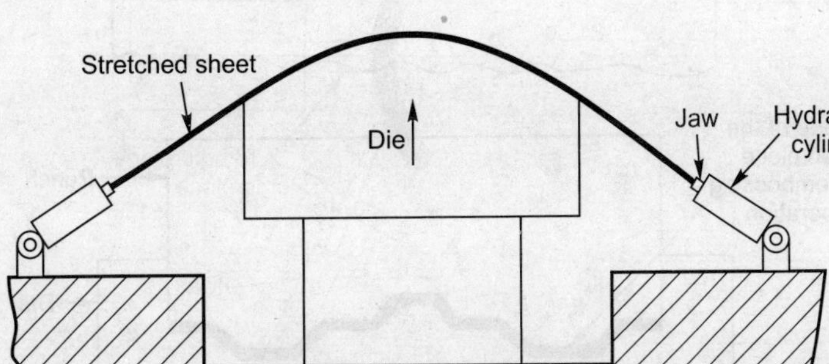

Fig. 6.22 Stretch forming.

- As in this process deformation is carried out in plastic region there is very little spring back and work piece closely conforms to the shape of tool.
- In stretch forming we can have component either single or double curved surfaces. The steel should have uniform thickness otherwise the thinner portions are likely to be over stretched. Also if the sheet is having any holes before the stretch forming they are likely to be enlarged.

- Stretch forming is quite popular in the air craft industry and frequently used to form aluminium and stainless steel into wing tips, scoops and other large panels.
- Sometimes mating male and female dies are used to shape the metal while it is being stretched. This process is known as stretch draw forming.

6.4.14 Embossing

Embossing is the operation in which projected or raised figures are made on sheet metals with corresponding relief on the other side. It basically involves drawing and bending operation. There may be the negligible change in the thickness of the metal.

- The die set consists of punch and die with required contours which are desired on the final product. The clearance between punch die at meeting is same as the thickness of required product.
- Embossing is used for providing rigidity to sheet metals and for decorative sheet work used in houses and religious places.

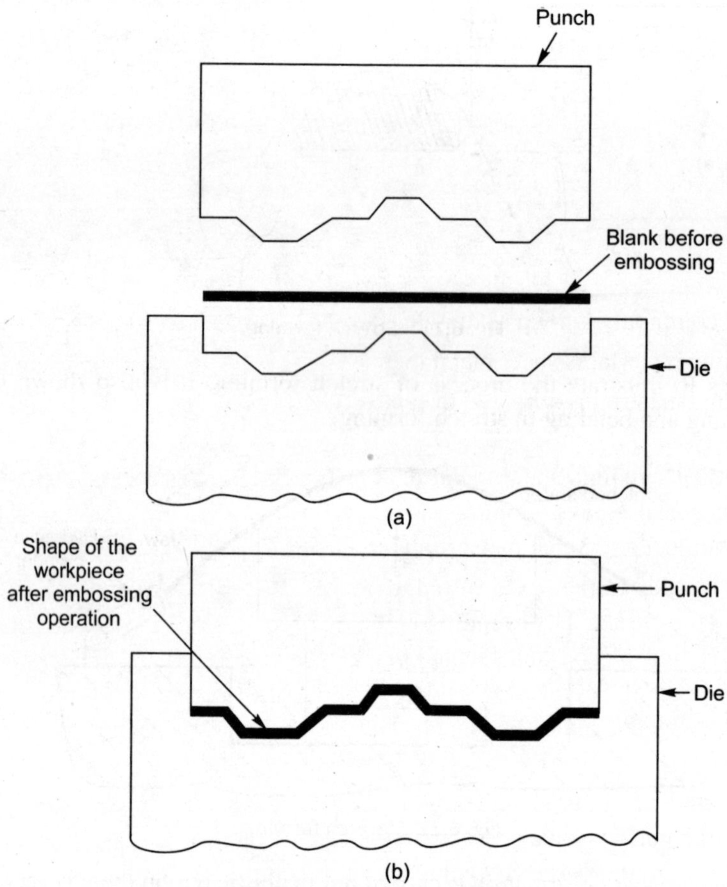

Fig. 6.23 Embossing operation.

6.4.15 Coining

Coining is basically a cold forging operation. The only difference is that here the flow of metal occurs only at top layers and not at the entire volume as in case of cold forging.

The process is used to produce coins, meals and similar products where exact size and fine details are required. The coining die consists of punch and die which are engraved with necessary details required on the both sides of final product. A blank is kept on the die and squeezed by it. The pressure required in coining is quite high as very fine details are requied in coining. For example, a pressure of 240,000 psi may be required to make a nickel coin of 50 paisa.

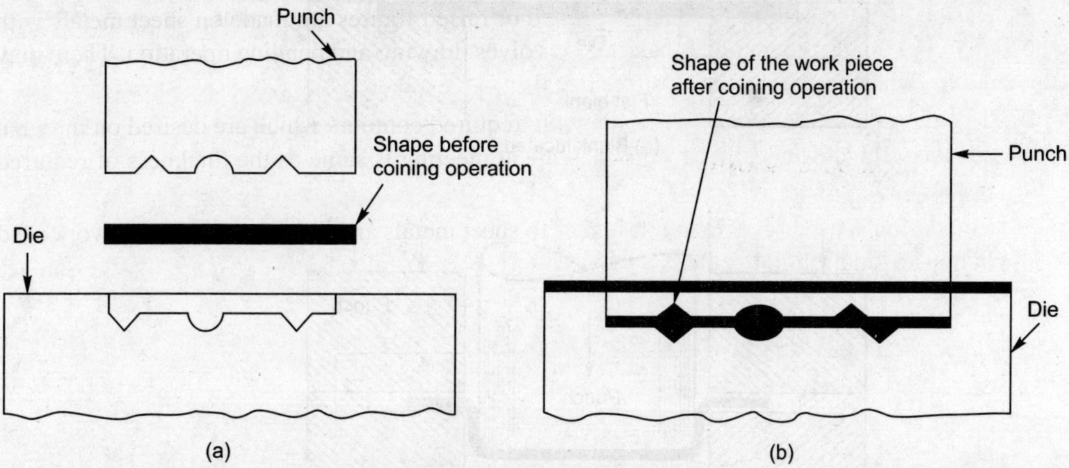

Fig. 6.24 Coining operation.

6.4.16 Cold Drawing

Cold drawing commonly known as drawing refers to two different operations.

(*a*) If the starting material is sheet metal then the operation is sheet metal drawing.

(*b*) If the starting material is wire, rod or tube then the operation is wire drawing rod drawing or tube drawing respectively.

(*a*) **Sheet Metal Drawing.** Sheet metal drwing is defined as process for making cup shaped aticles from flat sheet metal blanks. Common example of such components are dishes, trays, brake drums, cylindrical container etc. Sheet metal drawing operation is shown in Fig. 6.25.

The set up is similar to that used in balnking except that punch and die are provided with necessary rounding at the corner. The rounding is provided to allow the smooth flow of metal during drawing. The blank is first placed and located above the die. The blank holder comes down and properly holds the edges of the blank. After this punch moves downwards to force the blank to take the shape of cup formed by the end of the punch. The punch and blank holder then returns upward to complete the cycle.

- The punch and die radius should be equal and sufficient large (about 5 times the thickness of metal). Large punch radius prevents fracture of metal near the end of punch where tensile forces are maximum. Large die radius prevents tearings of metal as it flows over the edge of the die. The radius should not be too large otherwise wrinkle may form.
- The depth of draw in sheet meal drawing may be shallow, moderate or deep. If the depth of formed cup is less than half its diameter the process is called shallow drawing. If the depth of formed cup exceeds the diameter, it is known as deep drawing. Bottle caps and automobile panels are the examples of drawing process.

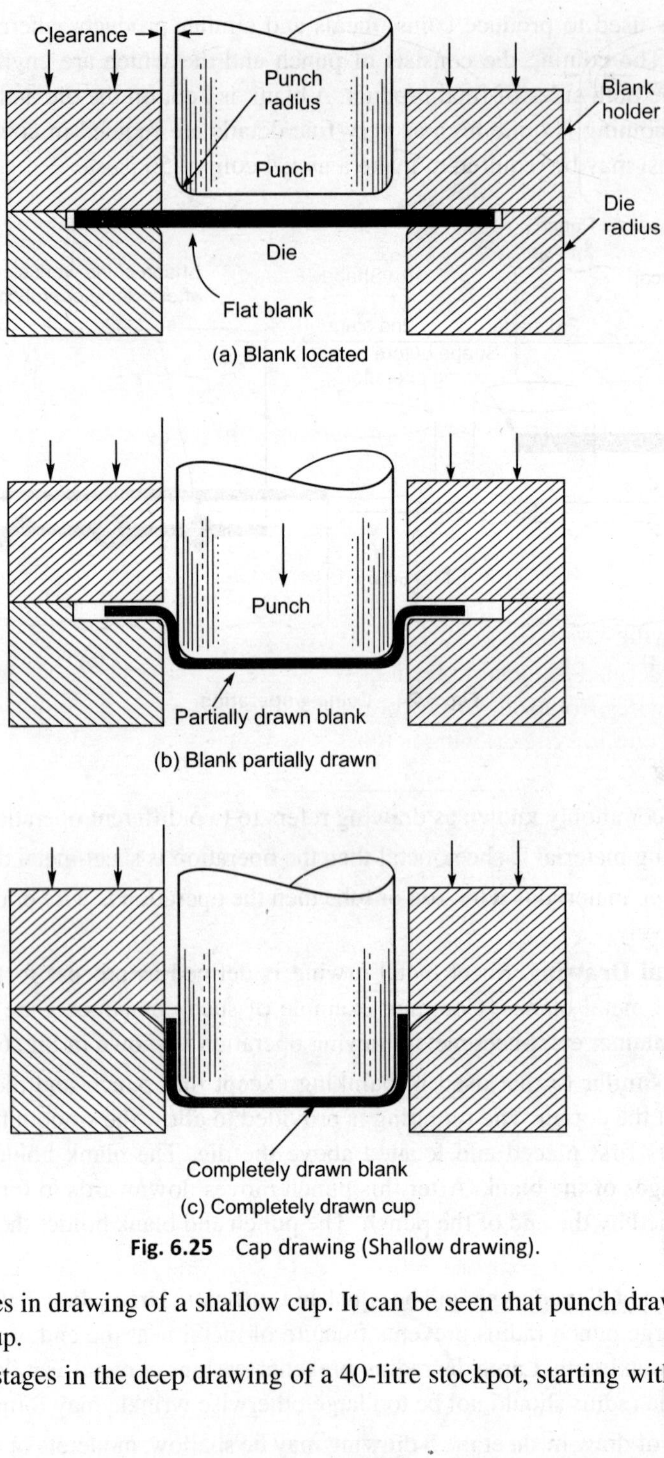

Fig. 6.25 Cap drawing (Shallow drawing).

Various stages in drawing of a shallow cup. It can be seen that punch draws the blank over the edges to form the cup.

Progressive stages in the deep drawing of a 40-litre stockpot, starting with a circular blank 2.5 mm thick.

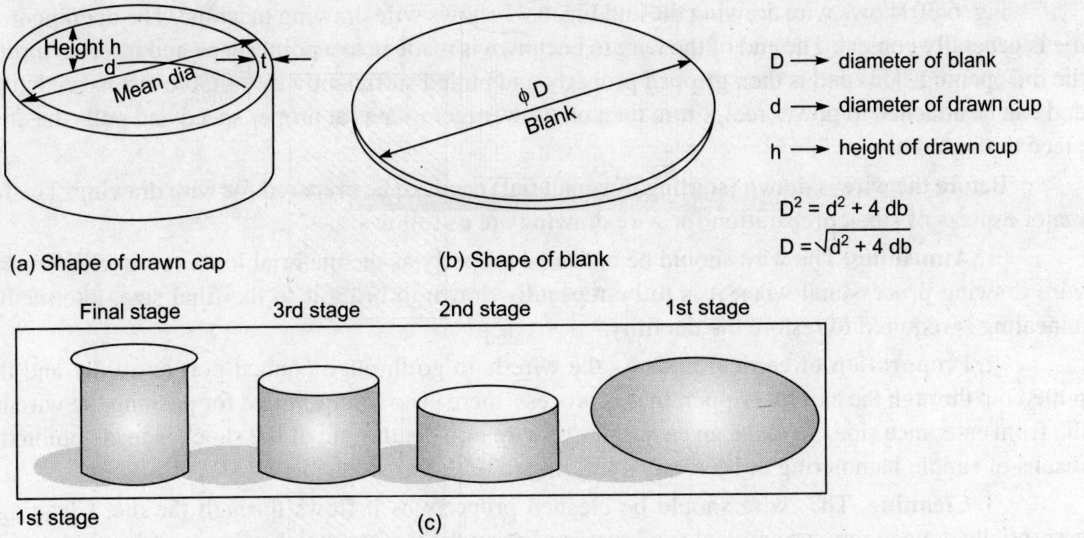

Fig. 6.26 Deep drawing.

(*b*) Wire Drawing

Wire may be defined as metal in form of thread or slender rod which is flexible. Wire drawing process is to obtain wires from rods of bigger diameter through wire drawing die. (Rods are the bars with circular cross section.) Wire drawing is mostly a cold working process.

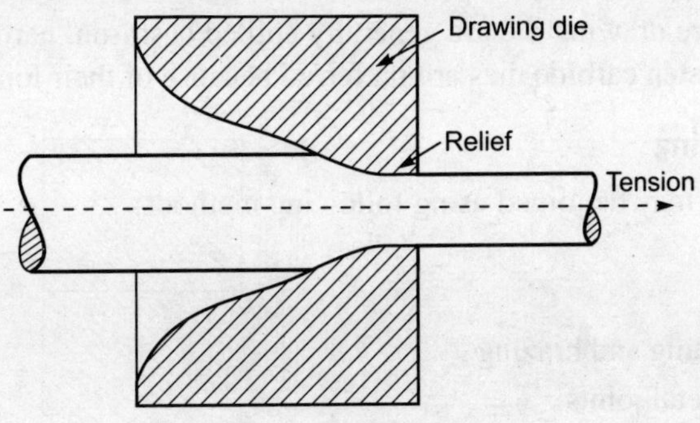

Fig. 6.27 Wire drawing die.

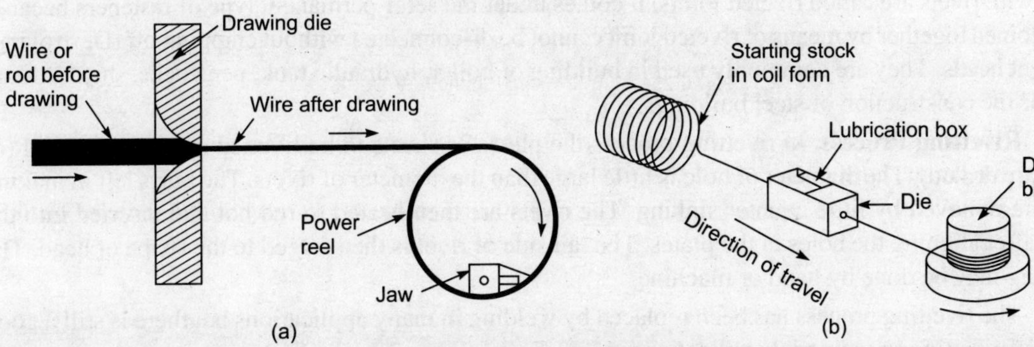

Fig. 6.28 Wire drawing machine.

Fig. 6.30 shows wire drawing die and Fig. 6.31 shows wire drawing machine. The opening in the die is generally conical. The end of the wire to be drawn is made in to a point shape and inserted through the die opening. This end is then gripped properly and pulled sufficiently by suitable means so that this end can be attached to power reel. From then on, power reel rotates at proper speed and pulls the entire piece through die.

Before the wire is drawn (starting raw material) needs to be prepared for wire drawing. The four major aspects of stock preparation for wire drawing are as follows:

1. **Annealing.** The wire should be annealed properly as the material looses its ductility during wire drawing process and when it is to be repeatdly drawn to bring it to the final size, intermediate annealing is required to restore the ductility.

2. **Preparation of conical end.** As the wire is to go through conical portion of die and then pulled out through the exit by gripper. In this process there is no force applied for pushing the wire into die from enterance side. To make an easy entry of wire into die the end of the shock is made pointed by means of simple hammering or by rotary swagging.

3. **Cleaning.** The wire should be cleaned properly as it flows through the die. Cleaning is essesntially done to remove any scale and rust present on the surface which may severely affect the die. Cleaning is normally done by acid pickling.

4. **Lubrication.** The pressure acting at the interference of metal and die is quite high and, therefore, lubrication of die is a serious problem. Two methods of lubrication are widely employed. In one method wire surface is cleaned, coated with lime and throughly dired. Before entering the die lubricant such as grease or soap is applied to surface of wire or rod. Second method is used for very thin wires. In this case Electrolytic coating of copper is used to reduce friction.

The material of wire drawing die are generally chilled cast iron, hardened alloy steel, tungsten carbide or diamond. Tungsten carbide dies are preferred because of their long life.

6.4.17 Sheet Metal Joining

Sheet metal pieces may be joined using following methods:

1. Rivetting
2. Screwing
3. Soldering, welding and brazing
4. Folded sheet metal joints.

1. Rivetting (Fig. 6.29). Rivets are used to join two or more plates or pieces of metals. Joints made with rivets are called riveted joints. It comes under the semi-permanent type of fasteners because sheet joined together by means of riveted joint cannot be disconnected without chipping off (Destroying) the rivet heads. They are commonly used in building of boiler, hydraulic tank, penstocks, ship building and for the construction of steel buildings.

Rivetting Process. In rivetting process the plates or sheets to be rivetted are first punched (or holes are drilled). The diameter of hole is little large than the diameter of rivets. The burrs left in making hole are removed by little counter sinking. The rivets are then heated to red hot and inserted into the holes after aligning the holes in the plates. The tail side of rivet is then forged to the shape of head. The rivetting may be done by hand or machine.

The rivetting process has been replaced by welding in many applications but there is still a good scope of rivetting in industrial applications.

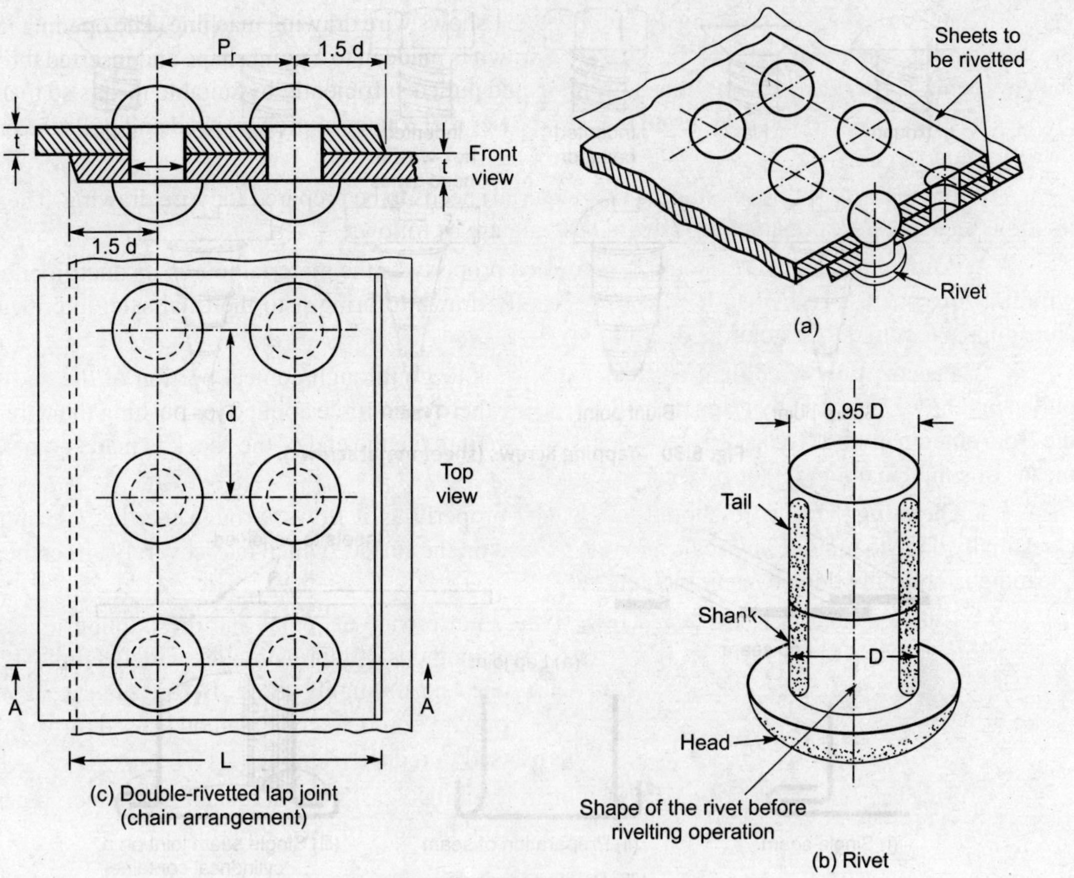

Fig. 6.29 Rivetting process.

There are basically two types of rivetted joints

1. Lap joints
2. Butt joints.

2. **Serewing (Fig. 6.30).** In case of joining very large parts especially when they are to be joined to other metallic or large non metallic bodies screwed fastening can be used. Tapping screws are designed especially for joining sheet metals.

Tapping screws can be categorised as thread cutting screw and thread forming screws:

➢ Thread cutting screws remove metal during the forming of threads in drilled or punched holes.

➢ Threads forming screws displaces the metal in drilled or punched hole when driven. They do not remove the metal.

Screwing procedure:

➢ While using sheet metal screw first.

➢ Layout of hole is prepared then holes are drilled or punched. The hole size should be equal to the root dia of the screw.

➢ The pieces of metal to be joined are held together firmly, their holes are aligned and then screw can be fastened in place with a screw driver.

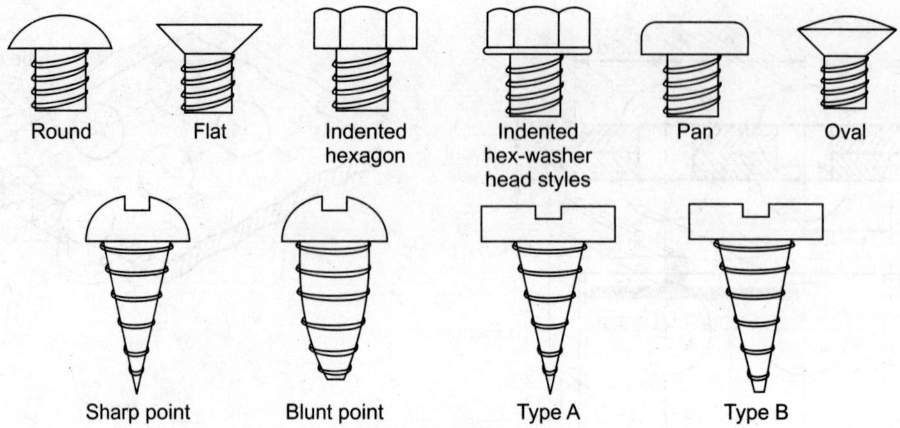

Fig. 6.30 Tapping Screws (sheet metal screws).

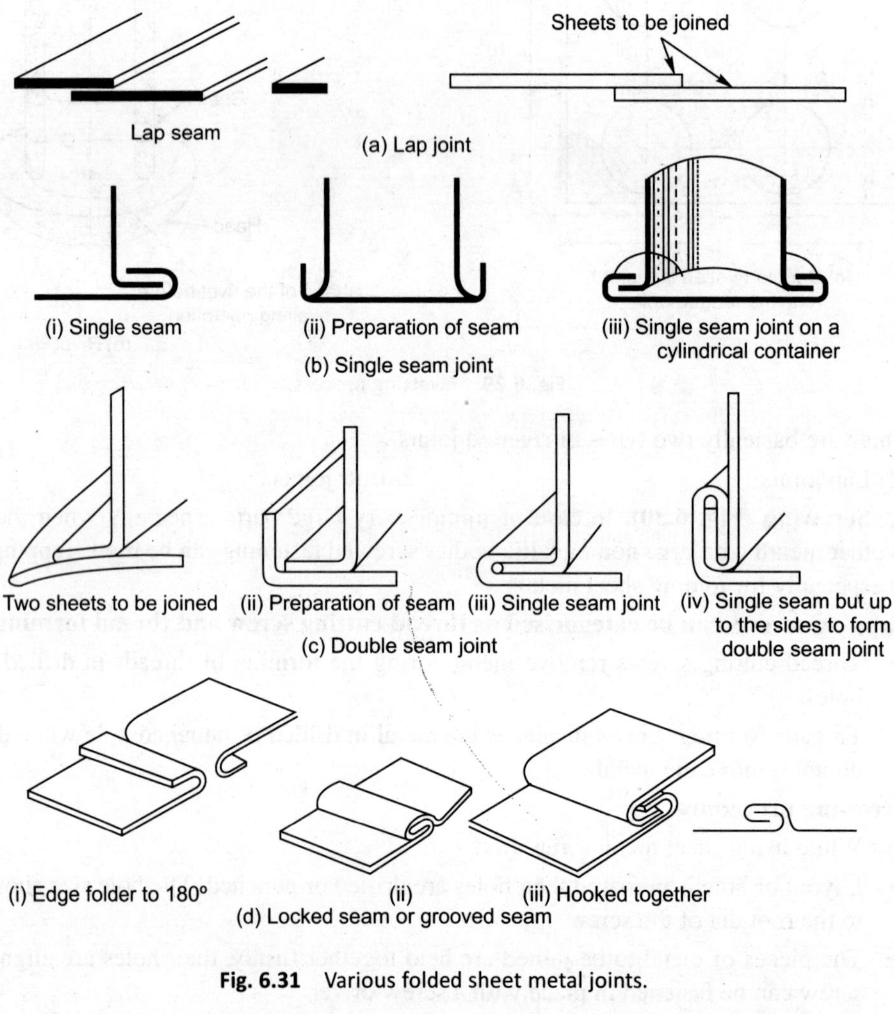

Fig. 6.31 Various folded sheet metal joints.

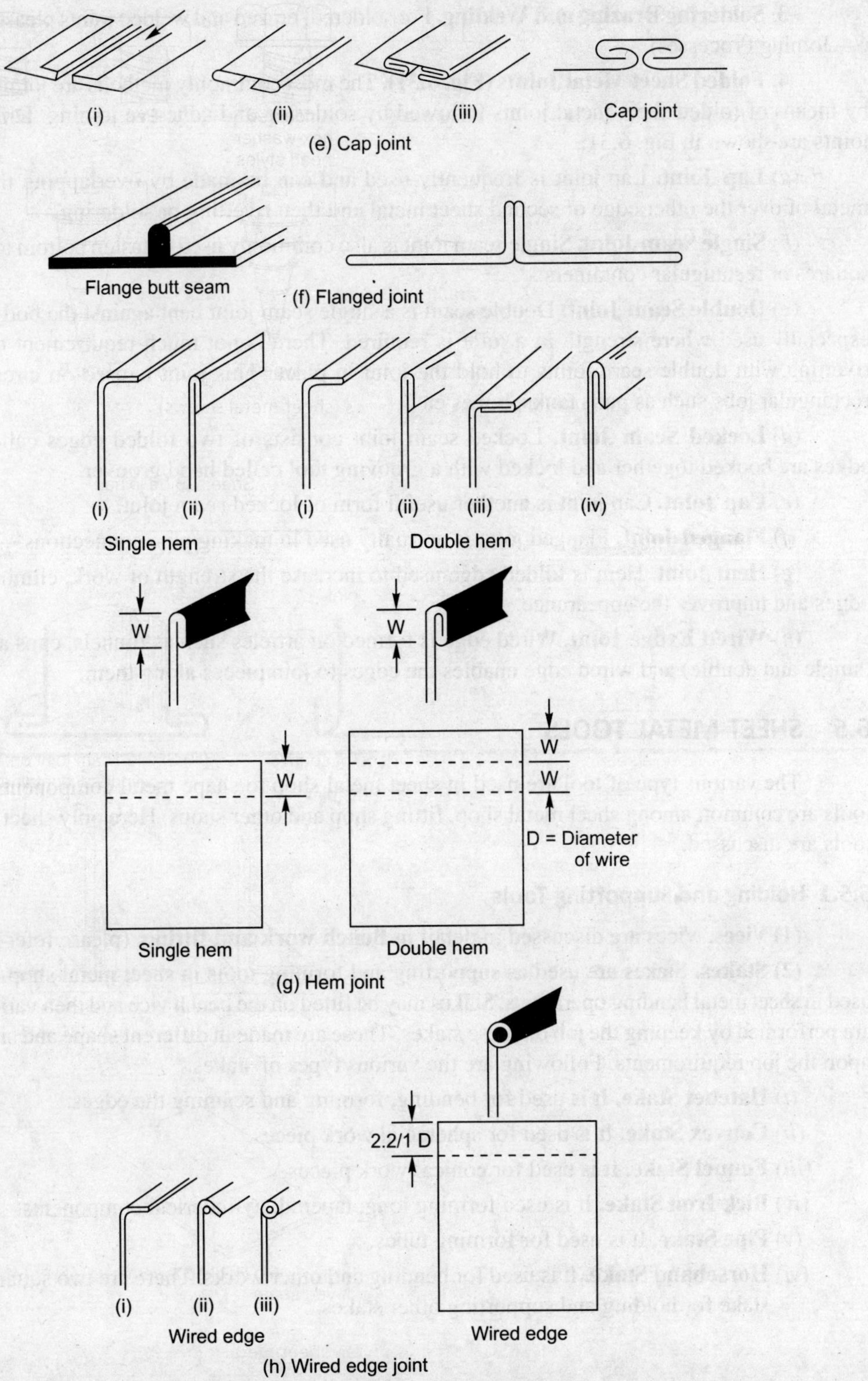

Fig. 6.31 Various Folded Sheet Metal Joints.

3. Soldering Brazing and Welding. For soldered, brazed and welded joints please refer (Chapter 6—Joining Processes)

4. Folded Sheet Metal Joints (Fig. 6.31). The most commonly methods are joining sheet metals by means of folded sheet metal joints followed by soldering and adhesive joining. Different types of joints are shown in Fig. 6.31.

(*a*) **Lap Joint.** Lap joint is frequently used and can be made by overlapping the one edge of metal of over the other edge of second sheet metal and then rivetting or soldering.

(*b*) **Single Seam Joint.** Single seam joint is also commonly used to fasten bottom to cyclinderical, squares or rectangular containers.

(*c*) **Double Seam Joint.** Double seam is a single seam joint bent against the body of object and especially used where strength in a joint is required. There is not much requirement of soldering or rivetting with double seam joints to hold the joint in place. This joint is used on circular, square or rectangular jobs such as pails tanks, boxes etc.

(*d*) **Locked Seam Joint.** Locked seam joint consists of two folded edges called locks. Two edges are booked together and locked with a grooving tool called hand groover.

(*e*) **Cap Joint.** Cap joint is another useful form of locked seam joint.

(*f*) **Flanged Joint.** Flanged joint is normally used in making pipe connections.

(*g*) **Hem Joint.** Hem is folded edge used to increase the strength of work, eliminates the sharp edges and improves the appearance.

(*h*) **Wired Exdge Joint.** Wired edge is formed on articles such as funnels, cups and trays. Hem (single and double) and wired edge enables the edges to join pieces along them.

6.5 SHEET METAL TOOLS

The various type of tool are used in sheet metal shop to shape metal components. Some of the tools are common among sheet metal shop, fitting shop and other shops. Here only sheet metal specific tools are discussed.

6.5.1 Holding and Supporting Tools

(1) Vices. Vices are discussed in detail in **Bench work and fitting.** (please refer 4.3.1)

(2) Stakes. Stakes are used as supporting and forming tools in sheet metal shop. They are also used in sheet metal bending operations. Stakes may be fitted on the bench vice and then various opeartions are performed by keeping the job on these stakes. These are made in different shape and sizes depending upon the job requirements. Following are the various types of stakes.

(*i*) **Hatchet Stake.** It is used for bending, forming and seaming the edges.

(*ii*) **Convex Stake.** It is used for spherecial work pieces.

(*iii*) **Funnel Stake.** It is used for conical work pieces.

(*iv*) **Bick Iron Stake.** It is used forming long, tapered, cylinderical components.

(*v*) **Pipe Stake.** It is used for forming tubes.

(*vi*) **Horseband Stake.** It is used for bending and other works. There are two square holes in this stake for holding and supporting other stakes.

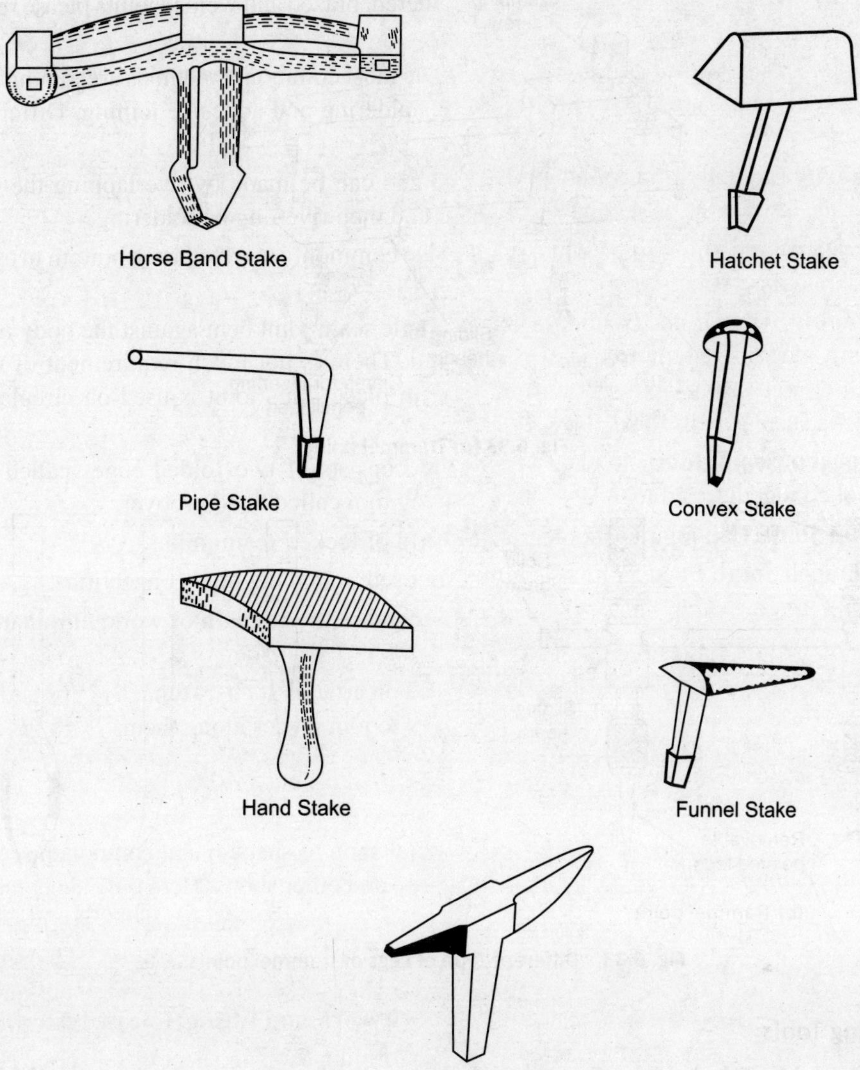

Fig. 6.32 Various types of stakes.

6.5.2 Measuring and Marking Tools

Various tools used in sheet shop for measuring and marking are steel rule, try square, vernier calipers, bevel protractor, combination set, scribers, divider, punches and trammel points etc. All these tools except trammel points are discussed in detail in bench work and fitting. (please refer 4.4)

(*i*) **Trammel Points.** It is used for drawing large arcs and circles which can not drawn with dividers. Trammel points has two straight removable legs having needle points. These legs are mounted on two separate heads which can slide on the beam as shown in Fig. 6.36. These heads can be locked at any position on beam with the help of thumb screws. A clamp is provided on one of the legs in which pencil can be inserted (for drawing arcs and circles).

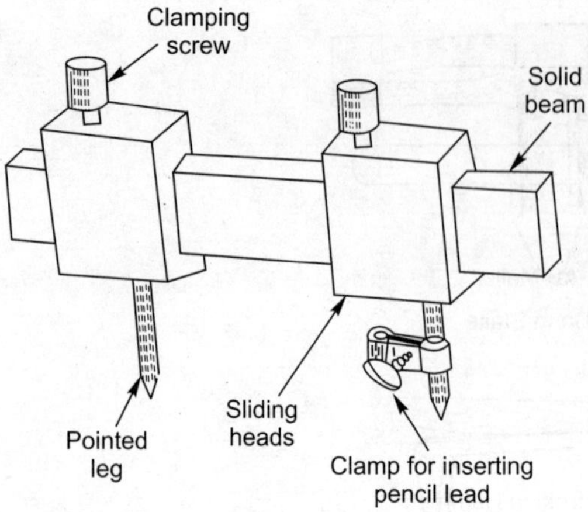

Fig. 6.33 (*a*) Trammel point.

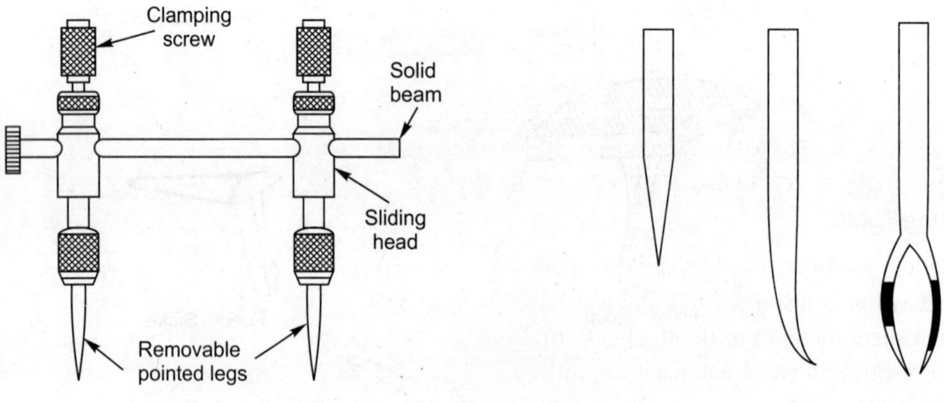

(b) Trammel point (c)

Fig. 6.33 Different Type of Legs of trammel points.

6.5.3 Striking Tools

Striking tools include hammers various hammers used in sheet metal shop are as follows:

(1) **Mallet.** Mallets are soft hammers made up of wood, plastic or fibrre. These are used whenever less force is required. Mallet is used for smoothening of sheet. The blow of mallet does not spoil the surface of sheet.

(2) **Setting Hammer.** This hammer has square flat face and is use for flattening the seams.

(3) **Raising Hammer.** It is used to form flat surface of sheet in to curved surface. It is used with raising block.

(4) **Riveting Hammer.** It is used for hammering rivets. The face of hammer is square and slightly curve with bevelled edges.

(5) **Ball peen, straight peen and cross peen hammers.** Hammers are discussed in detail in forging and smithy shop (please refer 5.4.12)

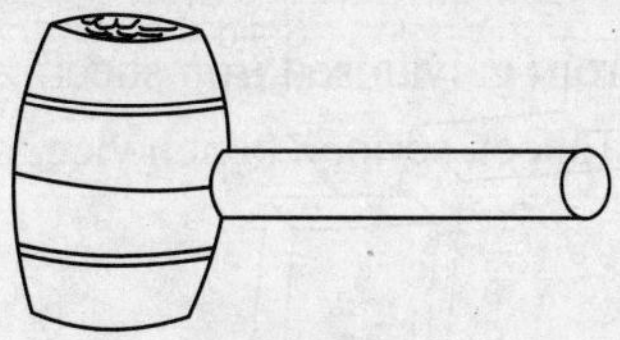

(a) Mallet

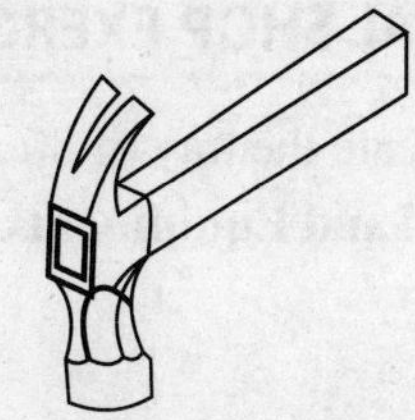

(b) Claw Hammer

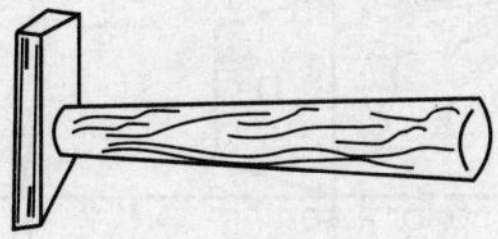

(c) Setting Hammer

(d) Raising Hammer

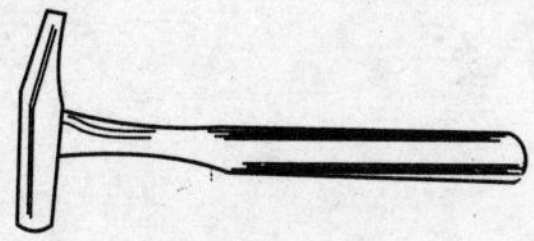

(e) Rivetting hammer

Fig. 6.34 Various types of hammers used in sheet metal shop.

6.5.4 Cutting Tools

The tools used for cutting sheet metal are called snips.

Hand Snips. Snips are also known as hand shear and are similar to a pair of seissors. These are usd to cut the sheet metal upto the thickness of 22 gauge. Most commonly used ship are straight snip and bent snips. Straight snips and used for cutting along a straight line whereas a bent snip is used for cutting along a curvature.

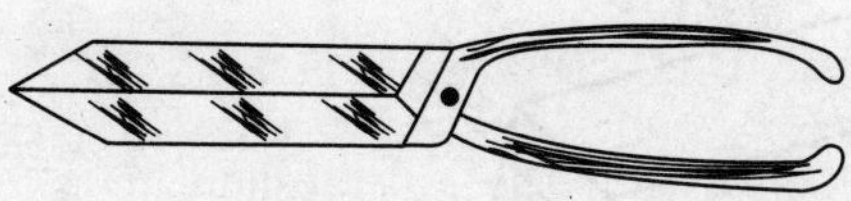

(a) Straight snip

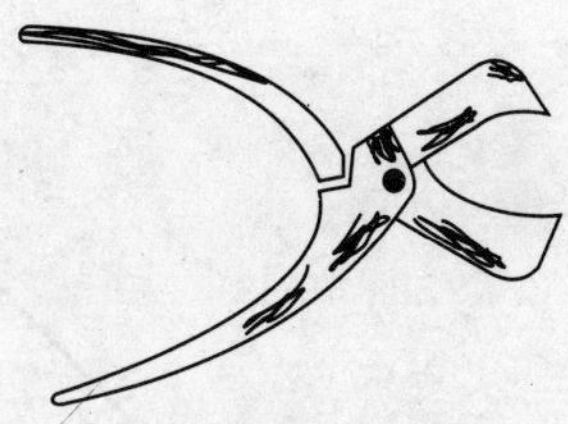

(b) Bent snip

Fig. 6.35 Various types of snips.

6.6 SHEET METAL SHOP EXERCISE

Aim. To fabricate the tray of 80 × 80 × 25 mm. from galvanised iron sheet.

Raw Material and Equipments. Galvanised Iron sheet, scriber, bench vice, marking punch, try square, soldering iron.

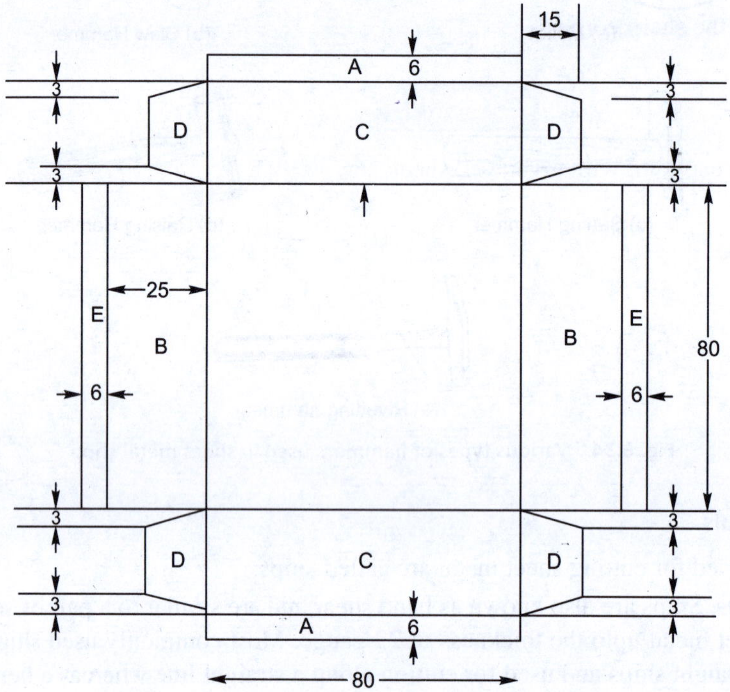

Fig. 6.36 Layout for fabrication of sheet metal tray.

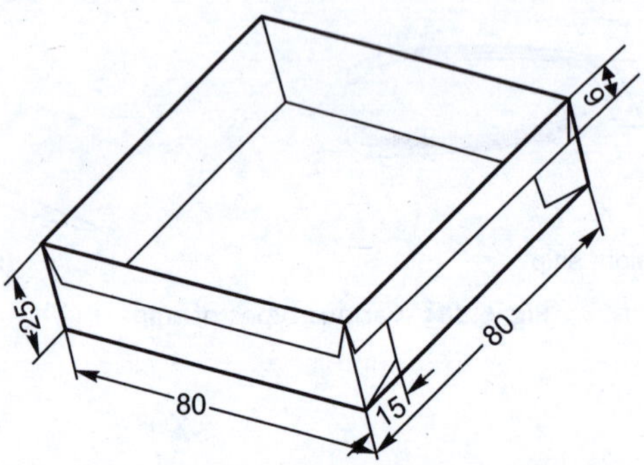

Fig. 6.37 Square tray.

Procedure:

(*i*) Cut a G.L. sheet of 142 × 142 mm with help of snips.

(*ii*) Draw layout with help of scriber as shown in Fig.

(*iii*) Cut the sheet along the marked line using a suitable snip.

(*iv*) Make closed fold on regions marked *A* away from lines.

(*v*) Make square bend on regions marked *B* towards the line.

(*vi*) Make square bend on region marked *C* toward the line.

(*vii*) Square bend four small region *D* upon the surface of *B*.

(*viii*) Fold the regions marked *E* away from lines to make single hem joint.

(*ix*) The corner lap joints sheet metal thus formed may be soldered if requierd.

(*x*) File all the sharp corners.

EXERCISE PROBELEMS

1. Make a cap joint with given G.I. sheet (Fig. 6.38).

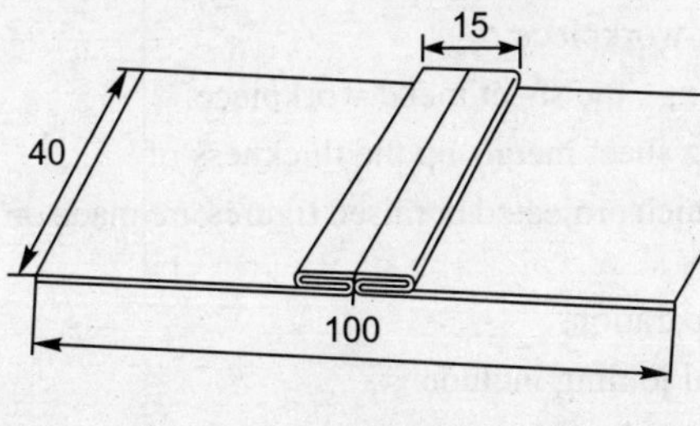

Fig. 6.38

2. Make a locked seam joint with given G.I. sheet (Fig. 6.39).

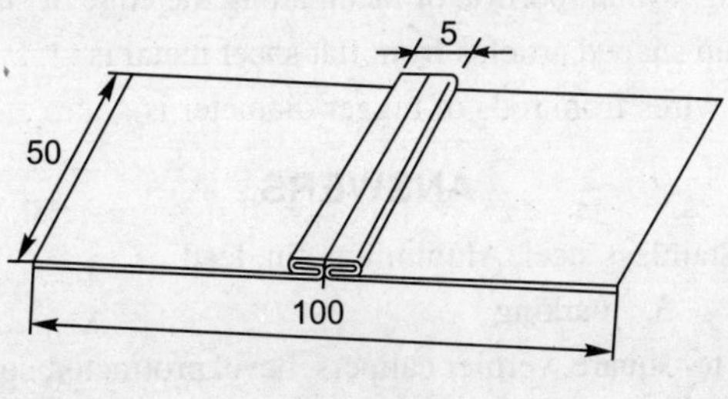

Fig. 6.39

6.7 SAFETY PRECAUTIONS WHILE WROKING IN SHEET METAL SHOP

(*i*) Make sure that square sheet 142 × 142 mm is cut and edges are perpendicular to each other.

(*ii*) Ensure that folding region *A* is first operations after laying out and folding region *E* is the last operation.

(*iii*) Sharp corners and burrs should be properly field and removed.

(*iv*) Use only shoft hammer (mallet) so that no hammering marks are visible.

(*v*) Make sure that square sheet is properly clamped in bench vice.

(*vi*) Be careful while using snips for cutting square sheet.

FILL IN THE BLANKS

1. Some of metals used for sheet metal work are ....
2. In sheet metal wire gauge as the gauge number increases, the thickness of sheet ....
3. Seriber are used for .... the sheets.
4. Various measuring and marking tools used in sheet metal work are ....
5. The soft hammer used in sheet metal work is known as ....
6. The process of coating iron sheets with thin layer of zinc is called ....
7. .... are used for drawing large circles which are not possible to draw with dividers.
8. Punching may also be termed as .... operation.
9. The thickness of sheet metal or dia of wire may be checked with ....
10. Stake are used as .... tools.
11. .... stake is used for conical workpiece.
12. Horseband slake is used for .... the sheet metal workpiece.
13. Hand snips are used for cutting sheet metals up the thickness of ....
14. The sheet metal operation in which projected or raised figures are made on sheet metals with corresponding relief on other side is known as ....
15. Coining is basically a .... operation.
16. Various method for sheet metal joining include ....
17. In punching operation sheet with the holes is .... product and removed piece is .... product.
18. In blanking operation the shapes cutout from sheet is .... product and the sheet left with holes is .... product.
19. The operation of removing a small portion of metal along the edge of sheet is ....
20. The process of making cup shaped articles from flat sheet metal is ....
21. The process of obtaining wires from rods of bigger diameter is ....

ANSWERS

1. galvanised iron, copper, stainless steel, Aluminium, tin, lead.
2. decreases **3.** marking
4. steel rule, trammel point, try square, vernier calipers, bevel protractor, combinationset, divideder punches etc.
5. mallet **6.** Galvanization **7.** Tramme points
8. piercing **9.** sheetmetal wire gauge **10.** supporting and forming
11. Funnel **12.** bending **13.** 0.8 mm
14. Embossing **15.** cold forging
16. rivetting, screwing, soldering, welding, brazing and folded sheet metal joints
26. required, waste **27.** required, waste **28.** notching
29. sheet metal drawing **30.** wire drawing

MULTIPLE CHOICE QUESTIONS

1. Scribers in sheet metal work are used for
   (*a*) Marking (*b*) Cutting (shearing)
   (*c*) bending (*d*) forming

2. The most accurate tool of measuring the thickness of sheet metal is
(*a*) Vernier Caliper (*b*) Steel scale
(*c*) Outside spring caliper (*d*) mircometer

3. Stakes are
(*a*) workpiece told tools (*b*) cutting tools
(*c*) supporting tools (*d*) supporting and forming tools

4. The tools like try square, divider, vernier caliper, V block are
(*a*) marking and measuring tools (*b*) working holding tools
(*c*) filing tools (*d*) threading tools

5. Funnel stake is used for
(*a*) conical components (*b*) bending forming and seaming the edges
(*c*) spherical work pieces (*d*) forming tubes

6. Trammel point is
(*a*) workholding tool (*b*) work supportin tool
(*c*) marking tool (*d*) cutting tool

7. Mallet is
(*a*) striking tool (*b*) marking tool
(*c*) cutting tool (*d*) work supporting and holding tool

8. Punching is the
(*a*) operation of cutting out flat pieces of various shapes from sheet metal
(*b*) operation of making holes of various shapes in sheet metal
(*c*) operation of forming the sheet metal to desired shape
(*d*) operation of bending and joining sheet metal to desired shape

9. Blanking is the
(*a*) operation of cutting out flat pieces of various shapes from sheet metal
(*b*) operation of making holes of various shapes in sheet metal
(*c*) operation of forming the sheet metal to desired shape
(*d*) operation of bending and joining sheet metal to desired shape

10. In punching operation shear is provided to
(*a*) punch (*b*) die
(*c*) either on punch or die (*d*) both on punch and die

11. In blanking operation shear is provided to
(*a*) punch (*b*) die
(*c*) either on punch or die (*d*) both on punch and die

12. Blanking and punching operation can be simultaneously performed in a
(*a*) combination die (*b*) compound die
(*c*) progressive die (*d*) simple die

13. Notching is the operation of
(*a*) cutting out flat piece from sheet metal
(*b*) making holes in sheet metal
(*c*) removing small portion of metal along the edge of sheet
(*d*) cutting out flat piece from sheet metal and folding it.

14. Process of making cup shaped articles from flat sheet metal blank is known as
    (*a*) bending (*b*) sheet metal drawing
    (*c*) punching (*d*) blanking
15. Sheet metals may be joined using
    (*a*) rivetting (*b*) folded sheet metal joints
    (*c*) soldering (*d*) all of above
16. Sheet metal may be defined as
    (*a*) plate of thickness about 6 mm (*b*) plate of any thickness
    (*c*) plate of thickness less than 2 mm (*d*) none of above

ANSWERS

1. (*a*)	**2.** (*d*)	**3.** (*d*)	**4.** (*a*)	**5.** (*a*)
6. (*c*)	**7.** (*a*)	**8.** (*b*)	**9.** (*a*)	**10.** (*a*)
11. (*b*)	**12.** (*b*)	**13.** (*c*)	**14.** (*b*)	**15.** (*d*)
16. (*a*)				

VIVA VOCE QUESTIONS

1. How a sheet metal can be defined?
2. What are the various metals used in sheet metal work?
3. What is meant by layout of a sheet metal?
4. Explain the difference between punching and blanking operation.
5. What is the difference between embossing and coining?
6. What is meant by sheet metal drawing?
7. What is various folded sheet metal joints?
8. Explain the difference between single hem and double hem joints.
9. What is the use of wire gauge?
10. What is various shearing tools used in sheet metal work?
11. What are the striking tools used in sheet metal work?

"Keep away from people who try to brittle your ambitions. Small people always do that, but the really great make you feel that you too can become great".

CHAPTER

7

CARPENTARY SHOP

INSIDE THIS CHAPTER

7.1 INTRODUCTION

Wood is available in nature in form of trees. Various forms of wood have been used as engineering material since thousands of years.

Timber

The wood obtained from fully grown trees is cut by means of jaws into various commercial sizes. This wood is known as timber and is widely used for engineering or building or purposes. Timber has the following advantages over other materials.

1. It is very easy to provide desired shape to the timber by using tools.
2. It is lighter than most of engineering material and have reasonable strength too.
3. In timber work cost or material as well as tools is less compared to other materials.
4. Timber is bad conductor of heat and is favorable material for construction of houses timber houses will remain cool in summer and warm in winter.
5. Timber has unique combination of properties like strength, light weight, durability, and economical.

"Successful often comes to those who dare and act; it seldom goes to timid who are ever afraid of the consequences."

6. Timber is very useful for making sound proof houses.
7. The polished and painted timber has very good appearance.

7.2 TYPES OF WOOD

The wood can be classified into following categories:

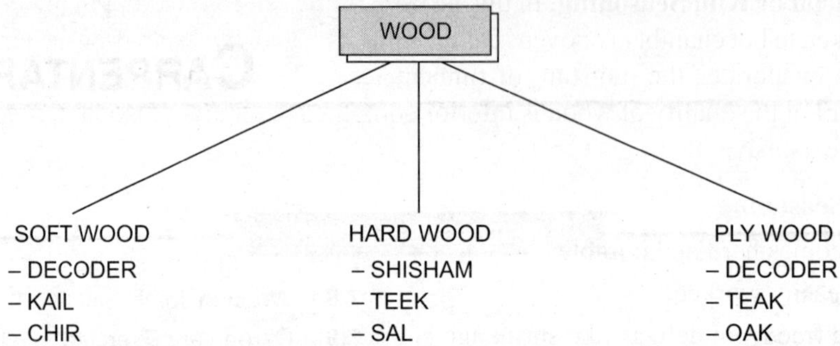

Fig. 7.1 Types of wood.

Characteristics of Hard Wood

1. The hard wood has closed structure.
2. It is heavier and dark in colour.
3. It is stronger and more durable.
4. It is difficult to work on it.
5. It has both good tensile as well as shear strength.
6. It can withstand higher temperatures.
7. It does not split quickly.

Characteristics of Soft Wood

1. Soft wood carries straight fibre and have fine texture.
2. It is lighter in weight and colour.
3. It is relative week and less durable.
4. It is easy to work on it.
5. It has good tensile strength but shear strength is comparatively less.
6. It cannot withstand higher temperatures.
7. It gets spoliated quickly.

7.3 SEASONING OF WOOD

The process of removing moisture from freshly cut down tress is known as seasoning. If the excess moisture is not taken out of wood, its presence may cause many problems like uneven shrinkage, wrapping etc. and wood will become unsuitable for use.

The commonly used seasoning methods are:

1. **Natural Seasoning.** In natural seasoning, timber boards are stacked in a shed such that they are not directly exposed to sun and rain, but free circulation of air takes place through them.

These are allowed to remain under this shed for considerably long period. Due to circulation of air moisture constants is evaporated and timber gets dried.

2. **Water Seasoning.** In water seasoning timber pieces are immersed in flowing water for a period of two to three weeks. The timber is then taken out and air seasoned in usual way. The seasoning time is less but the strength is reduced as flowing water takes away the sap from timber.
3. **Artificial or Kiln Seasoning.** In this process timber pieces are kept in big boxes. These box are taken to hot chambers or ovens called Kilns. Hot and dry steam is pushed into the chamber, which evaporates the moisture of timber pieces. This is the fastest way of seasoning the timber but the quality of wood is inferior compared to natural seasoning. Also this method is more expensive.

Advantages of Seasoning

1. Wood becomes hard and durable.
2. It can be easily worked.
3. It become free from defects like shrinkage and wrapping.
4. It becomes more tough.
5. Its resistance to fire is increased.
6. Density of wood is reduced.
7. Its ability to take up paint and polish is improved.

7.4 STRUCTURE OF WOOD

1. **Pith (or Medulla).** It is the dark central part of tree. It feeds sap from roots to leaves.
2. **Heart Wood.** The portion surrounding the pith is known as heart wood. It carries less moisture as compared to pith. This portion is harder, darker and stronger then remaining wood in tree.
3. **Annual Rings.** The concentric rings all around the pith are known as Annual ring. Every year one such ring is added to pith. That is why these are known as annual rings.

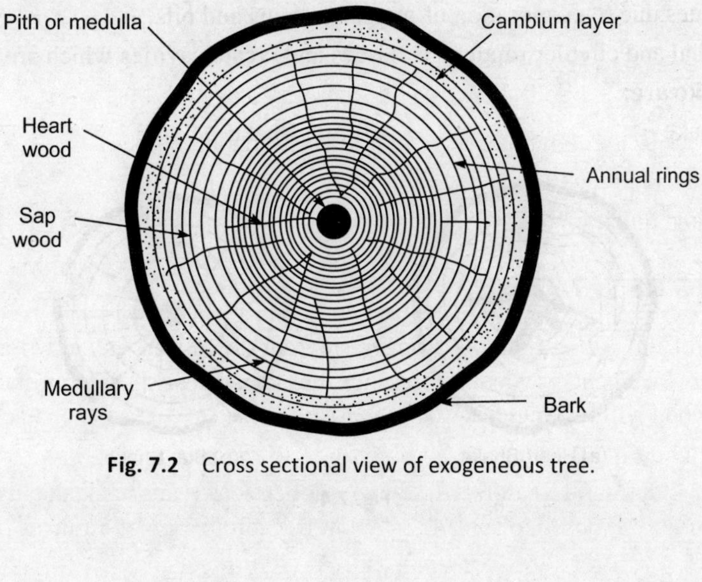

Fig. 7.2 Cross sectional view of exogeneous tree.

4. **Sap Wood.** The portion of wood between heart wood and cambium layer is known as sap wood. It is light in weight softer and weaker part. It carries high percentage of moisture and is not used for commercial and engineering purposes. Its mostly used as fuel wood.
5. **Cambium Layer.** The annular ring adjacent to bark is known as cambium layer. It has cellular structure and with passage of time is gradually converted into sap wood.
6. **Bark.** The outer surface of tree which acts as cover for inner portions is known as bark. It protects cambium layer from insects and frosts. It has fibrous construction.
7. **Medullary Rays.** These are the radial thin layers between pith and cambium layer. They are of cellur structure and help in binding annular rings together to provide a solid structure to tree. The sap food to the growing parts of the tree is supplied through them.

7.5 DEFECTS IN WOOD

Defects in wood can be divided into three categories:

(1) Natural defects caused in tree due to abnormality in the growth.

(2) Defects caused during seasoning operation.

(3) Defects caused due to insects or fungi.

(1) Natural Defects

(*i*) **Knots:** Knot is the centre of irregular grown part of the tree. The fibre cells of tree are wrapping around it and the normal shape of the internal structure is disturbed. In presence of knots strength of wood is reduced and is also difficult to work on wood. There are two types of knots:

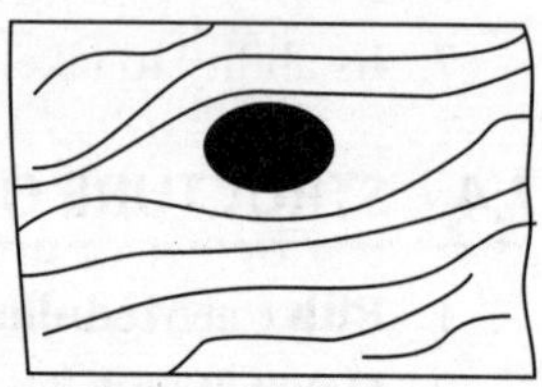

Fig. 7.3 Circular knot.

- **Dead Knots.** When the separation of branches takes place before the tree is cut.
- **Live Knots.** When the separation of branches takes place after the tree is cut. Live knots are not harmful to wood except offering difficultly in working.

(*ii*) **Shakes:** When the tree is not cut even after full maturity, the bonding material among the wood grain evaporates due to evaporation of moisture resins and oils.

It causes radial and circular ruptures is tissues and create cavities which are known as shake.

Type of shake are:

(*i*) Heart Shake (*ii*) Star Shake

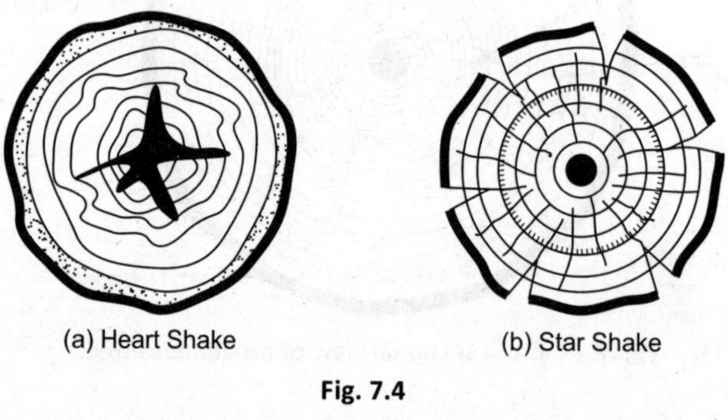

(a) Heart Shake (b) Star Shake

Fig. 7.4

(*iii*) **Ring Shake**

(1) **Heart Shake.** When cavities are in direction from pith toward sapwood they are known as heart shakes.

(2) **Star Shake.** When cavities originate from side of bark and extend toward the pith they are known as star shakes.

(3) **Ring Shake.** These are cavities appear in between the annual rings. These are caused due to excessive wind pressure.

(4) Shake is also caused by improper seasoning of the wood.

3. **Twisted Fibers.** This type of defect occur due to twisting of wood grain because of blowing wind. The fibre of the wood have different inclinations with respect to axis. It is very difficult to work on wood and a smooth surface cannot be obtained.
4. **Distortion.** Due to an even moisture inside the wood. The shape of the wood is diotorted. Even crack may occur inside the wood due to distortion. Twisting and lowing may also occur. This defect is also caused due to improper seasoning of tin wood.
5. **Case Hardening.** This defect is caused because uneven drying during seasoning. As the outer part of tree dries faster than inner portion stresses are developed on outer surface of wood. This causes hardening of the outer surfaces. As a result machining cannot be done easily.
6. **Honey Combing.** This type of defect occur during seasoning of wood. In this case. Inner portion of wood dry faster than the outer portion because of hygroscopic substances. Stresses are developed in inner surface of wood. This cause circular and radial crack in wood which intersect to produce the honey comb structure in the wood.
7. **The Defect Caused due to Insects and Fungi.** These are dry rot, wet rot and termites. These dry rot are caused by attack of fungi on dry wood whereas wet rot are caused by attack of fungi on wet wood.

Termites affect in life of timber e.g. white termite make the timber hollow from inside. Termite attack can be controlled by anti-termite treatment of live timber.

7.6 PRESERVATION OF WOOD

To protect the wood from attack insects and fungi and to increase its life chemical preservatives are used.

These preservatives can be classified into three broad categories.

1. **Tar Oils.** Most commonly used preservation are coal tar and wood tar. Coal is most effective and is applied on external surfaces.
2. **Water Soluble Chemical Salt.** The commonly used preservations are sodium floride, copper sulphate, zin chloride and mercuric chloride. These are used for providing good appearance to the wood. The wood is painted or polished after applying these chemical. These are cheaper than tar oils.
3. **Organic Solvent Chemicals.** The nepthol and phenol are commonly used preservatives before applying these chemicals. They are dissoled in spirit of volatile oil. Wood has to be painted after the application of these chemicals. These are costlier oils and provide very good results. Timber treated by them does not wrapper shrink after drying.

7.7 CARPENTARY TOOLS

The efficiency and accuracy of workmen depend upon the type of tools in the carpentary shop. The good quality of tools make the work easy). There are large number of tool used in carpentary shop. These tools can be broadly classified as

(*i*) Holding device
(*ii*) Marking and measuring tools
(*iii*) Cutting tools
(*iv*) Planning tools
(*v*) Drilling and boring tools
(*vi*) Striking tools
(*vii*) Filing tools

7.7.1 Holding and Supporting Devices

(*i*) **Work Bench.** It consists of heavy table of rigid hard wood. The size of commonly used table is 6 feet × 3 feet. The height of table is about 2½′ to 3′ feet. Generally 4 carpenters vice (bench vice) are fitted on 4 corners of table.

(*ii*) **Carpentary Vice.** Carpentary vice (bench vice) are discussed in detail in bench work and fitting shop (please refer 4.3).

(*iii*) **C-Clamp.** C-clamp is used for holding small work pieces. It is also used to hold and align wooden pieces after joining them with paste or glue.

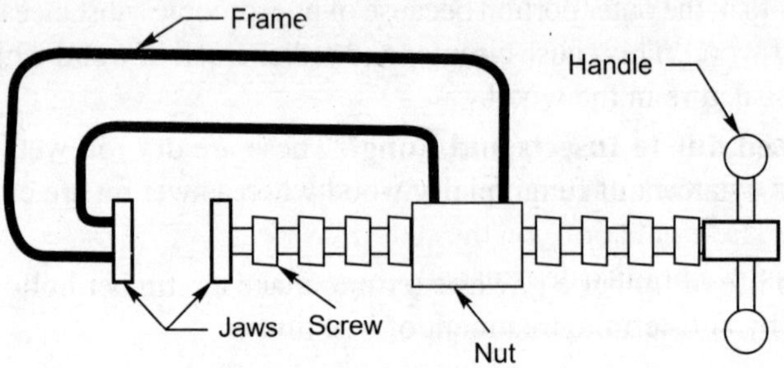

Fig. 7.5 C-Clamp.

(*iv*) **Pincer.** Pincer is very useful tool and is used for pulling out nails from wood. It consists of two steel forged arms which are hinged. The jaws of arms are hardened and tempered. One of the arm has a ball end and other arm has claw end.

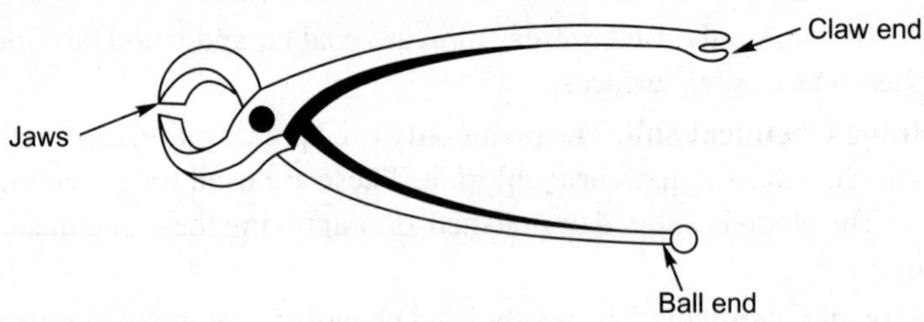

Fig. 7.6 Pincer.

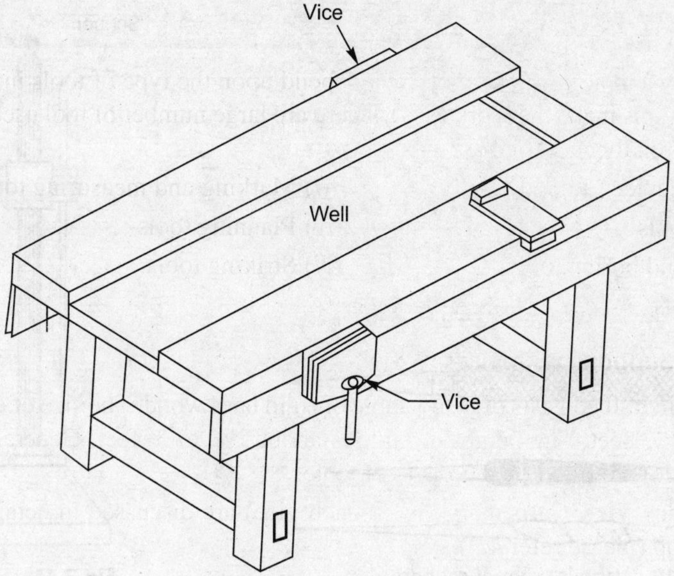

Fig. 7.7 Carpenter's bench.

7.7.2 Marking and Measuring Tools

(*i*) **Steel Rule.** This is a very basic measuring instrument which is widely used among school students as well as engineers for measuring length. Length may be expressed as distance between two lines (line measurement) or as distance between two faces (end measurement). Scale is the most common example of line measurement. Steel rules are available from ½ feet to 2 feet. The graduations on the steel rule are in mm on one edge and in inches on other edge. The least count of scale is usually 1 mm.

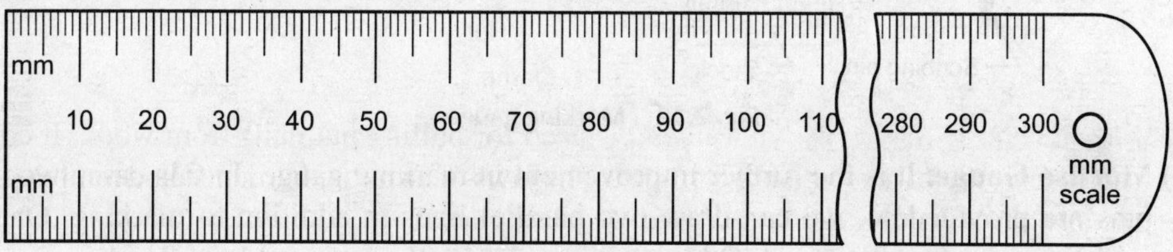

Fig. 7.8 Steel rule.

(*ii*) **Inch Tape.** It is made of a thin flexible steel strip. It can be folded into a compact case attached with a small handle. It is generally graduated in feet and centimeters along the edges of steel strip.

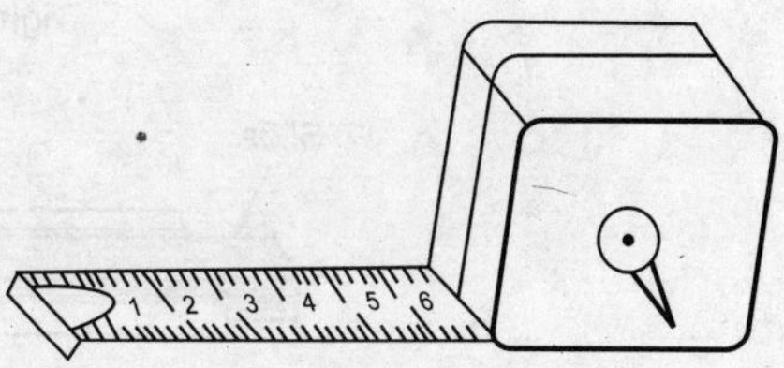

Fig. 7.9 Inch tape.

(*iii*) **Scriber and Surface Gauge.** Scriber is used for marking lines. It is made of high carbon steel and is having hardened front edge. Scriber mounted on a vertical bar with a base is known as surface gauge. Surface gauge is also used for marking and drawing parallel lines.

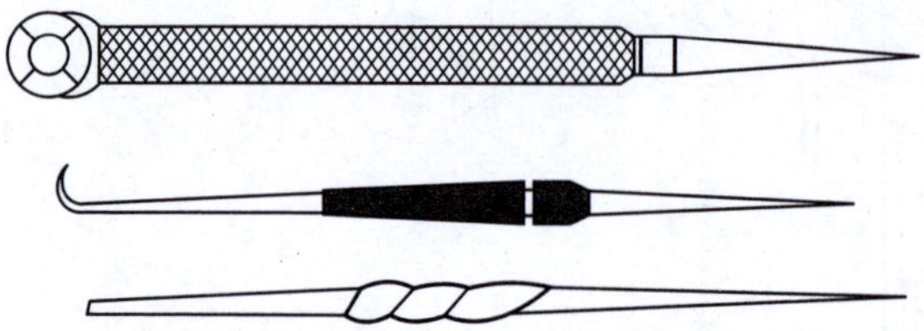

Fig. 7.10 Various types of scribers.

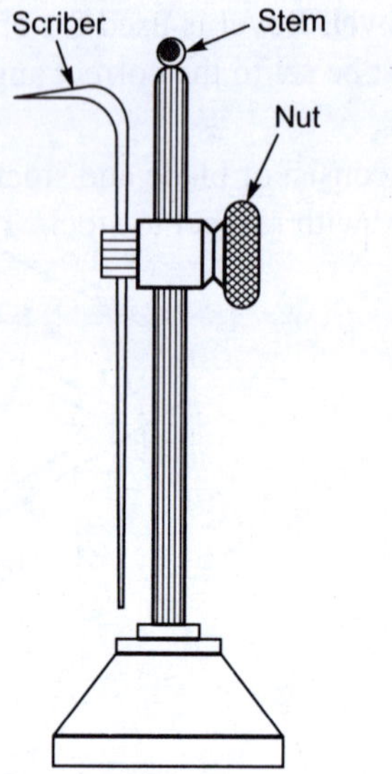

Fig. 7.11 Surface gauge.

(*iv*) **Marking Gauge.** It is used to draw parallel lines for marking. It has two main parts stem and stock. The stem is provided with graduations and a scribing pin for marking. The stock is sliding wooden piece on stem. Which can be blocked at any position on stem with help of thumb screw. One face of stock remains in contact with workpiece while marking.

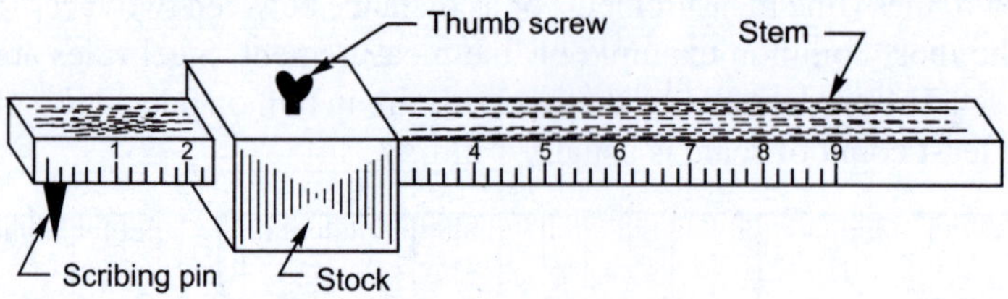

Fig. 7.12 Marking gauge.

(*v*) **Mortise Gauge.** It is the further improvement in marking gauge. In this case, two scribing pins are provided so, we can draw two parallel lines in addition to marking. One of the scribing pin is fixed as usual. Other scribing pin is either adjustable of fixed.

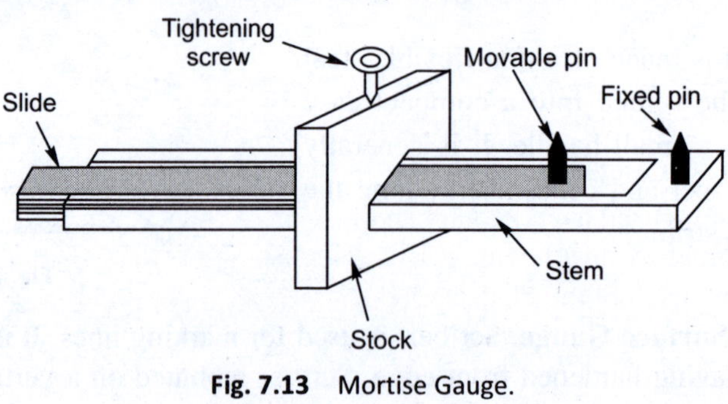

Fig. 7.13 Mortise Gauge.

Bevel. Bevel is used for checking inclination of two surfaces at any angle other than 90°. The level must be set to the correct angle before use. It may also be used for duplicating the angle on other surface.

It consist of blade and stock. The blade can slide along the slot and can be set at any angle from 0° to 180° with respect to stock. The tightening screw is used to lock the blade at given angle.

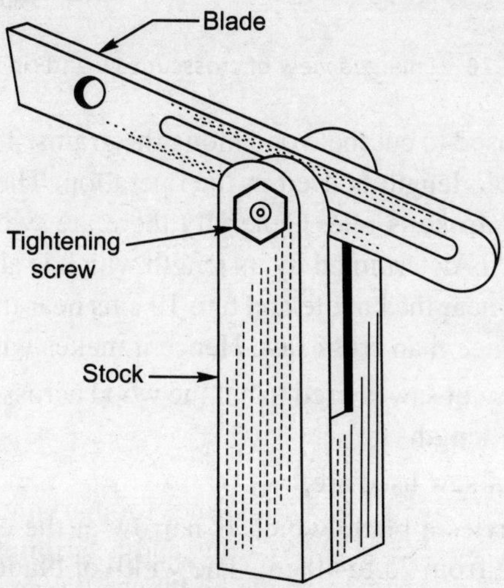

Fig. 7.14 Bevel.

Straight Edge. Straight edge is used for checking the flatness of the surfaces and edges. It is usually made of steel. When an edge is placed against a surface and held up to the light, any discrepancy can be detected by the appearance of light between two.

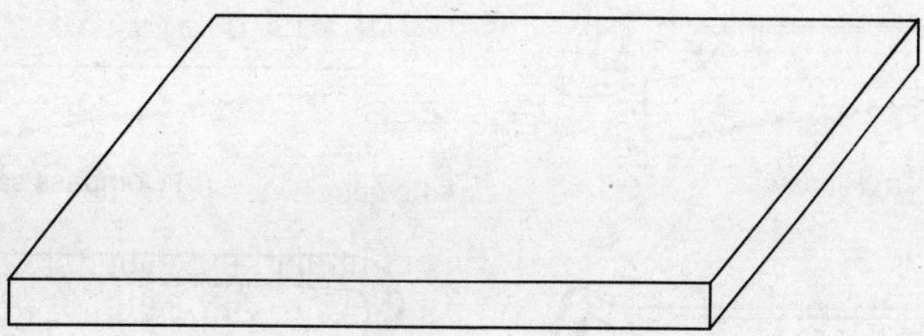

Fig. 7.15 Straight Edge.

7.7.3 Cutting Tools

The cutting tools in carpentary may be classified as

(1) Saws (2) Chiesels (3) Axe

1. **Saws.** All the saws used in carpentary shop consist of blade and wooden handle. The blade is made up of thin spring hardened steel having a series of sharp teeth on one edge. Each tooth cuts away small portion of wood. The teeth of the blade are bent alternatively to provide clearance to blade and to reduce friction.

The saws can be classified as (*i*) push saws and (*ii*) pull saws. Push saws perform cutting in forward stroke where as pull saws perform cutting in backward stroke. In carpentry shop push saws are more commonly used than pull saws. Please note that Hexsaw (used in fitting shop) is a pull saw.

One other way of classifying saw is according to the direction of wood grains are (*i*) rip saw and (*ii*) cross cut saw.

Fig. 7.16 Enlarged view of crosscut saw and rip saw.

(*i*) **Rip Saw.** Rip saw is used to cut the wood along the grains. The cutting action starts from the tip and gradually whole length is used in the operation. The cutting takes place in forward stroke and backward stroke is idle. Generally there are two teeth per centimeter length of blade. The size of saw is determined by its length which is about 75 mm. The width of blade is 12 to 20 centimeter near the handle and 6 to 10 cms near the tip. It should he noted that rip saw have more clearance than cross saw. Hence it makes wider cats.

(*ii*) **Cross Cut Saw.** Cross cut saw is used to cut the wood across the grain. Usually there are 3 to 4 teeth per centimeter length.

The various other types of saw used are:

1. **Compass Saw.** It carries a blade which is narrow at the end as shown in Fig. 7.17. The length of blade varies from 25 to 40 cm. The width of blade is about 2.5 cm at the end and 5 to 6 cm at the blade handle end. Compass saw is used for taking straight or curved cut inside the wood.
2. **Keyhole Saw.** It is used for cutting internal and intricate work where it is not possible to use other saws.

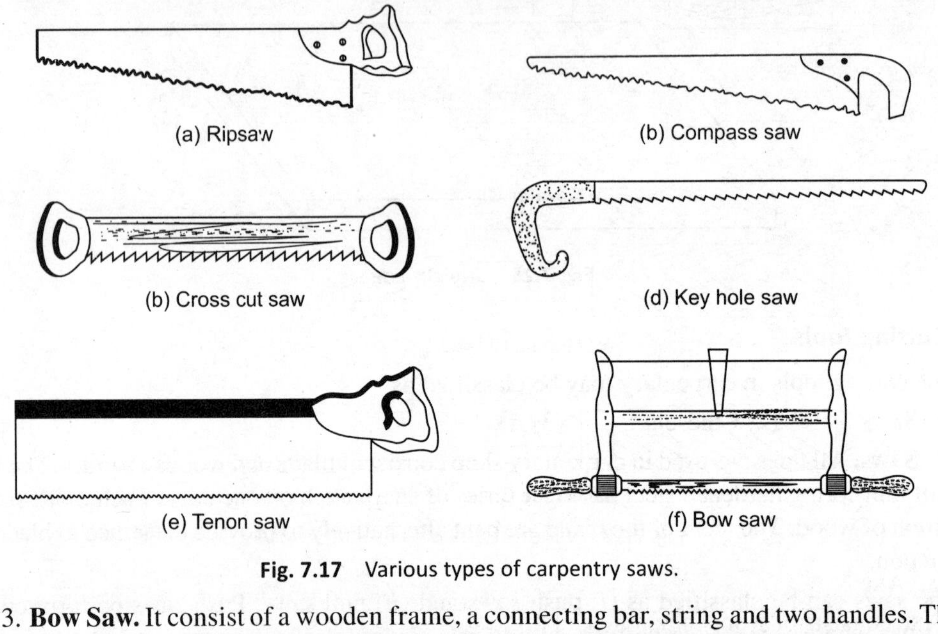

Fig. 7.17 Various types of carpentry saws.

3. **Bow Saw.** It consist of a wooden frame, a connecting bar, string and two handles. The blade is fastened at one end of frame. The tension to blade is provided by twisting the string by means of leaver. Bow saw is used for fine curves and profiles as compared to compass saw.

4. **Tanon Saw.** Tanon saw is for cutting the wood across the grains. It is used for finer work than rip saw or cross cut saw. It contains 4 to 6 teeth per centimeter length of blade. The blade is parallel and provided with reinforces strip at the top so that is does not bend. During cutting and straight cut is obtained.

Chisels

Chisels are used in carpentary shop to produced various shapes and cavities. Chisel consist of steel blade and wooden handle. The cutting edge of steel blade are ground to get finished surfaces.

1. **Firmer Chisel**

It is most widely used for taking wider cuts and finishing flat surfaces. The cross section of blade is rectangular.

2. **Dove tail Chisel**

It is used for finishing V grooves and sharp corners. The cross section of blade has bevelled shape as shown in Fig. 7.18. Because of reduced blade thickness on the sides. It can enter sharp corner and V grooves easily to finish them.

3. **Mortise Chisel**

It is used in mortise and tennon joint for producing rectangular or square holes and recesses in wooden piece. It is also used for taking heavy and deep cut. The maximum width of blade is upto 18 mm and thickness varies from 6 to 16 mm. Because of more thickness it is stronger than other chisels and is capable of bearing heavy blows.

4. **Gauge Chisel**

It is used for finishing curved surfaces. The blade of chisel is in hollow curved shape to fish curved surfaces.

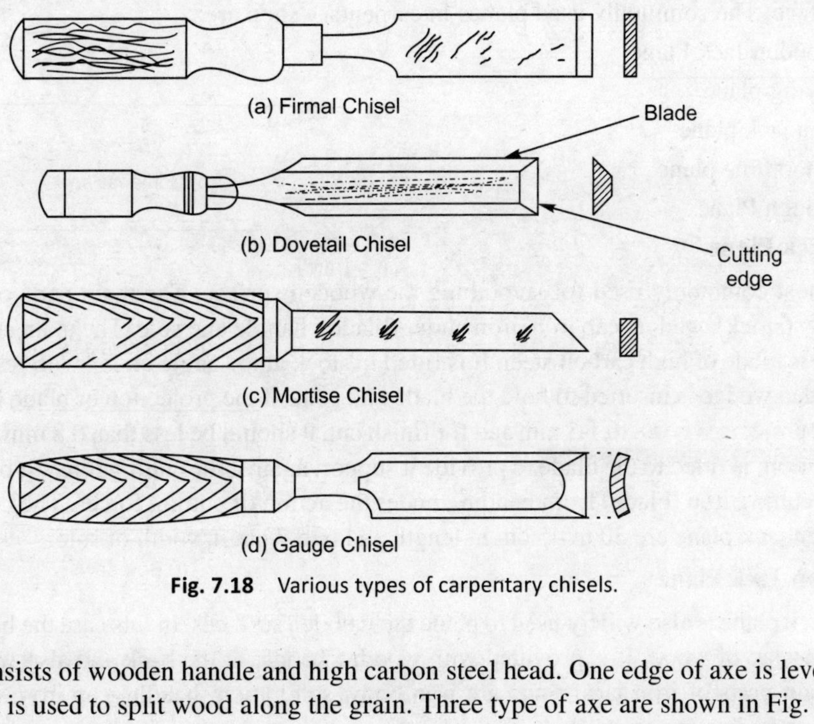

Fig. 7.18 Various types of carpentary chisels.

5. **Axe**

Axe consists of wooden handle and high carbon steel head. One edge of axe is levelled to form cutting edge. It is used to split wood along the grain. Three type of axe are shown in Fig. 7.19.

(*i*) **Axe.** In axe, cutting edge is along the length of handle.

(*ii*) **Side axe.** In side axe, cutting edge is perpendicular to length axis of handle.

(*iii*) **Adze.** It has outer face convex and inner face concave. It is used to produce concave surfaces.

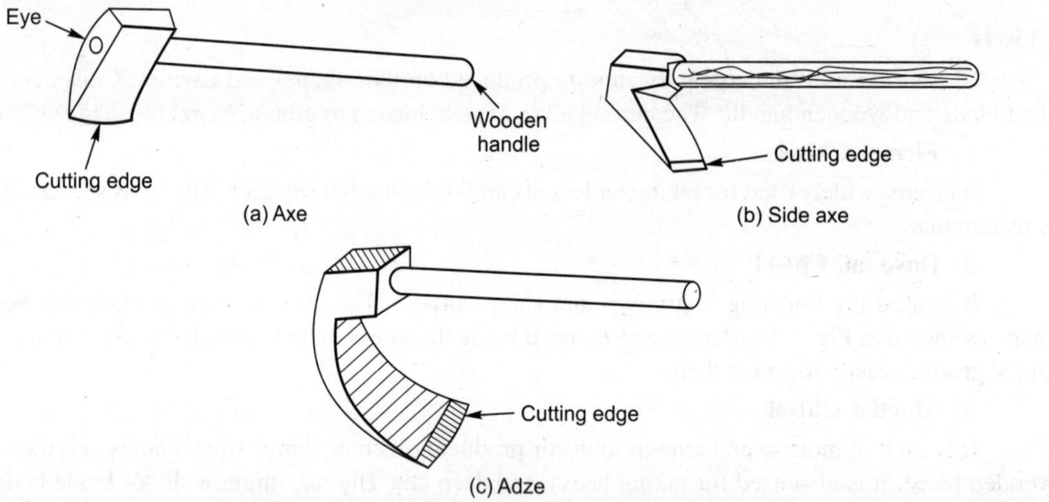

Fig. 7.19 Various types of axe used in carpentary shop.

7.7.4 Planning Tools

Planning tools are used in carpentary shop shaving or smoothing wooden surfaces. Planning tools may also be categorised in cutting tool as they remove the material in form of wooden chips to smith the surface. The commonly used planes in carpentary shop are:

(1) Wooden Jack Plane

(2) Trying plane

(3) Iron jack plane

(4) Smoothing plane

(5) Plough Plane

1. **Wooden Jack Plane**

It is most commonly used for smoothing the wooden surface. The main parts of wooden jack plane are body (stock), wedge, cap iron, iron cutter (blade), handle and sole. The main cutting part iron cutter (blade) is made of high carbon steel. It is fitted in stock at the angle of 45° with respect to sole. A Tapered wooden wedge is inserted to hold the blade and control the projection of blade below the sole. For rough cut projection is 0.8 to 1.5 mm and for finish cut, it should be less than 0.8 mm. An iron piece, known as cap iron, is fitted to the blade to provide it support against the cutting force. It provides rigidly and prevents cutting iron (blade) from bending under the action of cutting forces. The commonly used size of wooden jack plane are 30 to 45 cm in length and 6 to 7 cm in width of sole.

2. **Iron Jack Plane**

Iron jack plane is also widely used to plane the wooden surfaces. In this case the body is made up of cast iron instead of wood. It is provided with wooden handle at the back and also wooden knob at front. The main parts of iron jack plane are iron body, sole, knob, handle, cap iron, cap iron lever, cutting blade and adjusting screw.

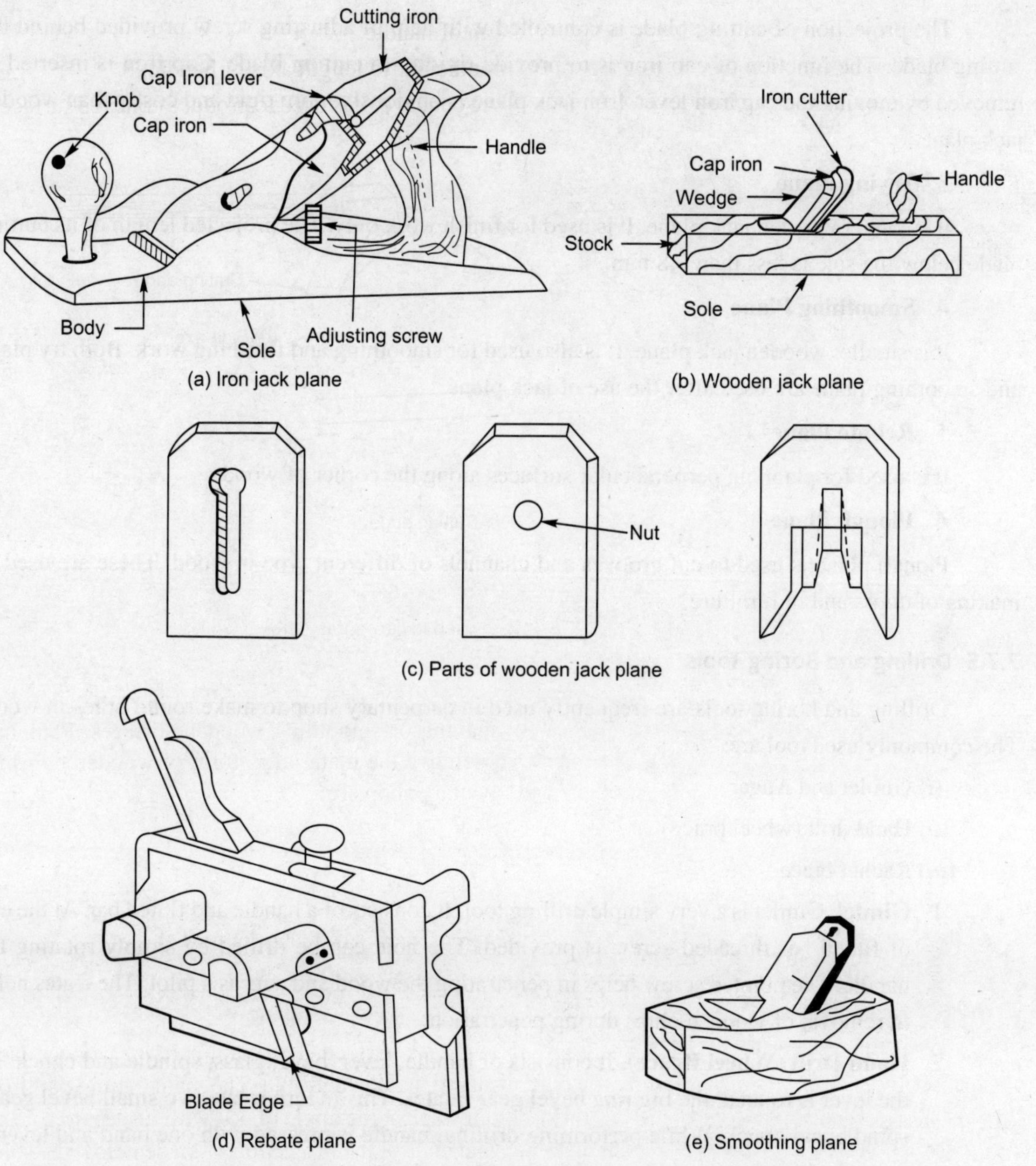

(a) Iron jack plane

(b) Wooden jack plane

(c) Parts of wooden jack plane

(d) Rebate plane

(e) Smoothing plane

Fig. 7.20 Various type of planes used in carpentry shop.

The projection of cutting blade is controlled with help of adjusting screw provided behind the cutting blade. The function of cap iron is to provide rigidity to cutting blade. Cap iron is inserted or removed by moving the cap iron lever. Iron jack plane is obviously more rigid and costly than wooden jack plane.

3. **Trying Plane**

It is longer wooden jack plane. It is used for finish work only. The projected length of its cutting blade below the sole is less than 0.8 mm.

4. **Smoothing Plane**

It is smaller wooden jack plane. It is also used for smoothing and finishing work. Both try plane and smoothing plane are used after the use of jack plane.

5. **Rebate Plane**

It is used for planning perpendicular surfaces along the corner of wood.

6. **Plough Plane**

Plough plane is used to cut grooves and channels of different type in wood. These are used in making of doors and in furniture.

7.7.5 Drilling and Boring Tools

Drilling and boring tools are frequently used in carpentary shop to make round holes in wood. The commonly used tool are:

(*i*) Gimlet and Augar

(*ii*) Hand drill (wheel brace)

(*iii*) Rachet brace.

1. **Gimlet.** Gimlet is a very simple drilling tool. It consists of a handle and fluted bar. At the end of fluted bar threaded screw is provided. The hole can be drilled by simply rotating the handle. The pointed screw helps in penetrating the wood and acts as a pilot. The flutes helps in removal of wooden chips during penetration.
2. **Hand Drill (Wheel Brace).** It consists of handle, lever, bevel glass, spindle and chuck. As the lever is rotated, the big ring bevel gear rotates. This in turn rotate two small bevel gears, spindle and chuck. While performing drilling, handle is pressed with one hand and lever is rotated with other hand.

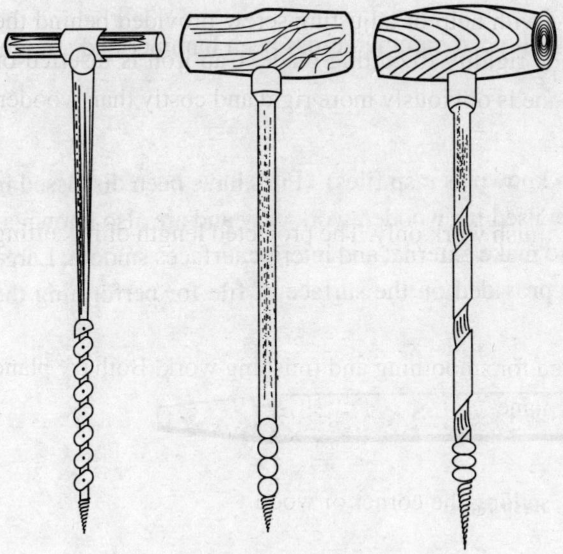

Fig. 7.21 Drilling tools.

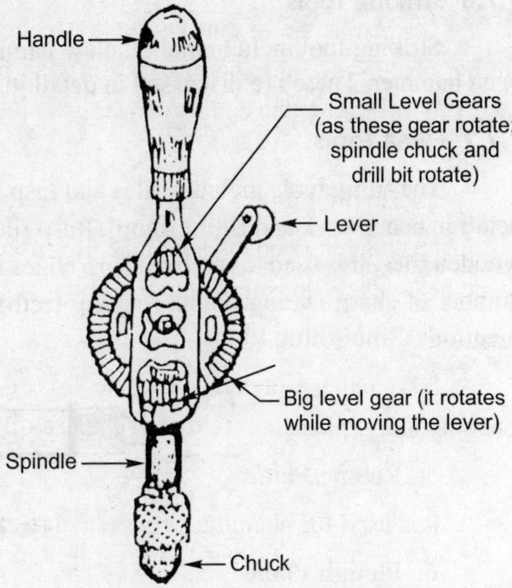

Fig. 7.22 Hand drilling machine.

3. **Rachet Brace.** It consists of head, crank handle, rachet and chuck for making the drill, head is pressed with one hand and crank is rotated by second hand. The rachet arrangement helps rotation of drill bit in one direction only. This arrangement is very useful where crank cannot complete the turn due to lack of space. When crank is rotated in reverse direction drill bit does not rotate. Various types and size of the bit can be accommodated in drill.

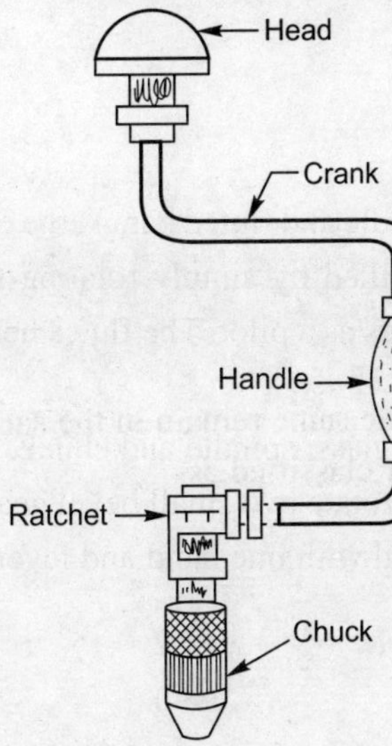

Fig. 7.23 Rachet brace.

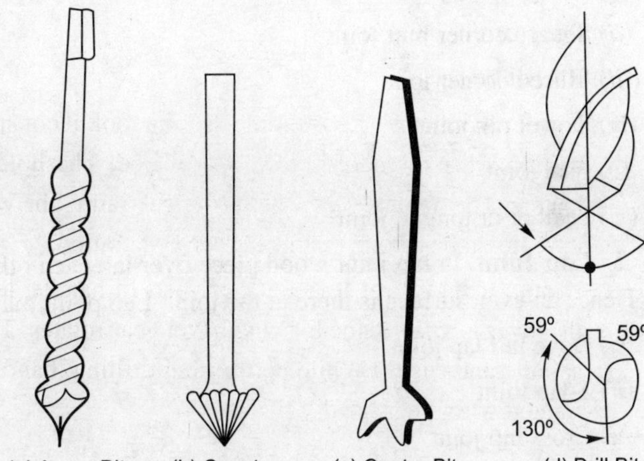

(a) Angur Bit (b) Counter Bore Bit (c) Centre Bit (d) Drill Bit

Fig. 7.24 Various type drill bits for rachet brace.

7.7.6 Striking Tools

Striking tool include mallet, claw hammer, ball peen hammer, cross peen hammer and straight peen hammer. These are discussed in detail in 5.4.12.

7.7.7 Filing Tools

The filing tools includes files and rasp (also known as rasp files). (Files have been discussed in detail in benchwork and fitting shop). Rasp files are used for wooden work only and are also known as wooden files. It is used to remove sharp edges and to make external and interior surfaces smooth. Large number of sharp cutting points (cutting teeth) are provided on the surface of file for performing the function.

Fig. 7.25 Rasp file.

7.8 WOODEN JOINTS

In carpentary shop various types of wooden joints are used to connect different wood parts together. These joints are joined by means of suitable faster like nails, screws, nuts and bolts, dowels or by adhesive glues. Joints can be probably classified in butt joints according to the method of joining.

1. **Butt Joint.** In butt joint one end (or one side) of the one piece is placed against the end of other, butt joints can be further classified as:

(*i*) Straight butt joint

(*ii*) Square corner butt joint

(*iii*) Milred corner joint

(*iv*) Dowel pin joint

(*v*) Fish joint

(*vi*) Feather or tongue joint

2. **Lap Joint.** In lap joint wood piece over lap each other and at the same remain in the same plane. Hence an even surface is there at the joint. Lap point may be further classified as:

(*i*) Dove tail lap joint

(*ii*) T-lap joint

(*iii*) Cross lap joint

(*iv*) Corner lap joint

(*v*) Briddle joint

(*vi*) Mortise and tenon joint

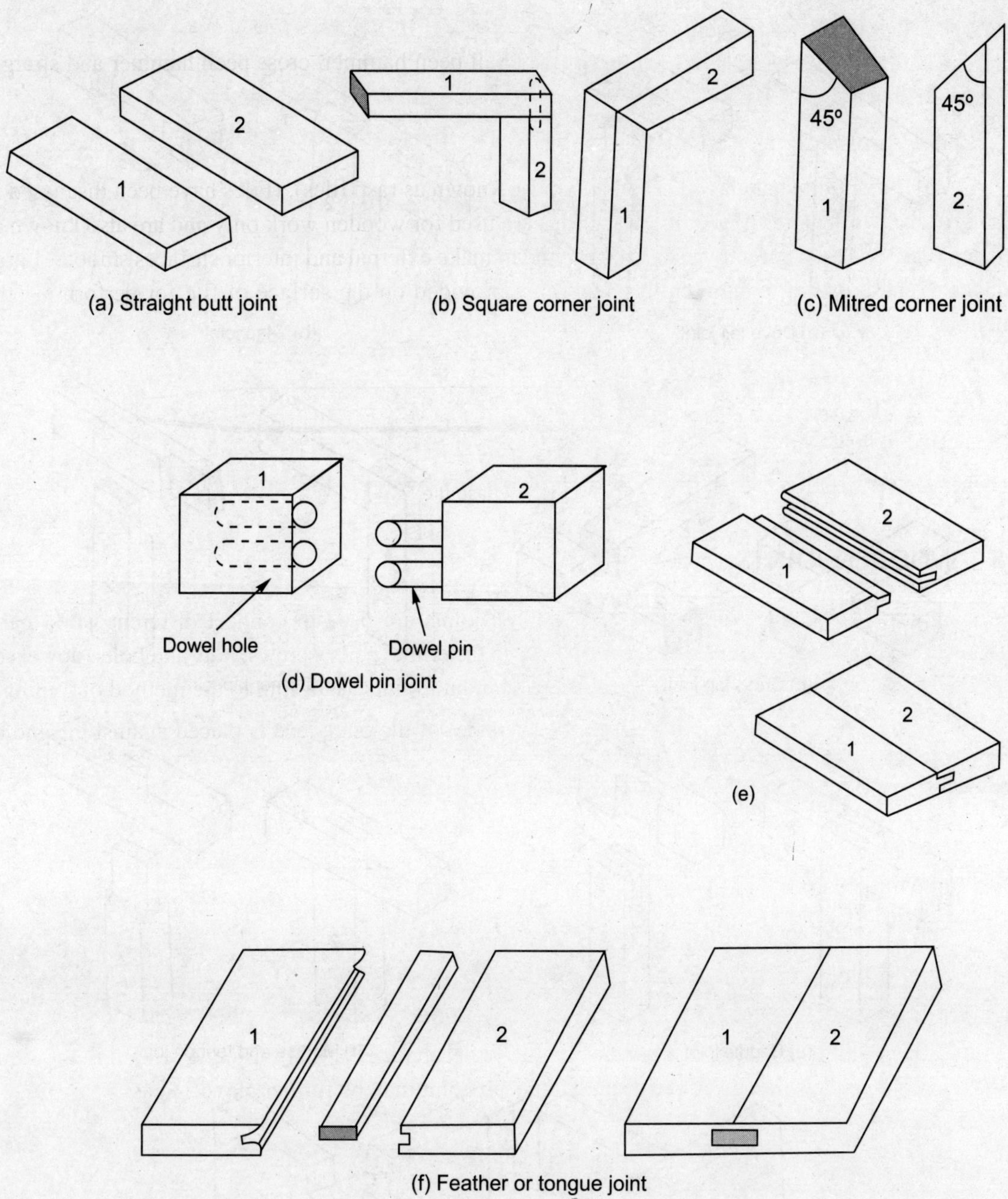

Fig. 7.26 Various type of butt joints.

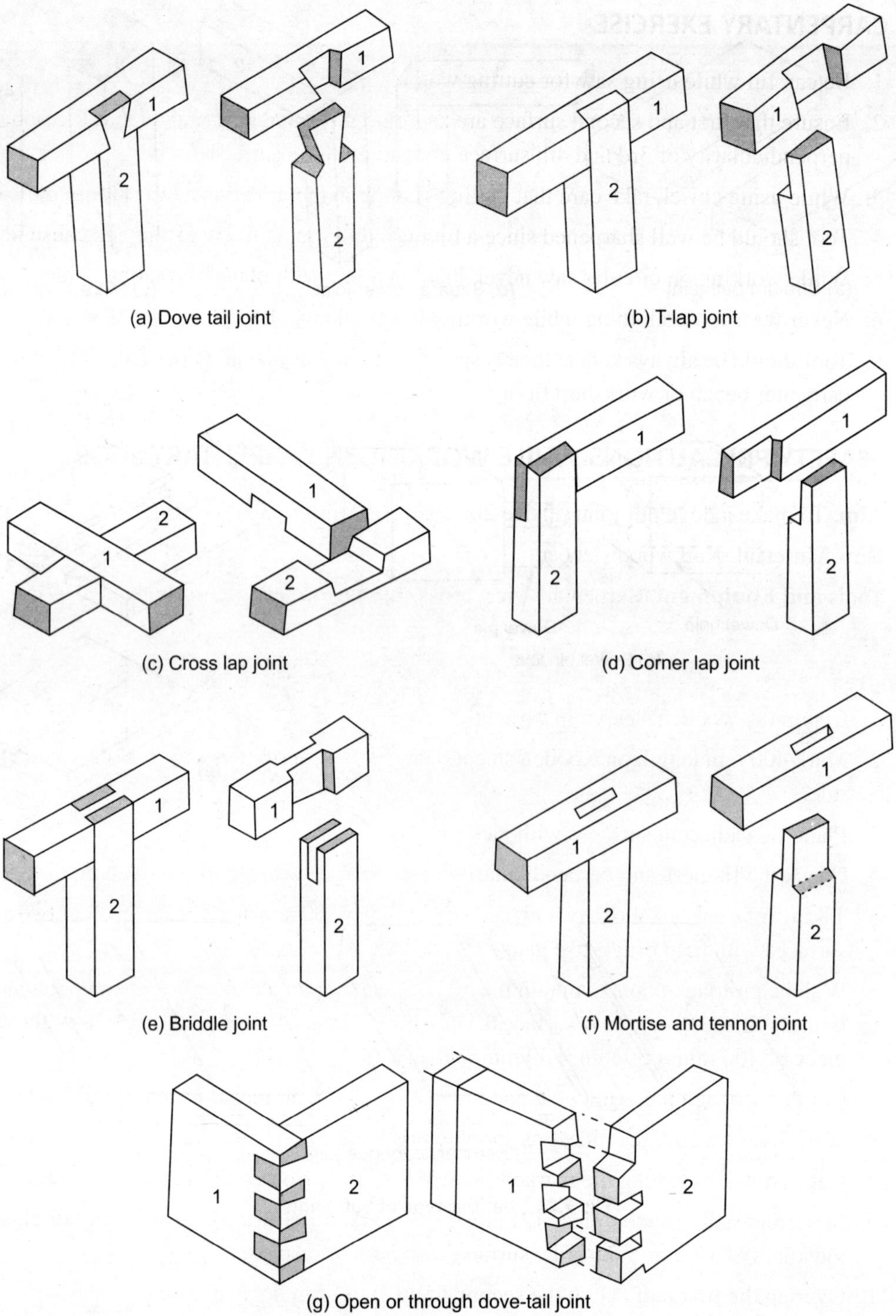

Fig. 7.26 Various types of lap joint.

7.9 CARPENTARY EXERCISE

1. Be careful while using saw for cutting wooden piece.
2. Ensure that first and second surface are accurately flat and perpendicular as the flatness and perpendicularity of 3rd and 4th surface depend on these surfaces.
3. While using chisel, take care that cutting is done in the directions away from your body.
4. Tool should be well sharpened since a blunt or dull tool is likely to slip and cause injury.
5. While working on circular saw never stand in a line with plane of rotating blade.
6. Never wear loose clothing while working in carpentary shop.
7. Tool should be always kept at their respective place. They should not be allowed to scatter on carpenter bench or workshop floor.

7.10 SAFETY PRECAUTIONS WHILE WORKING IN CARPENTARY SHOP

Aim. To make a dove tail joint of size 200 × 50 × 30 mm.

Raw Material. Kail wood sleeper.

Tools and Equipment. Carpentary vice, cross cut saw, rip saw, measuring and marking gauges, planner, chisel.

Procedure:

1. Clamp the wooden sleeper in carpentary vice.
2. Mark 400 mm length on wooden sleeper and cut it with rip saw with help of pencil and steel rule.
3. Plane two adjacent surfaces with jack plane.
4. Check the flatness and perpendicularity of these two surfaces with try square.
5. Taking reference of above two planed surfaces mark 50 mm and 30 mm on other two adjacent surfaces with help of mortise gauge.
6. With help of jack plane maintain the 50 mm and 30 mm and check the perpendicularity and flatness of these new two surfaces. Verify the dimensions by steel rule. Now the wooden piece of 400 mm × 50 mm × 30 mm is ready.
7. Cut this wooden piece in two equal halves (i.e. 200 × 50 mm × 30 mm) with the help of
8. Mark lines on both the pieces as shown in figure 7.27.
9. Cut first half (1) along the marked lines with help of rip saw upto the thickness of 15 mm.
10. In second half wooden piece (2) along the marked line with help of dove tail chisel upto thickness of 15 mm. Finish the surface with help of rasp cut file.
11. Overlap the first half (1) on the second half (2) to form dove tail joint.

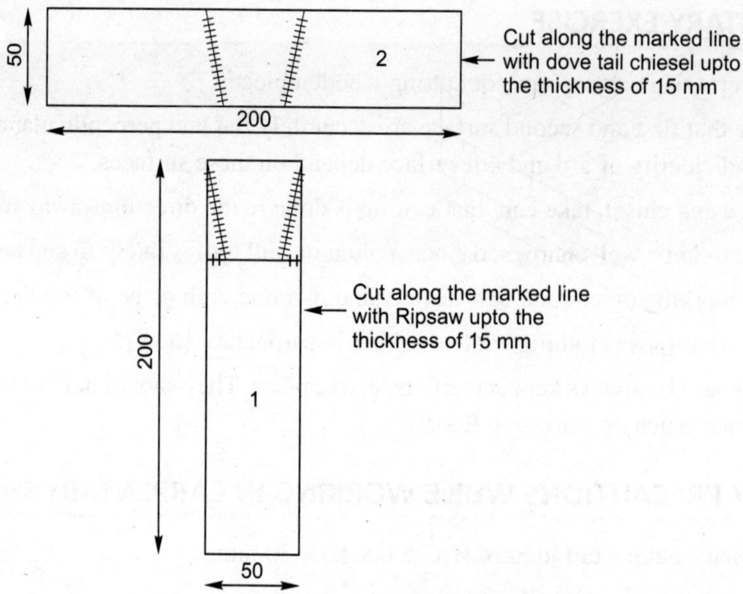

Fig. 7.27 Markings on dove Tail Joint.

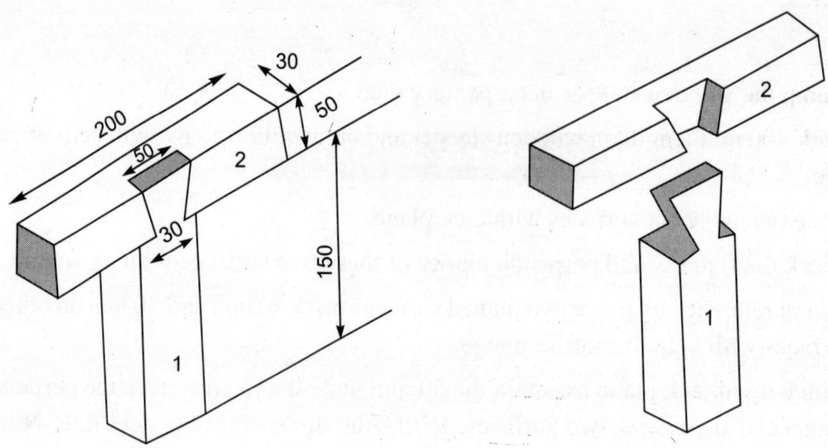

Fig. 7.28 Dove tail joint.

EXERCISE PROBLEMS

1. Make line dove tail joint using given wooden piece (Job 1).
2. Make the mortise and Tenon Joint using given wooden piece (Job 2).
3. Make the bridle joint using the given wooden piece.
4. Make the cross lap joint-using given wooden piece (Job 4)

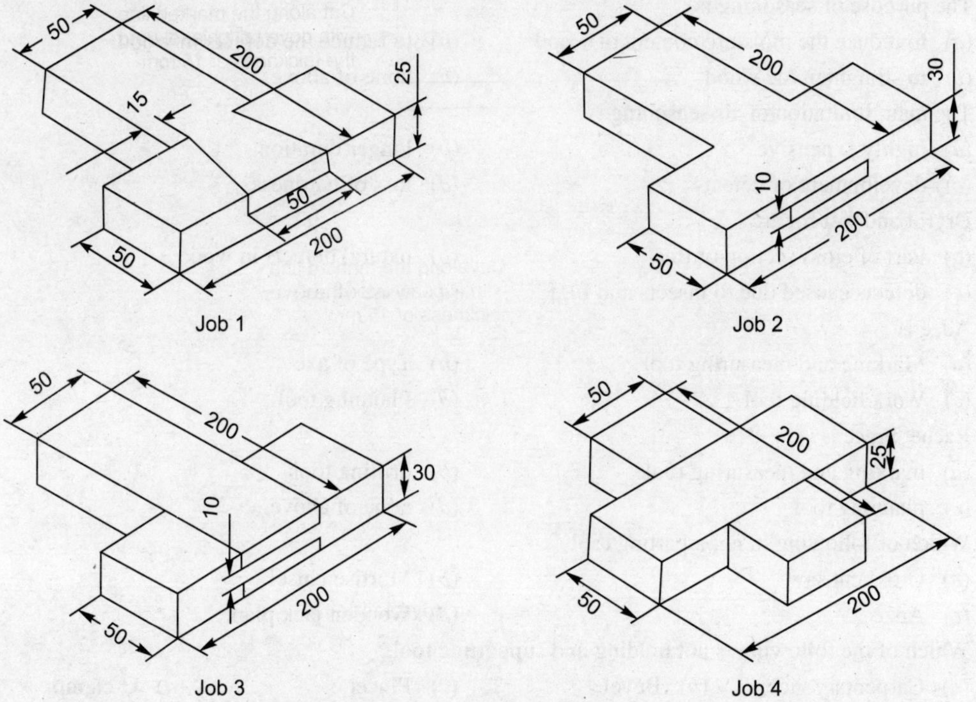

Fig. 7.29 Exercise problems in carpentary shop.

OBJECTIVE QUESTIONS

1. Timber is widely used because

(*a*) It possess high tensile strength (*b*) It has god life

(*c*) It is easily available (*d*) It can withstand the effect of whether.

2. Compressive strength of wood is

(*a*) equal in all directions (*b*) high along the grains but low across the grains

(*c*) less along grains and more across the grains (*d*) none of above.

3. The life of tree is estimated on the basis of

(*a*) number of annual ring all around the pith (*b*) visual condition of the tree

(*c*) diameter of trunk (*d*) height of the tree.

4. Which of following tool is holding and supporting tool.

(*a*) Pincer (*b*) Mortise gange (*c*) Adze (*d*) Angar.

5. Mortise gauge is

(*a*) holding and supporting tool (*b*) cutting tool

(*c*) marking tool (*d*) planning tool.

6. Which of the following is the hardest wood.

(*a*) Kail (*b*) Chir (*c*) Babul (*d*) Shisham.

7. Which of the following is a natural defect of wood.

(*a*) Knots (*b*) Cadmium layer (*c*) Pith (or medulla) (*d*) bark.

8. The purpose of seasoning is

(*a*) to reduce the moisture content of wood (*b*) to reduce the defects in wood.

(*c*) to straighten the wood (*d*) none of above.

9. The main limitation of air seasoning is

(*a*) highly expensive (*b*) longer duration

(*c*) development of defects (*d*) loss of hardness.

10. Dryrot and wet rot are

(*a*) Part of cross section of tree (*b*) natural defects in wood

(*c*) defects caused due to insects and fugi (*d*) none of above.

11. Adze is

(*a*) Marking and measuring tool (*b*) Type of axe

(*c*) Work holding tool (*d*) Planning tool.

12. Rachet brace is

(*a*) marking and measuring tool (*b*) drilling tool

(*c*) planning tool (*d*) none of above.

13. Which of following in not a parting tool

(*a*) Cross cut saw (*b*) Mortise chisel

(*c*) Adze (*d*) Wooden jack plane.

14. Which of the following is not holding and supporting tool

(*a*) Carpentary vice (*b*) Bevel (*c*) Pincer (*d*) C-clamp.

15. Which of following is not a drilling and boring tool.

(*a*) Angar (*b*) Gimlet (*c*) Ratchet brace (*d*) Adze.

ANSWERS

1. (*c*)	**2.** (*b*)	**3.** (*a*)	**4.** (*a*)	**5.** (*c*)
6. (*d*)	**7.** (*a*)	**8.** (*a*)	**9.** (*b*)	**10.** (*c*)
11. (*b*)	**12.** (*b*)	**13.** (*d*)	**14.** (*b*)	**15.** (*d*).

FILL IN THE BLANKS (CARPENTARY)

1. The process of removal of moisture from freshly cut trees is known as .... of wood.

2. The dark central part of tree is known as ....

3. The life of the tree is estimated on the basis of ....

4. Preservation of wood is done by ....

5. Straight edge is used for checking ....

6. The .... is used for checking Inclination of two surfaces at any angle other than 90°.

7. .... saw is used to cut the wood along the grains.

8. .... saw is used to cut the line wood across the grains.

9. .... saw has more clearances than .... saw and hence it makes wider cuts.

10. The tools used in carpentary shop for shaving or smoothing wooden surface one known as ....

11. Various types of drilling and boring tools used in carpentary shop are ....

12. The files used for wood work are also known as ....

13. Feather or tongue joint is .... Joint.

14. Dove tail joint is .... joint.

15. The main limitation of air seasoning of wood is ....

ANSWERS

1. seasoning **2.** Pith (or medulla)

3. number of annular rings all around the pith.

4. Tar oils, water soluble chemical salts and organic solvants.

5. checking the flatness of the surfaces and edges. **6.** Bevel

7. Rip **8.** Crosscut **9.** Ripsaw, crosscut

10. planning tools **11.** Gimlet and Augar, hand drill, and rachet brace etc.

12. rasp files **13.** Butt **14.** lap

15. that seasoning process takes longer time

VIVA-VOCE QUESTIONS

1. Why seasoning of wood is done.

2. Name different type of seasonings.

3. Name type of hammers used in carpentary shop.

4. Which are the work holding devices in carpentary shop?

5. What are the difference between straight peen and cross peen hammer?

6. Name various type of chisels used in carpentary shop.

7. Explain the uses of firmer and mortise chisel.

8. Explain the difference between marking gauge and mortise gauge used in carpentary shop.

9. Name the various parts of iron jack plane and wooden jack plane.

10. Name various joints made in carpentary shop.

11. Explain various types of saws used in carpentary shop.

12. Explain the function of gimlet drill and hand drill and ratchet brace.

Chapter

8

WELDING PROCESSES (WELDING SHOP)

INSIDE THIS CHAPTER

8.1 INTRODUCTION TO WELDING PROCESSES

Welding is a metal joining process in which fusion is obtained by application of heat and/or pressure. Combination of these two variables may be high temperature with no pressure, high pressure with no rise in temperature or both high temperature and pressure. Welding is an atomic bonding process and metallurgical bond is accomplished by the attracting forces between atoms. One beauty of welding process in comparison to other processes is that by this process we can have strength of joint equal and sometime greater than the parent metals. This makes welding one of the most extensively used manufacturing process.

Welding is extensively used in automobile industry, aircraft machine frames, structural work, repair work, ship building, etc. It is also useful method for repairing of broken castings and defective metal parts.

"Success often comes to those who dare and act; it seldom goes to timid who are ever afraid of the consequences."

8.2 CLASSIFICATION OF WELDING PROCESSES

Welding may be broadly classified into fusion welding and solid state welding. The welding classification is shown in Fig. 8.1 and Fig. 8.2.

1. Fusion Welding
2. Solid state welding

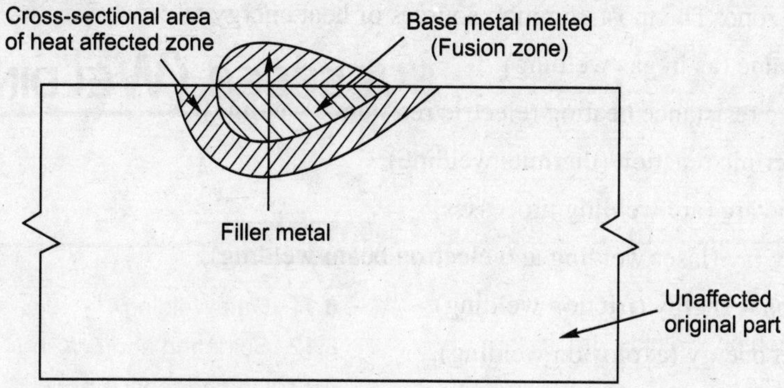

Fig. 8.1 Cross-sectional area of a bead on plate showing fusion zone, heat affected zone and unaffected original part.

Welding Process

Solid state welding

Name of the welding		Heat source
1. Forge welding	→	Smiths forge
2. Friction welding	→	Frictional Energy
3. Diffusion welding	→	Diffusion
4. Explosive welding	→	Impact Energy
5. Ultrasonic welding	→	Ultrasonic Waves

Fusion welding

Name of the welding		Heat source
1. Metal Arc welding	→	Electric Arc
2. Carbon Arc welding	→	Electric Arc
3. Tungsten Arc welding	→	Electric Arc
4. Resistance spot welding	→	Electric Resistance
5. Resistance seam welding	→	Electric Resistance
6. Resistance Butt welding	→	Electric Resistance
7. Gas welding	→	Gas Flame
8. Electro slag welding	→	Electric Arc to Electric Resistance
9. Laser Beam welding	→	Energy of Rays
10. Electron Beam welding	→	Energy of Rays

Fig. 8.2 Welding heat sources for various welding processes.

Fusion Welding. In fusion welding process joint is completely melted of obtain the joint and if necessary a molten filler material is also added. The fusion weld can also be described as minicasting surrounded by heat treated zone. In a fusion welded part there are three distinct zones (*i*) fusion zone (*ii*) heat affected unmelted zone around the fusion zone (*iii*) unaffected original part.

Solid State Welding. These are the welding processes where the metal is brought to the plastic state without actual melting and joining is obtained by application of large forces. In few cases, solid state welding may also be carried out at room temperature, but it does not allow recrystallisation and thus reduces the ductility. Solid state welding at high temperatures not only increases the ductility but also eliminates some of the welding defects.

Welding Heat Sources. A heat source suitable for welding should release the heat in a sharply defined coolated zone. The most common sources of heat energy are:

1. Gas flame (as in gas welding)
2. Electric resistance heating (electric resistance welding)
3. Exothermic reaction (thermits welding)
4. Electric arc (arc welding processes)
5. Energy ray (laser welding and electron beam welding)
6. Frictional energy (friction welding)
7. Impact energy (explosion welding).

8.3 TYPES OF JOINTS

Different types of joints used in welding processes are: lap, butt, edge, T joint, fillet and corner joints. These joints are shown in Fig. 8.3. Lap joints and butt joints are normally used with resistance welding processes, gas welding and arc welding process. The joints of forge welding do not resemble to those shown in Fig. 8.3.

Type of Joints

Lap Joint. In lap joint the pieces to be welded over lap each other.

Butt Joint. The butt joint, the pieces to be welded are held together side by side.

Cornor Joint. The corner joint the pieces to be welded are joined by corner edges.

T-Joint. In *t* joint both the pieces are welded in T-position.

Fillet Weld. Fillet weld is used to join two pieces which are approximately at 90°. The metal is deposited into the edge made by the mating surfaces of joining pieces. Fillet weld is used in T joint, corner joint and lap joint.

Plug Weld. Plug weld is used to join two pieces in lap joint. The upper piece is essentially having a hole or slot. The filler metal is deposited in this hole (or slot) to make the joint.

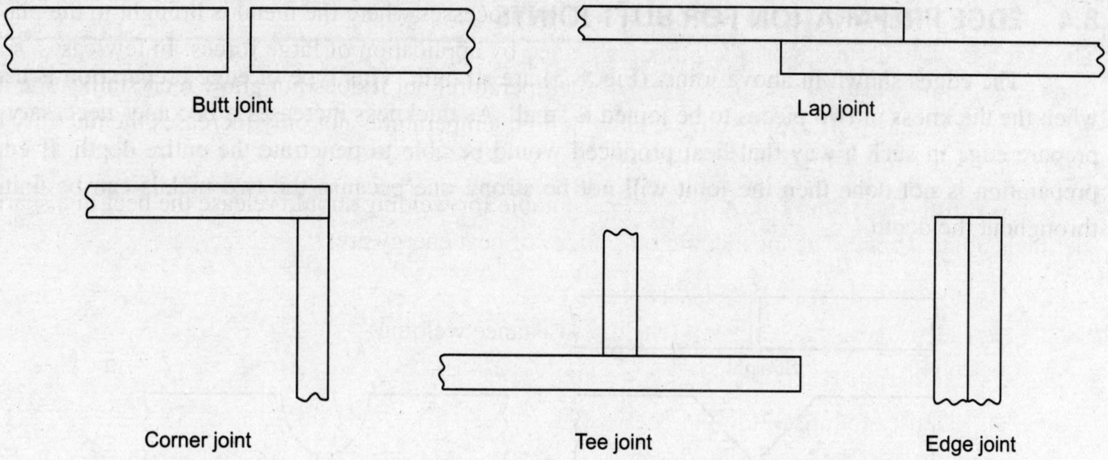

Fig. 8.3

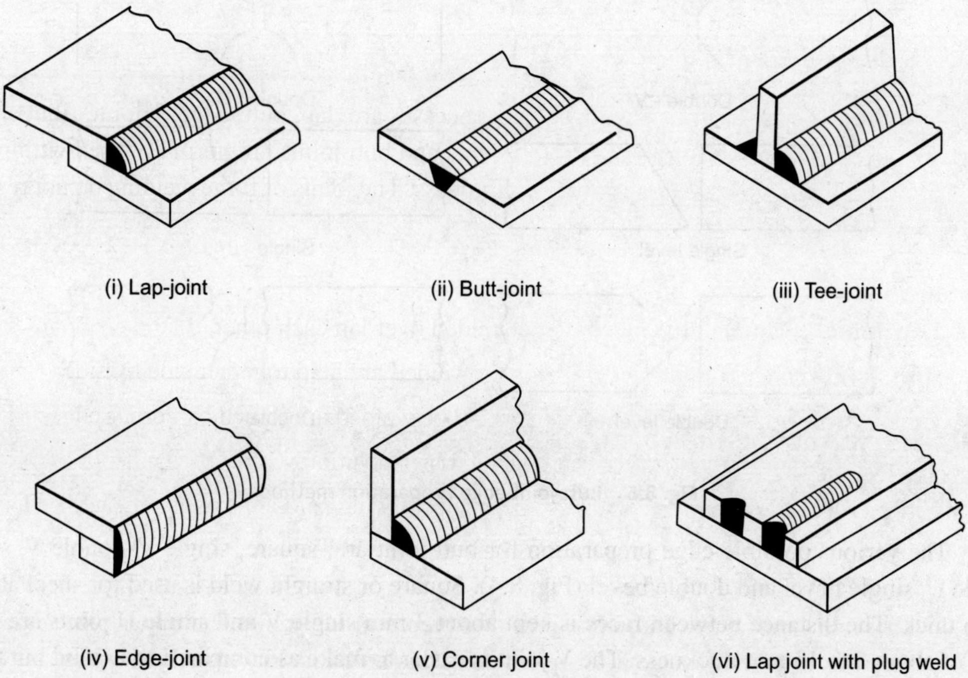

Fig. 8.4 Different types of welded joints.

8.4 EDGE PREPARATION FOR BUTT JOINTS

The edges shown in above joints (Fig. 8.3) are straight. This type of edge preparation is used when the thickness of two pieces to be joined is small. As thickness increases it becomes necessary to prepare edge in such a way that heat produced would be able to penetrate the entire depth. If edge preparation is not done then the joint will not be strong one because the two metals can be united throughout the depth.

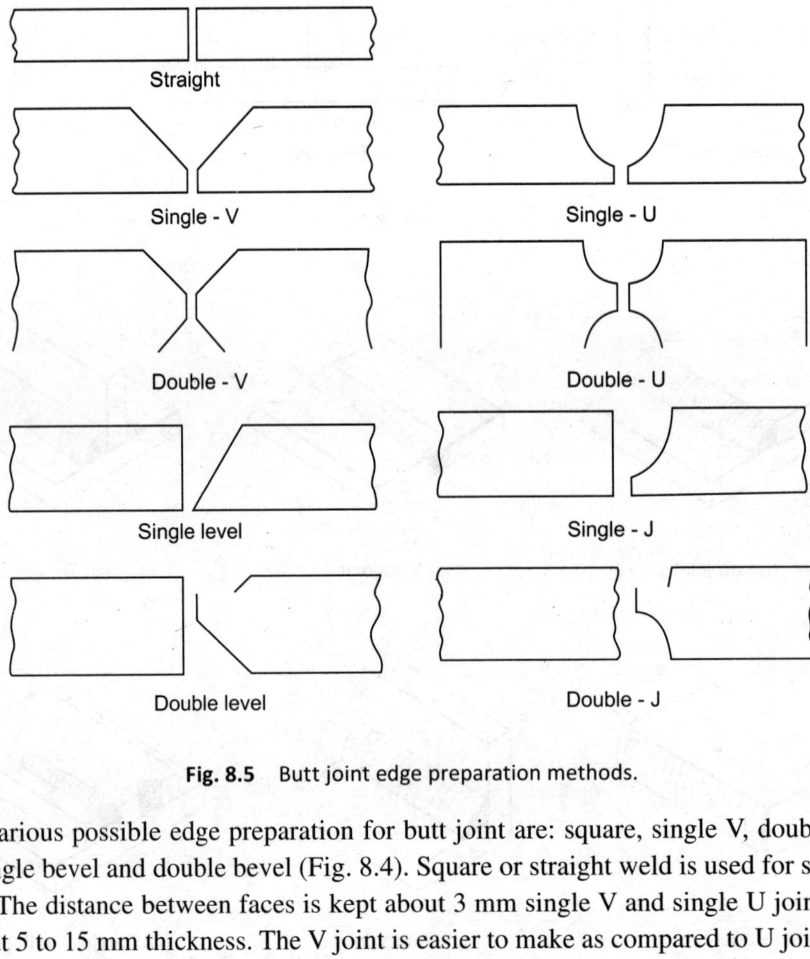

Fig. 8.5 Butt joint edge preparation methods.

The various possible edge preparation for butt joint are: square, single V, double V, single U, double U, single bevel and double bevel (Fig. 8.4). Square or straight weld is used for sheet about 1 to 5 mm thick. The distance between faces is kept about 3 mm single V and single U joints are used for sheet of about 5 to 15 mm thickness. The V joint is easier to make as compared to U joint but the filler metal increases greatly with increase in thickness. That is why some times U joint is preferred but this joint is difficult to make. Double U (above 15 mm) joint is preferred for joining very thick material because of less requirement of filler material.

The gap between plates should be kept at an optimum value since penetration will be insufficient with small gap and with wide gap molten pools will fall through and welding will be difficult to control.

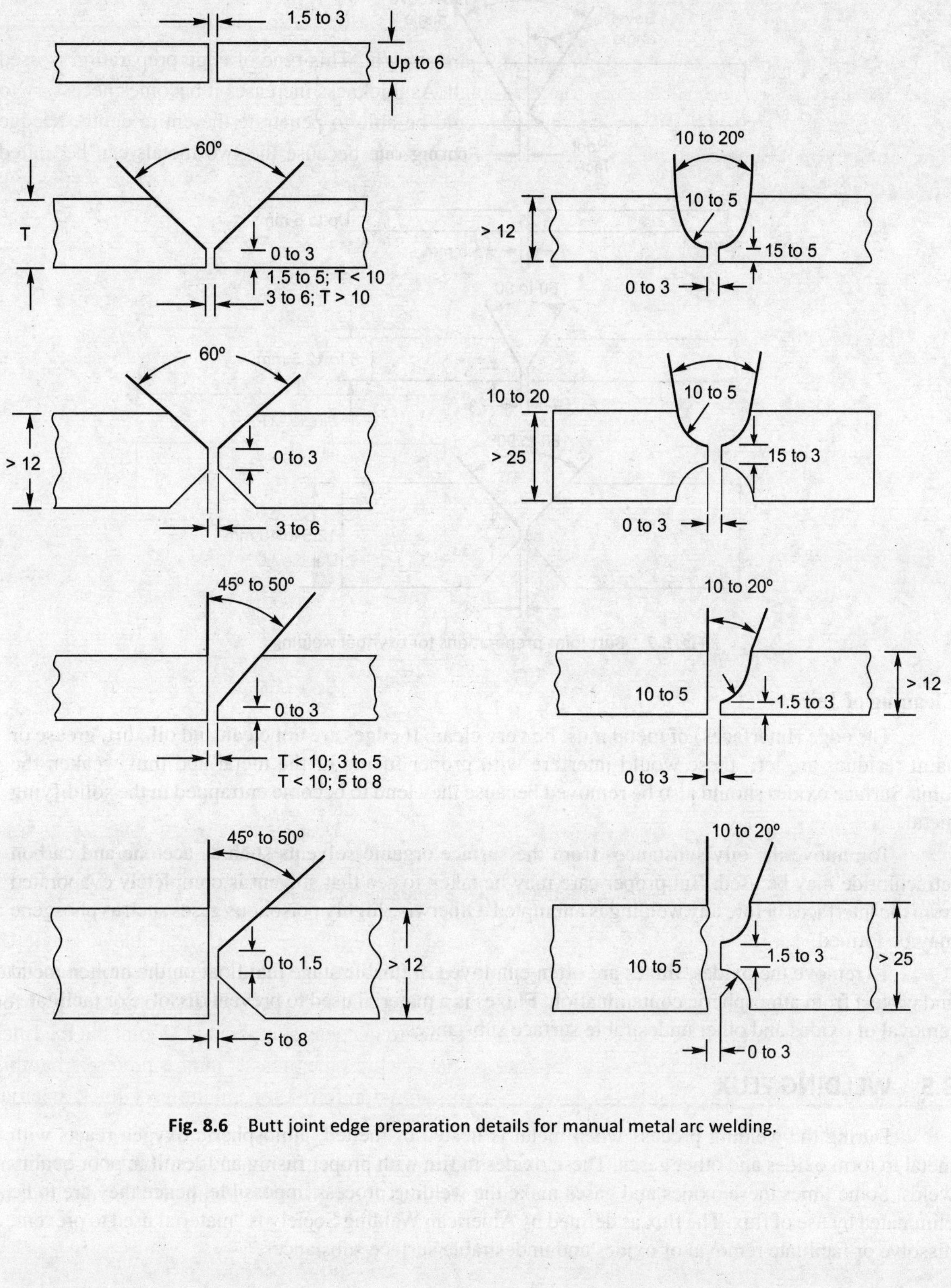

Fig. 8.6 Butt joint edge preparation details for manual metal arc welding.

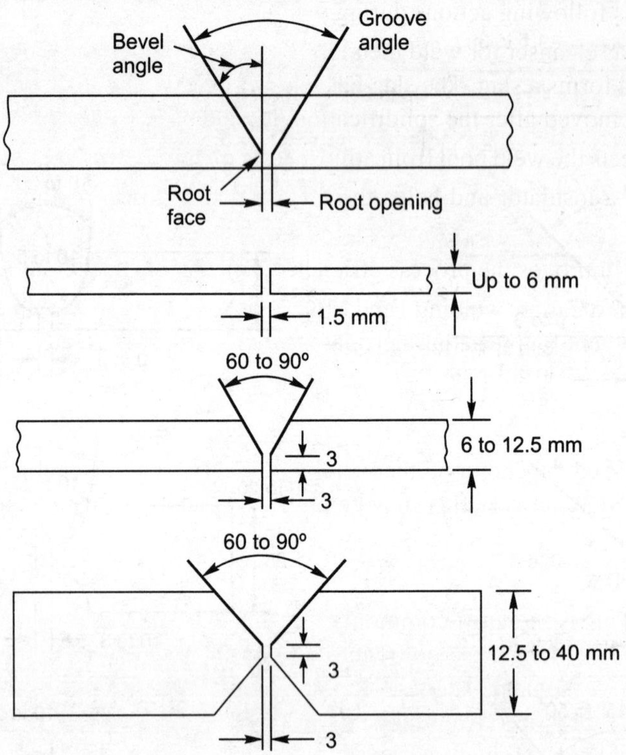

Fig. 8.7 Butt joint preparations for oxy-fuel welding.

Cleaning of Joints

The edge (Interfaces) of metal must be very clean. If edges are not clean and oil, dirt, grease or paint residues are left, these would interfere with proper fusing of the metal and thus weaken the joint. Surface oxides should also be removed because they tend to become entrapped in the solidifying metal.

To remove the oily substances from the surface organic solvents such as acetone and carbon tetrachloride may be used. But proper care may be taken to see that solvent is completely evaporated from the interfaces before any welding is attempted. Otherwise highly poisonous gases such as phosgene may be formed.

To remove the oxides, fluxes are often employed in fusible stage that float on the molten metal and protect from atmospheric contamination. Fluxes is a material used to prevent dissolve or facilitates removal of oxides and other undesirable surface substances.

8.5 WELDING FLUX

During the welding process when metal is heated or melted, atmospheric oxygen reacts with metal to form oxides and other gases. These oxides in fire with proper fusing and result in poor quality welds. Some times these oxides and gases make the welding process impossible, hence they are to be eliminated by use of flux. The flux as defined by American Welding Society is "material used to prevent, dissolve or facilitate removal of oxides and undesirable surface substances."

The flux performs following actions during welding.

(*i*) The flux acts as cleanser for weld metal. It reacts with oxides and other impurities present in weld pool and forms a slag. The slag has a lower density and floats on the top of weld pool. Slag can be removed after the solidification of weld bead by chipping hammer.

(*ii*) The slag protects the weld pool from atmospheric oxygen during the weld bead solidification.

(*iii*) The slag acts a insulator and helps the metal to cool slowly and prevent the formation of brittle weld.

(*iv*) The flux also improves the process efficiency by reducing spatter and burning losses.

In TIG (tunsten inert gas arc welding) and MIG (metal inert gas arc welding). Welding processes the purpose of flux is solvent by means of inert gases whereas in SMAW the electrode coating serves the purpose of flux.

Composition of Fluxes

The type of flux used depend upon the welding operation and the characteristics of base metal that is being welded. In SAW process flux may be physically submerged arc welding classified as:

(1) Fused flux

(2) Agglomerate flux

1. **Fused flux.** This is the most commonly use flux. The flux is manufacturing from minerals like sand (SiO_2), Manganese ore ($MnSiO_3$), Chalk ($CaCO_3$), dolomite $CaMg\ (CO_3)_2$, and fluorspar etc. As the name suggests flux is prepared by fusing the ingredients together in a furnace. This flux is free from moisture. The composition of these fluxes and their grains have less mechanical strength because of which flux cannot be used repeatedly.

Table 8.1. Ingredients of Agglomerate Fluxes

Ingredient	Composition (by weight)
Marable	57.9
Rutile	15.0
Fluorospar	20.0
Ferromanganese	0.5
Ferro-silicon	0.8
Ferro-titanium	6.0
Sodium silicate (1.35 density)	20.0

Table 8.2. Ingredients of Fused Flux

Name of the ingredient	Quantity
Silica	41.00–43.5%
Magnesium oxide	5.5–7.5
Manganous oxide	34–38%
Calcium flouride	4–5%
Calcium oxide	5.5%
Alumina	4.5%
Sulphur	0.15%
Phosphorus	0.12%

2. **Agglomerated flux.** These also are included ceramic fluxes and are prepared by mixing together the ingredients and bonding the grains with sodium silicate. These fluxes contain ferro manganese, ferro silicon ferro-titanium alloys.

The composition of one such flux is shown in Table 10.2. Drawback of ceramic flux is that it quickly absorb moisture.

8.6 WELDING POSITIONS AND TECHNIQUES

Welding Position (Fig. 8.8)

Depending upon the position during welding each of the joint mentioned earlier can have any of the four welding positions. All the welding positions are shown in Fig. 8.8.

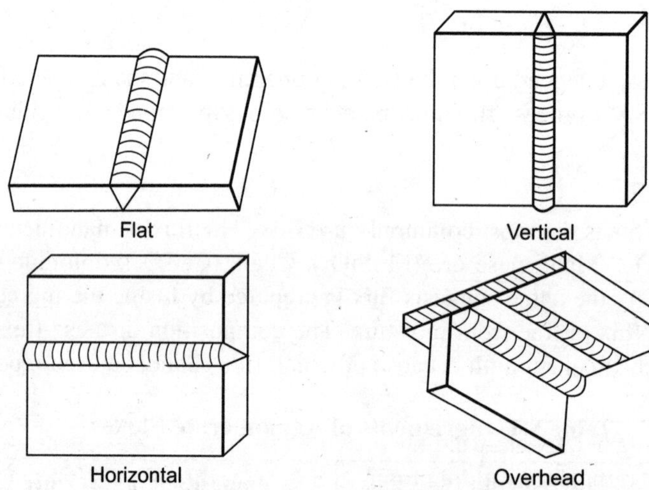

Fig. 8.8 Various welding positions.

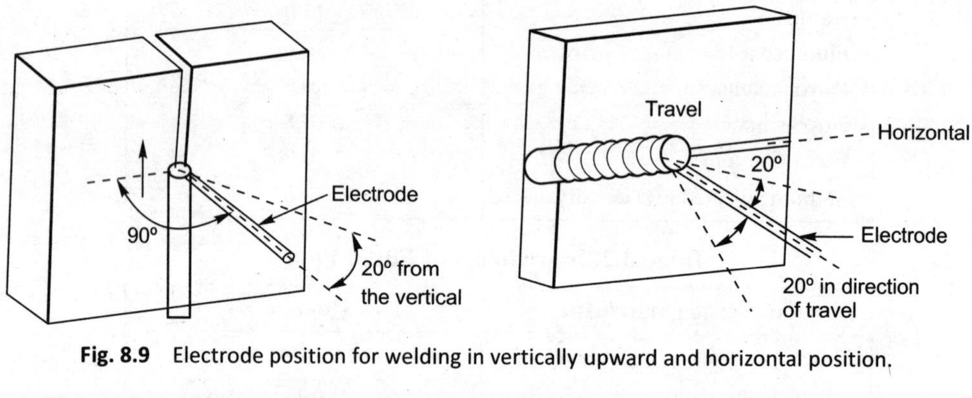

Fig. 8.9 Electrode position for welding in vertically upward and horizontal position.

1. **Flat or down hand welding.** According to international norms weld is a weld in any direction on horizontal surface.
2. **Horizontal welding.** Horizontal weld is one running horizontally on the vertical surface.
3. **Vertical welding.** A vertical weld is that which runs vertically on a vertical surface.
4. **Overhead welding.** An overhead weld is one that is deposited above the operator's head.

Welding Techniques

Welding techniques can be classified as forehand welding (left ward welding) and backhand welding (rightward welding).

(*a*) **Forehand welding.** In case of forehand welding the torch tip should be positioned above the metal plate at an angle of 30° to 45° from the horizontal plane. The torch movement should be either oscillating or circular. In case of forehand welding the torch is moved in the direction of the welding torch tip. This tends to preheat the metal.

(*b*) **Backhand welding.** In backhand welding the torch move backhand and it allows the joint to be continuously annealed and thus, relieving the welding stresses. It is generally used for thicker materials as it forms bigger weld bead.

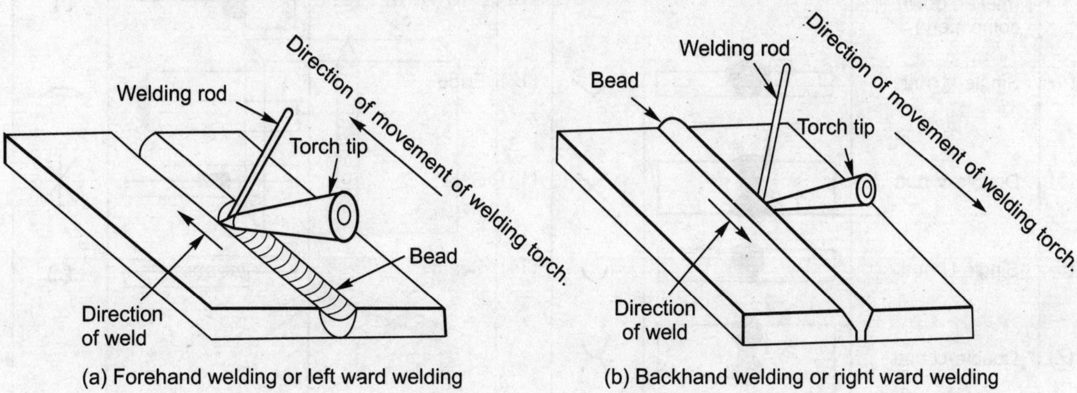

Fig. 8.10 Welding techniques.

In forehand welding or leftward welding torch is moved in the direction of torch tip [Fig. 8.10 (*a*)] whereas in back hand or right ward welding torch moves backward.

8.7 WELDING SYMBOLS

Welding symbols provides all the information relating to welding process. The welding symbols provides information regarding type of weld, joint preparation, weld size, weld contour, root gap and degree of penetration etc. welding symbols may be assumed as short hand language about information of welding processes.

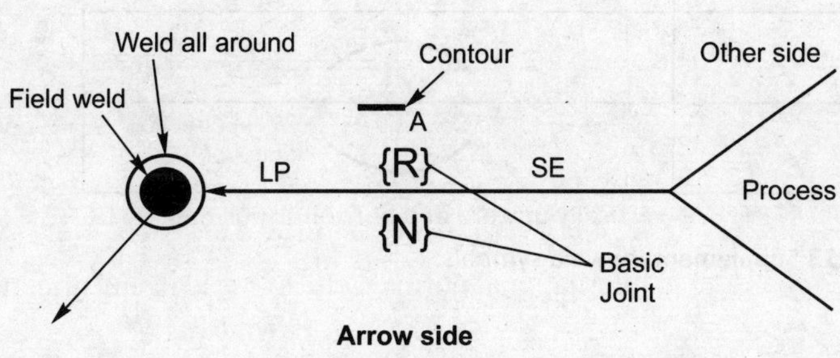

L = Length of weld
P = Pitch of weld (center to center spacing of welds)
A = Groove angle (included angle of countersink for plug weld)
R = Root opening (depth of filling for plug and spot weld)
S = Depth of penetration (size or strength for certain welds)
E = Effective throat
F = Finish of weld
N = Spot weld number

Fig. 8.11 Welding symbols.

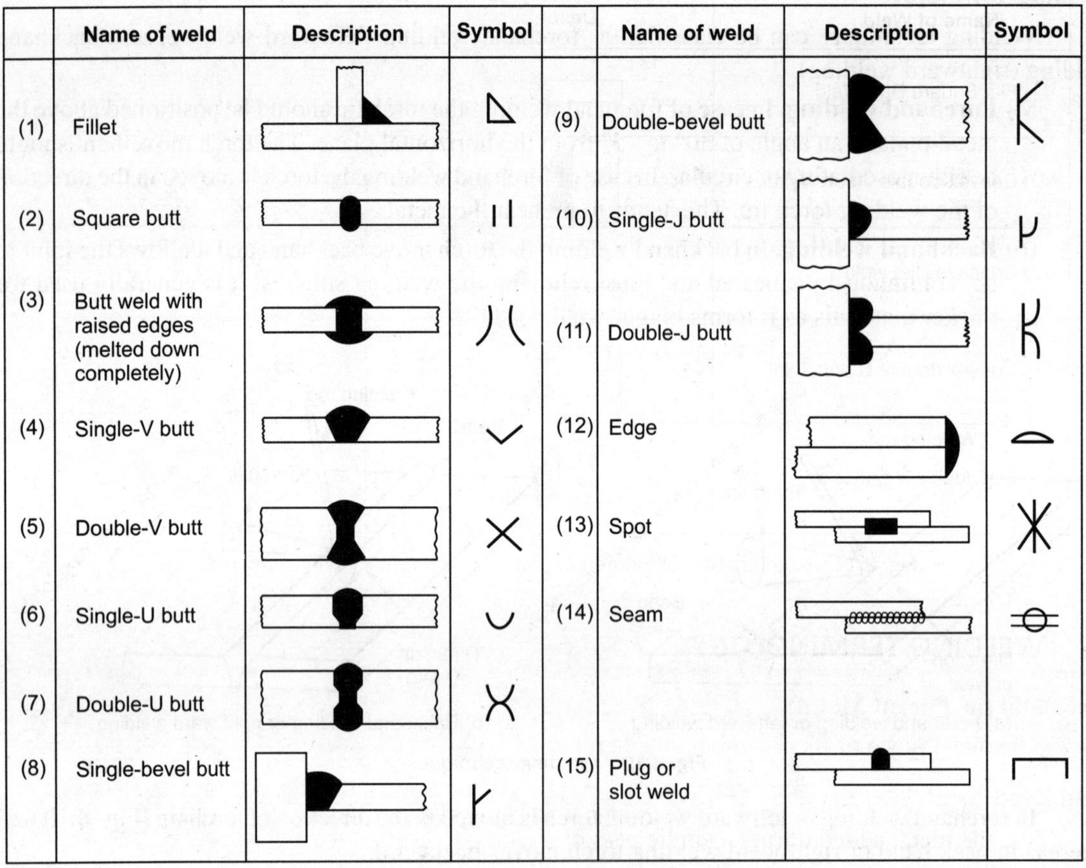

Fig. 8.12 Elementary weld symbols as recommended by bureau of indian standards.

Shape of Weld Surface	Symbol
(1) Flat (usually flash)	
(2) Convex	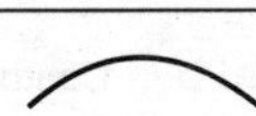
(3) Concave	

Fig. 8.13 Supplementary weld symbols.

Name of Weld	Description	Symbol
(1) Flat Square Butt Weld		
(2) Convex double J butt weld	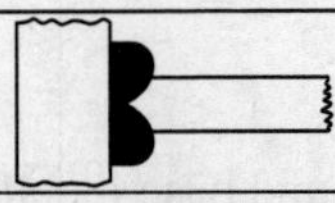	
(3) Concave fillet weld	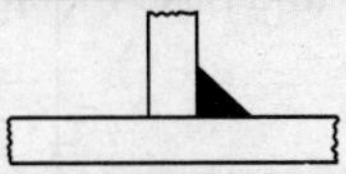	
(4) Convex double U butt Joint		
(5) Flat double V butt joint		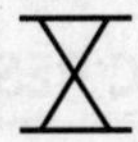

Fig. 8.14 Combination of Elementary and Supplementary Weld Symbols.

8.8 WELDING TERMINOLOGY

Basemetal (or Parent Metal)

The metal piece or pieces to be joined are termed as base metal.

Weld Bead

It is the metal added during a single pass of welding.

Weld Metal Pool

The portion of joint which is melted by heat of welding is known as weld pool or puddle.

Crater

In Arc welding crater is the depression in the weld metal pool at the point where arc forces the molten metal out of molten metal pool (weld pool). It appears as small cavity.

Contour

It is the shape of weld bead. It may be flat, convex or concave.

Deposition Rate

The rate at which weld metal is deposited per unit time. It is normally expressed as kg/hour.

Penetration

It is the depth upto which weld metal penetrates and combined with base metal. It is **measured from top surface of joint.**

Tacking or Tack Weld

It is the small weld used to temporarily hold two piece in alignment just before making final weld.

Root

It is the point at which two pieces to be joined by welding are nearest.

Weld Face

The exposed surface of weld.

Toe of Weld

It is the junction between weld face and base metal.

Weld Pass

It is the single movement of welding torch or electrode along the length of joint.

Weld Bead

The metal deposited during single pass is known as bead.

8.9 ELECTRIC RESISTANCE WELDING

Electric resistance welding is a fusion welding process where both, heat and pressure are applied on the joint but no filler material or fluxes are added. Resistance welding is based on well known principal that heat is generated by the resistance offered to flow of current. Resistance welding has wide applications in welding of sheet metal of all engineering material (except Cu, Ag), automobile and aircraft industries, pipes and tupes production.

Principle

In resistance welding process a low voltage and high current is passed through the joint for a very short time. Due to the contact resistance at the joint high heat is generated which melts the joint. At the same time pressure is maintained on the joint. The heat generated can be expressed as:

$$H = I^2 Rt \qquad \text{(equation 10.1)}$$

$H \rightarrow$ Heat generated in workpiece in joules

$I \rightarrow$ Electrical current in amperes

$R \rightarrow$ Resistance of joint in ohms

$t \rightarrow$ Time for which electrical current is passed through the joint in seconds.

Here because of square of the current in above equation I needs to be properly controlled for any proper joint. The main requirement of the process is low voltage and high current which is obtained by means of a step down transformer.

8.9.1 Resistance Spot Welding

Resistance spot welding is the most common resistance welding process. It is essentially done to join sheet metal jobs in essentially lap joint. The workpieces are clamped between two (water cooled) copper cylindrical electrodes. On the passage of high transient current, the interface melts over a spot and forms the weld. The high current is needed for a fraction of second to complete the weld. The cooling at the electrode limits the size of spot.

Spot welding machine is shown in Fig. 8.15. It consists of two electrodes out of which one is fixed. The upper electrode is fixed to a rocker arm (to provide mechanical advantage) for transmitting the mechanical force from a pneumatic cylinder or hydraulic cylinder. The proper alignment of the electrodes should be maintained to obtain a good quality joint.

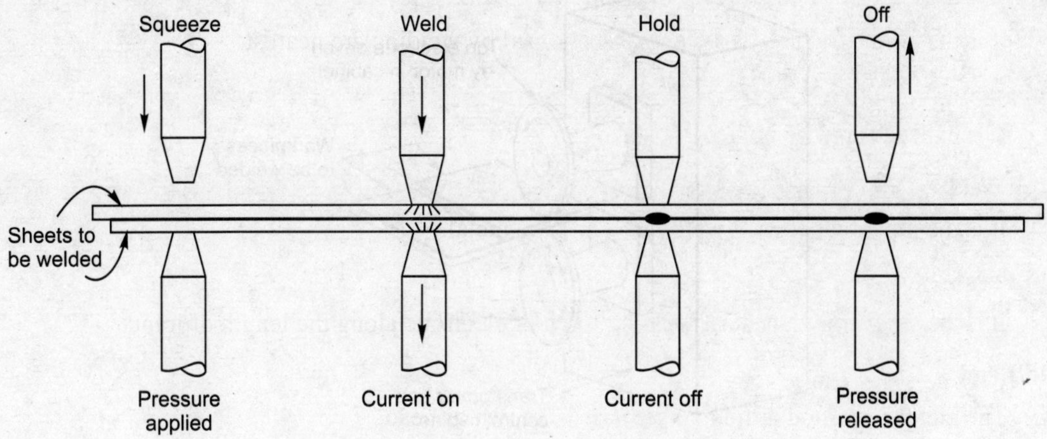

Fig. 8.15 Various stages while making a spot weld.

8.9.2 Seam Welding

Seam welding is a continuous welding process wherein spot welds overlap each other to desired extent. Here disc electrodes are used instead of cylindrical electrodes. As the electrode rotates the work is moved between them and at the same time pressure on the joint is maintained. Electrodes need not to be separated at any time. The current is applied through electrodes in a series of pulses at proper interval and timing are so adjusted that the pulses overlap each other and form a continuous seam joint.

Seam welding is used for producing leak proof joint. Welds should overlap 15% to 20% of each nugget for producing leakproof joint. Seam welding has wide applications in sheet metal containers like petrol tank for automobiles.

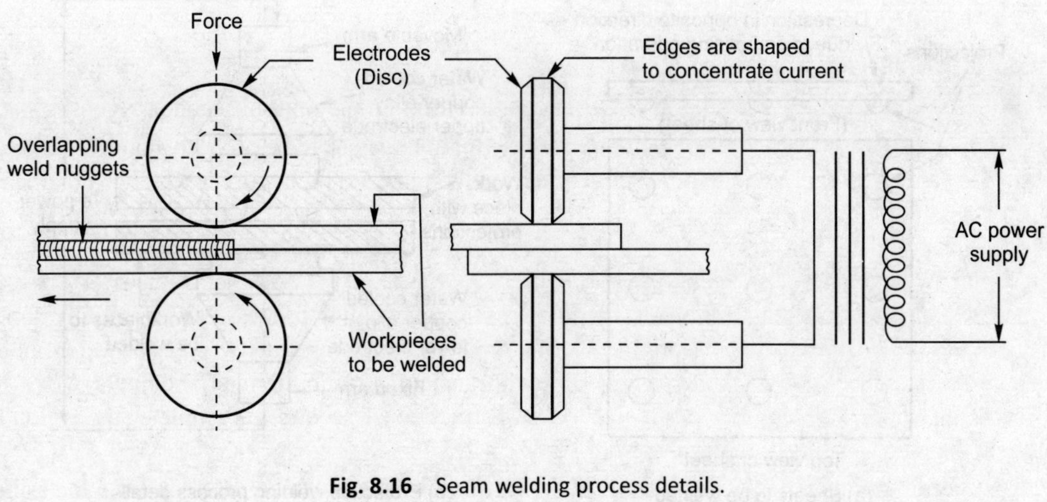

Fig. 8.16 Seam welding process details.

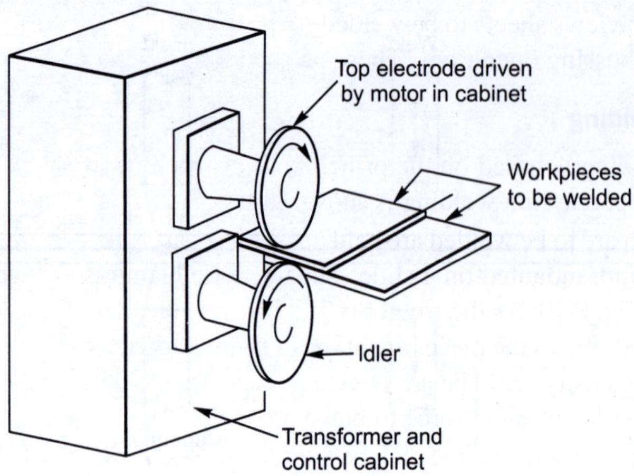

Fig. 8.17 Seam welding machine.

8.9.3 Projection Welding

Projection welding is another variation of resistance spot welding where one of the sheets to be welded is provided with number of projections. The welding spots are predetermined because of the number of projections. The projections are generally obtained by embossing or coining process. As the welding current is passed through projections they get melted and fusion joint is obtained while at the same time pressure on joint is maintained by electrodes.

One of the attractive features of projection welding is that it gives longer electrode life with less wear and maintenance. It is no more necessary to use small sized electrode. It is also possible to weld more than one spot simultaneously. Upto five projections can be successfully welded at a given time that come under the tip of the electrode.

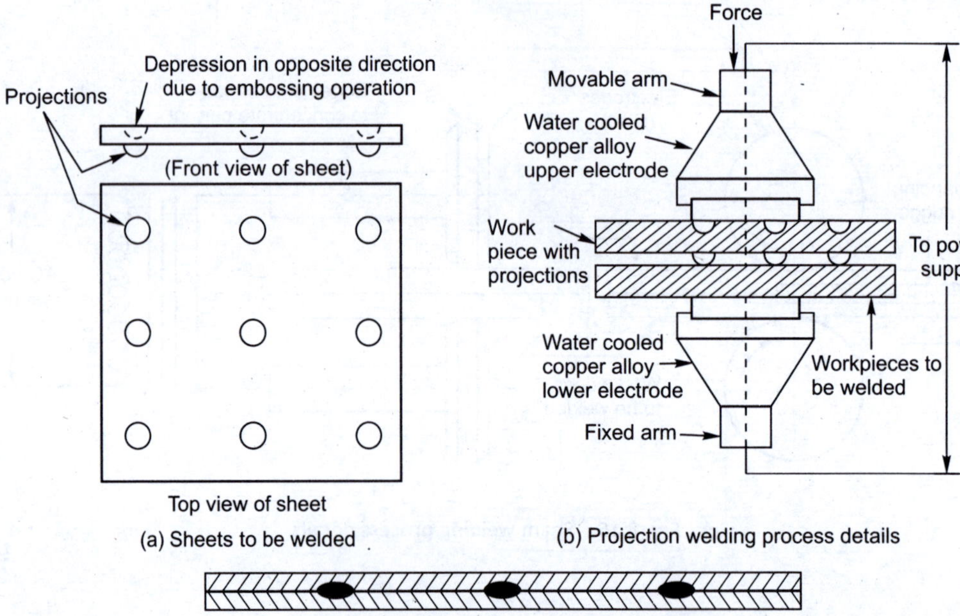

Fig. 8.18 Projection Welding.

Front views top views sheets to be welded is shown in Fig. 8.13 (*a*). The sheet may be provided with projection by embossing operations. These projections predetermine the welding spot.

8.9.4 Percussion Welding

Percussion welding is based on the principle of arc welding, for heating rather than resistance heating. The setup for percussion welding is shown in Fig. 8.19.

The parts which are to be welded are held apart. One part is held in stationary holder and other is held in a movable clamp, mounted on a slide. The movable clamp is backed up against heavy spring pressure as shown in Fig. 8.19. As the trigger is pulled, movable clamp is moves rapidly carrying with it the piece to be welded. When the pieces are about (1.5 mm) apart, there is sudden discharge of electric energy, causing high intensity arc. The arc is extinguished by the percussion blow (rapid blow) of two part coming together with sufficient force to make the weld.

The electric energy for discharge may be built up in two ways; high voltages breakdown and low voltage arc drawn method. In both methods energy source are capacitor, which are changed by variable voltage transformers cum rectifiers.

— The process can be successfully used for joining dissimilar metals combinations like copper and stainless steel and copper and nichrome.

— The major limitation of process is that only small area upto 625 mm^2 of regular sections can be welded.

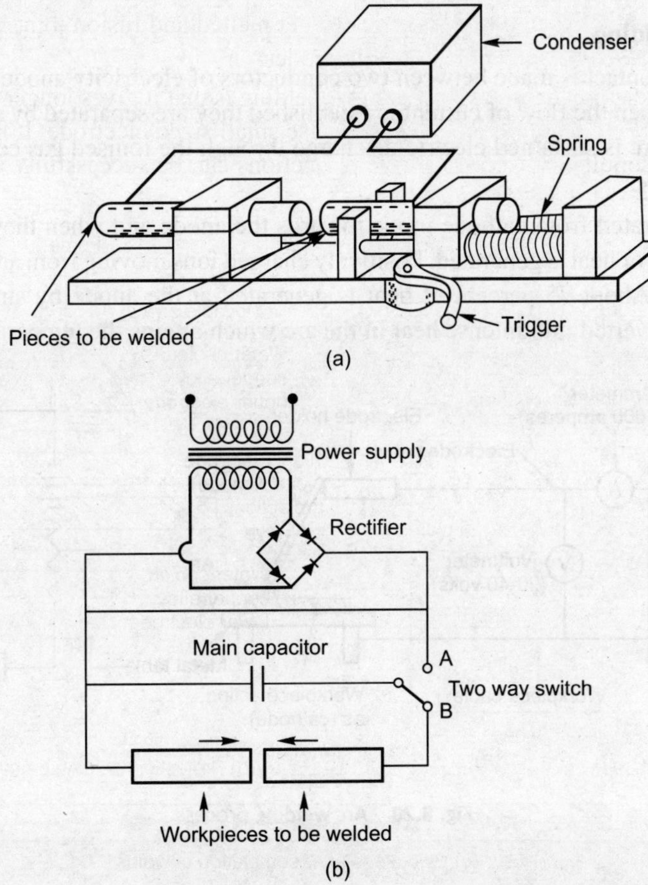

Fig. 8.19 Setup for percussion welding.

8.10 ELECTRIC ARC WELDING

Introduction

Electric arc welding is one of the most versatile joining process and is extensively used all over the world. One of the attractive feature of arc welding is ease of use and high production rate that can be achieved economically.

Arc welding is a fusion welding process where joining is obtained by heat produced by an electric arc. The arc may be obtained:

(*a*) Between workpiece and consumable electrode (acts as filler materials also) as in shielded metal arc welding (SMAW), MIG welding etc.

(*b*) Between workpiece and non-consumable electrode as in tungsten inert gas arc welding. (TIG welding)

(*c*) Between two non-consumable electrodes as in carbon arc welding.

(*d*) Or between metal pieces which are to be welded together.

In arc welding process when consumable electrodes are used it also acts as filter material which is consumed continuously and deposited into the joint to make the weld. When non-consumable electrodes are used separated filler metal rods are used.

Principle of Arc Welding

Initially, the contact is made between two conductors of electricity anode and cathode to create an electric circuit. When the flow of current is established they are separated by small distance and arc is formed. Actually arc is sustained electric discharge through the ionised gas column (called plasma) between two electrodes.

Electrons liberated from cathode move towards the anode and when they strike anode at high velocity large amount of heat is generated. Positively charged ions moving from anode are strike cathode thus liberating heat. About 75 percent of heat is generated at the anode by striking electrons. Thus electric energy is converted into intense heat in the arc which attains the temperature around 5500°C.

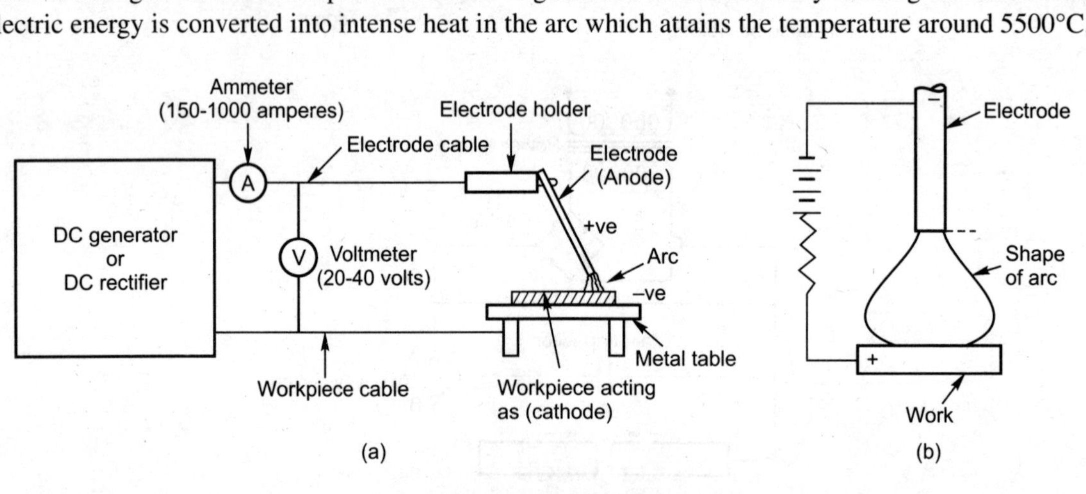

Fig. 8.20 Arc welding process.

Fig. 8.20 (*a*) Basic circuit for arc welding (DC arc welding with reverse polarity). Fig. 10.23 (*b*) Basic shape of arc between electrode and workpiece.

The potential difference between the electrodes and air gap between electrodes are two significant parameters which are to be controlled efficiently. If the air gap becomes too large for voltage the arc may be extinguished. Larger air gap requires higher potential differences.

Fig. 8.20 (*a*) shows basic circuit for arc welding. The sources of current may be alternating current or direct current. The workpiece is kept on the metallic table. One cable from power supply is connected to the electrode holder into which the electrode is gripped. Other lead is connected to the metallic table on which workpiece is kept. When the electrode is brought into contact with workpiece arc generates and welding takes place.

Arc Welding Equipment

The arc welding machines can be classified into two categories depending upon the source of electric power.

1. Alternating current machines.
2. Direct current machines.

Either alternating current or direct current can be used for electric arc welding.

Direct current arc welding is more expensive than AC are welding, even then DC arc welding is preferred because of the control of heat input and easy arc maintenances.

8.10.1 Polarity in Arc Welding

As discussed earlier in case of DC arc welding about 60–70% of heat is generated at anode and rest of the heat is generated at cathode. So if more heat is required on the workpiece side, say for welding thick materials and materials having high thermal conductivity. The work piece may be made as anode (*and electrode in the electrode holder of welding machine is working as cathode*) liberating large heat near it. This arrangement is termed as straight polarity or direct current electrode negative (DCEN).

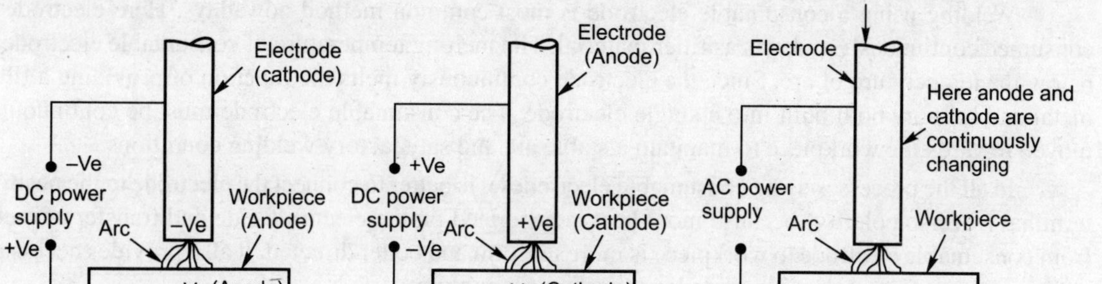

(a) Straight polarity (D C E N) (b) Reverse polarity (D C E P) (c) AC Arc welding

Fig. 8.21 Polarity in arc welding.

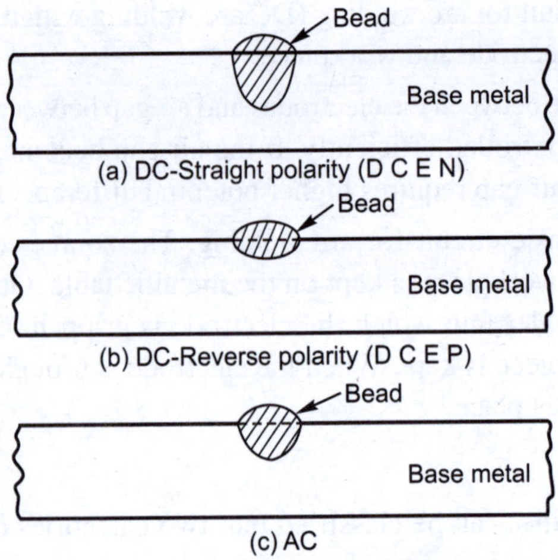

Fig. 8.22 Illustrates how the penetration is affected by the polarity of welding.

For thinner materials where less heat is required on work piece side, the work piece may be made as cathode. This is termed as reverse polarity or direct current electrode positive (DCEP). DC are welding is preferred for difficult tasks such as overhead welding as it maintains stable arc.

For AC arc welding there is no question of polarity because anode and cathode will interchange after every half cycle, so heat generated at each electrode is same.

The penetration obtained in straight polarity is large. In reverse polarity penetration is small. In AC arc welding penetration is somewhere in between above two types as shown in Fig. 8.22.

8.10.2 Arc Welding Electrodes

The electrodes used for providing heat are of two types:

1. Consumable electrodes.
2. Non-consumable electrodes.

Consumable Electrodes

Welding using a consumable electrode is most common method nowadays. Here electrode is consumed continuously and acts as filler material. The melting temperature of consumable electrode is below the temperature of arc. Since the electrode continuously melts the function of providing a filler metal and heat are both built into a single electrode. The consumable electrode must be continuously moved towards the workpiece to maintain a stable arc and satisfactory welding conditions.

In all the processes using consumable electrodes it is better to connect the electrode to the positive terminal (reverse polarity) because more heat is generated on the electrode side and transfer of metal from consumable electrode to workpiece is more uniform and better directed. It also provides necessary cleaning action on metals with oxide layer such as aluminium.

Consumable electrodes are made of various materials depending upon the purpose and chemical composition of metals to be welded. They may be made of steel, cast iron, aluminium, copper, brass or bronze. When consumable electrode is used welding process is called metal arc welding.

The consumable electrode may be either bare or coated type. The coated electrodes most commonly known as stick electrodes are used in manual arc welding process. The stick electrodes are normally available in diameters 3.2, 4, 5, 6, 8, and 9 mm and the length is 350 or 400 mm.

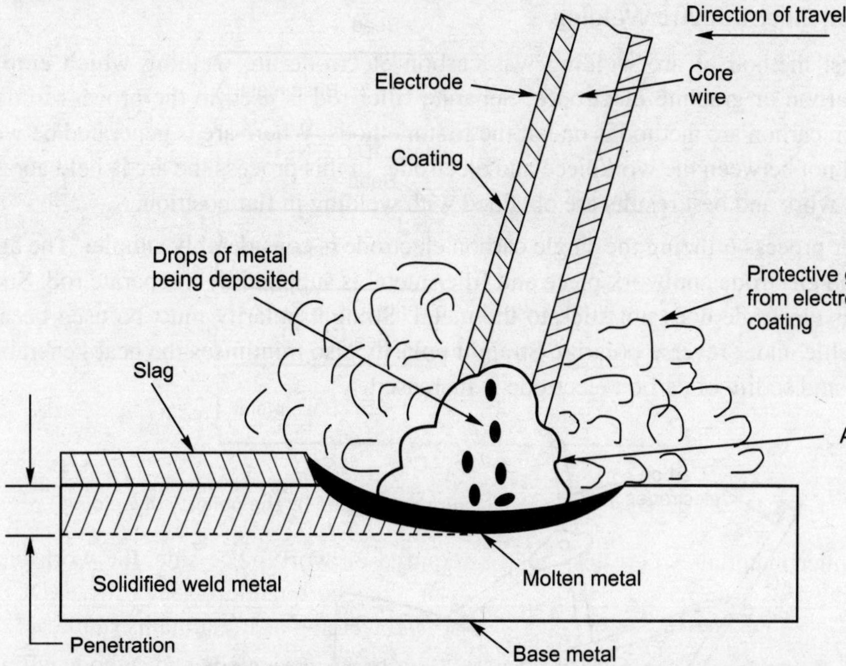

Fig. 8.23 Important features of arc welding processes using coated electrodes.

Characteristics of Coated Electrodes

Fig. 8.23 shows the important features of arc welding process using coated electrodes. Coating on the electrodes add many desirable characteristics and serve following purpose:

(*i*) Coating on the electrode provides a gas shielding around the arc to eliminate the formation of undersirable oxides and nitride, which would otherwise form by reacting with atmospheric oxygen and nitrogen.

(*ii*) Coating forms a slag which being lighter than molten metal floats on the top of weld pool and protects it against the surrounding air during weld bead solidification. It also helps metal to cool slowly, preventing the formation of brittle weld.

(*iii*) Elements for stabilisation of arc are added to these coatings.

(*iv*) Alloying elements can also be added through these coatings to improve the strength and physical properties of weld metal.

Moisture is one of the major problem with coated electrodes. Therefore, it is preferable that electrodes should be kept in an oven to keep them thoroughly dry.

Non-Consumable Electrodes

When non-consumable electrodes are used, separate filler metals rods are used. We may say that in case of welding processes using non-consumable electrodes heat source and filler metal deposition can be separately controlled. For all processes using non-consumable electrodes it is better to connect the electrode to the negative terminal (straight polarity) to keep the heat losses minimum. However, in case of welding aluminium and magnesium it is preferable to use AC power supply.

Non-consumable electrodes are usually made of tungsten, graphite or carbon. When non-consumable electrodes are used processes are named after the electrode material like tungsten arc welding (tungsten is used as electrode) or carbon arc welding (carbon is used as electrode).

8.10.3 Carbon Electrode Arc Welding

The first method of arc welding was carbon electrode arc welding which employed non-consumable carbon or graphite electrodes. Separate filler rod is used in the process to deposit filler metal. The twin carbon arc method is one of the first methods. Where arc is generated between the two electrodes and not between the workpiece and electrode. In this process the arc is held about 6.5 to 9.5 mm above the work and best results are obtained with welding in flat position.

Another process utilizing the single carbon electrode is considerably simpler. The arc is created between carbon electrode and work piece and filler metal is supplied by a separate rod. Such an arc is easy to start as electrode does not stick to the metal. Straight polarity must be used because carbon becomes unstable under reverse polarity. Straight polarity also minimises the heat generation near the electrode side and so life of carbon electrode is increased.

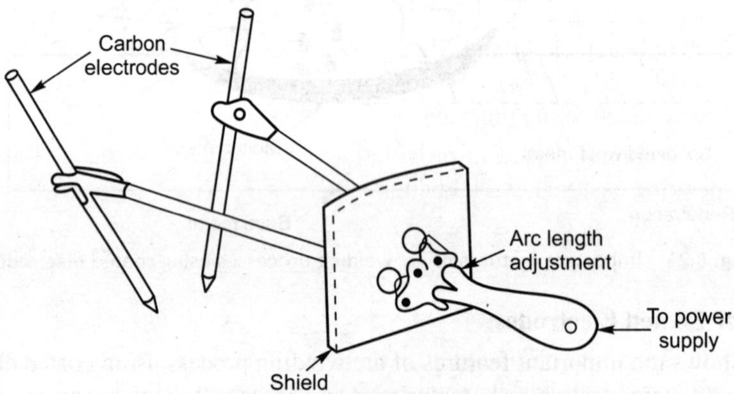

Fig. 8.24 Carbon electrode arc weld.

Twin electrode carbon arc welding. Here arc is generated between two carbon electrodes.

8.10.4 Shielded Metal Arc Welding (Fig. 8.20)

It is the most common and most versatile joining process and extensively used all over the world. Metal arc welding is done with stick electrodes. An arc is created by striking the workpiece with an electrode and quickly raising it a short distance. The equipment for shielded metal arc welding is least expensive of all the arc welding processes and weld can be made in any position. Job of any thickness can be welded by shielded metal arc welding but generally 3 mm to 20 mm thickness is preferred. Both AC and DC power source may be used. Generally AC power source is preferred in welding heavy plates or fillet welding because of less problem of “arc blow” with AC welding. DC power source is normally used with reverse polarity. The main disadvantage is the slow speed of process. The weld metal deposition rate is in the range of 1 to 8 kg/hr.

8.10.5 Tungsten Inert Gas Welding (TIG)

Tungsten inert gas welding or GTAW (gas tungsten arc welding) is the welding process where arc is maintained between a non-consumable tungsten electrode and workpiece in a protective inert gas atmosphere. Filler material needed is supplied externally. For welding steel, cast iron, copper alloys and stainless steel, direct current with straight polarity is used. Reverse polarity is rarely used.

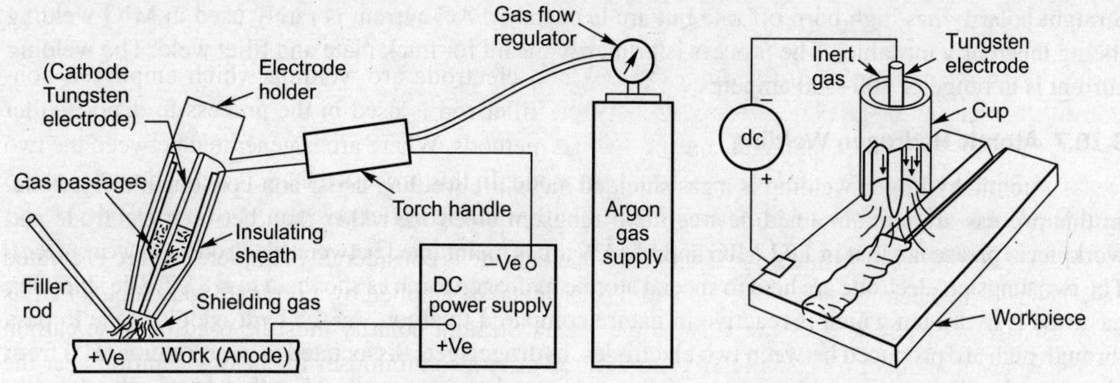

(a) Tungsten inert gas (TIG) welding process (b) Pictorial view of (TIG) welding process

Fig. 8.25 Tungsten inert gas welding setup.

For welding metal such as Al and Mg if straight polarity is used, oxide layer persists which prevents the formation of weld pool so an AC arc is used for such material. To avoid melting of electrode thorium or zirconium is added to the tungsten.

Argon is most commonly used to provide inert atmosphere. Nitrogen is sometimes used for welding copper. Operating cost makes TIG better suited for welding of light work rather than welding heavier materials.

8.10.6 Metal Inert Gas Arc Welding

Figure 8.26 represents a metal inert gas arc welding process. It is also known as gas metal arc welding (GMAW). In this case arc is maintained between a consumable electrode and workpiece in an inert gas atmosphere. The electrode is bare wire consumable electrode and is fed continuously by drive rolls. Because filler material is transferred through the protected arc, greater efficiency is obtained resulting in more rapid welding. The metal is deposited in an atmosphere that prevents contamination. Direct current reverse polarity provides stable arc. This is the major difference between MIG and TIG.

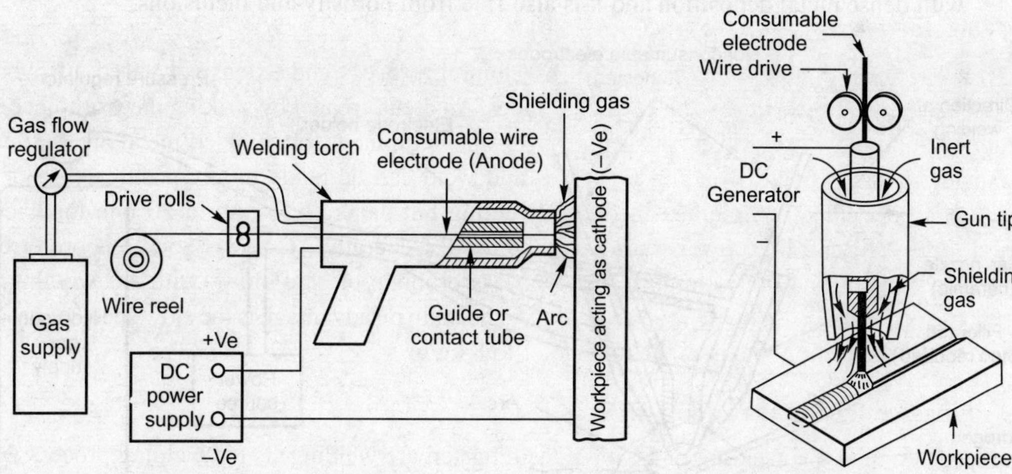

(a) Metal inert gas arc welding process (b) Metal inert gas arc welding process (Pictorial view)

Fig. 8.26 Metal inert gas welding setup.

Straight polarity has high burn off rate but arc is unstable. AC current is rarely used in MIG welding (being inherently unstable). The process is primarily meant for thick plate and fillet weld. The welding current is in range of 100–300 amperes.

8.10.7 Atomic Hydrogen Welding

Atomic hydrogen welding is a gas shielded metal are welding using non-consumable electrode. In this process arc is maintained between two tungsten electrode rather than between electrode and workpiece (please not that in TIG, MIG and SMAW arc is maintained between electrode and workpiece). The two tungsten electrode are held in special atomic hydrogen torch as shown in Fig. 8.27. The shielding gas used is hydrogen which is reactive in nature compared to argon. As the hydrogen is made to pass through high arc produced between two electrodes, hydrogen gets dissociated in single atoms (H^+) from H_2 molecules. The single hydrogen atoms have strong tendency to reunite when they approach relatively cool surface of workpiece. When hydrogen atoms reunite, intense heat is released producing the temperature of about 3500°C and joint is melted.

Following are the salient features of the process:

(1) Because of the strong reactivity of hydrogen atom, it is able to break oxide on the base metal and clean weld is obtained.

(2) As arc is independent of workpiece, heat input to the welding zone can be controlled by varying the distances between electrodes.

(3) The normal process voltage ranges between 50 to 80 V and current varies from 15 to 1500 amperes.

(4) The process is used for flat positions only because of generation of high temperature.

(5) The process is applicable to all the materials which are welded by non-consumable electrode inert gas are welding process. In addition process is also used for repairing to steel moulds, dies and tools.

(6) AC power supply is used to reduce the wear of particular electrode.

(7) If the process is followed properly, atomic hydrogen welding gives extremely clean weld with dense metal deposition and it is also free from porosity and inclusions.

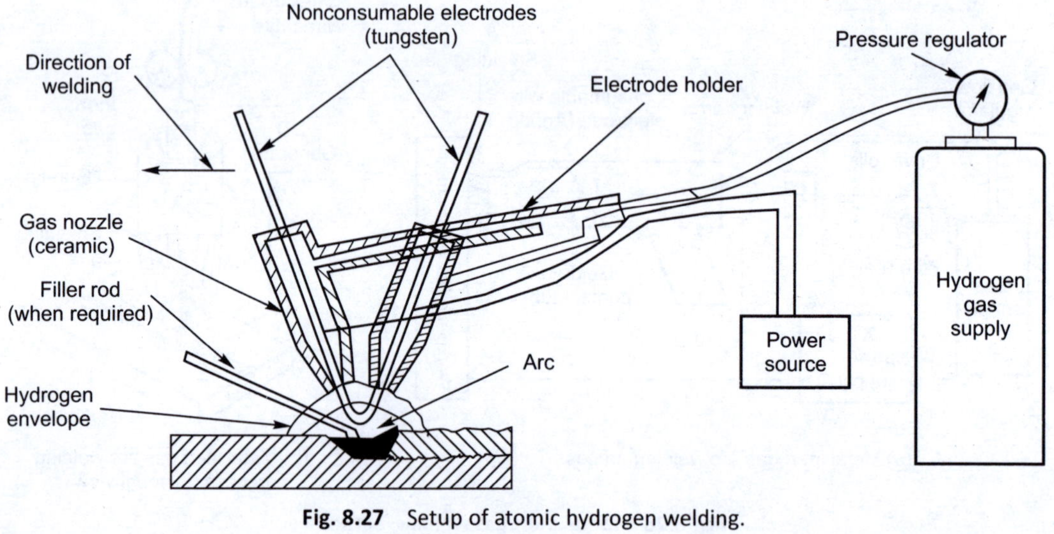

Fig. 8.27 Setup of atomic hydrogen welding.

8.10.8 Submerged Arc Welding

Submerged arc welding is used for doing faster welding jobs. High deposition rate, deep penetration and high cleanliness are the characteristics of submerged arc welding process. In this process arc is produced between bare (non coated) consumable electrode and workpiece. The welding zone and electric arc are completely covered with a granular, fusible material called flux during the welding. Flux is delivered ahead of the welding electrode by means of welding flux feed tube (Fig. 8.28). A portion of flux melts and removes impurities from large pool of molten metal, while the unmelted flux provides additional shielding. The molten flux acts as cleanser, absorbing impurities from molten metal, forms a slag. The slag being lighter floats on top of molten metal. The slag, alongwith flux which is not melted, provides a thermal insulation which slows down the cooling of weld metal and results in soft and ductile weld. The salient features of submerged arc welding are:

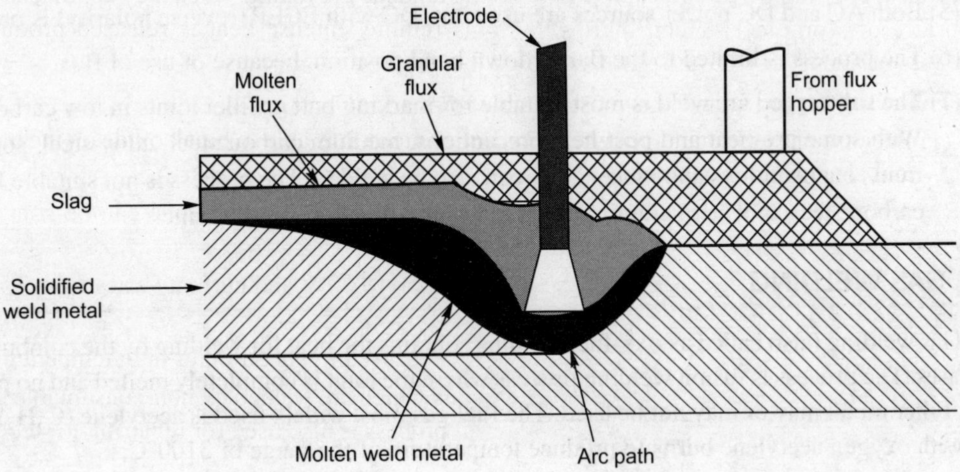

(a)

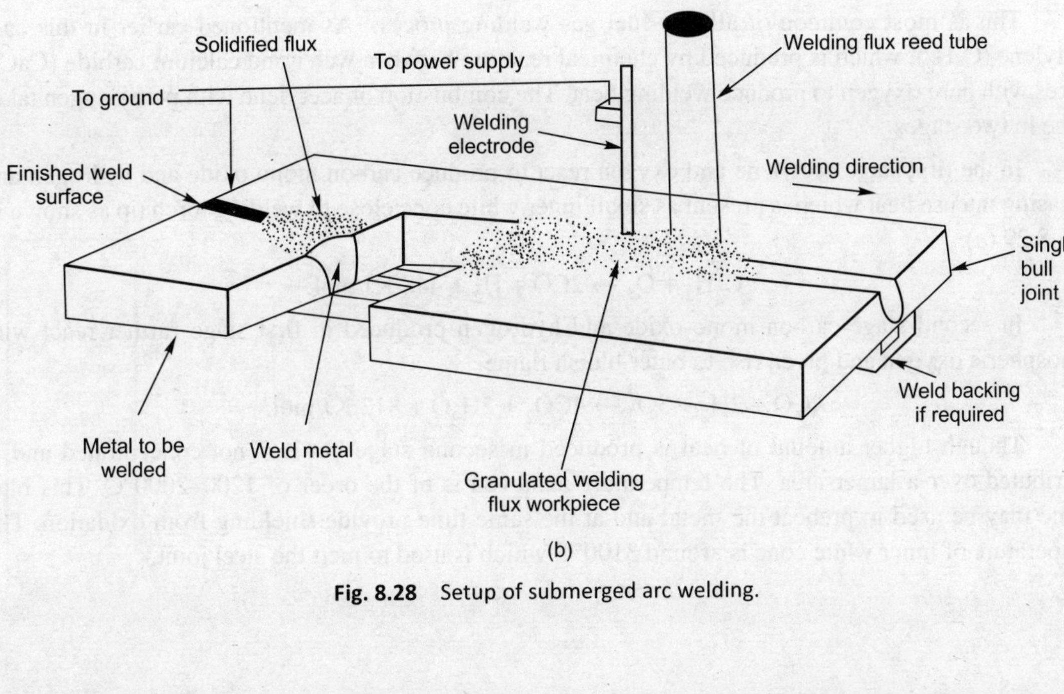

(b)

Fig. 8.28 Setup of submerged arc welding.

(1) High deposition rate, deep penetration and high cleanliness are the characteristics of submerged arc welding process.

(2) High current upto 4000 ampere may be used, so that high metal deposition rate of the order of 20 kg/hour can be achieved. Current density ranges from 18,000 to 70,000 amperes per square inch of electrode cross-sectional area. Welding speed of 0.8 m/min in 1 inch thick plate can be easily achieved.

(3) As the arc remain submerged underneath flux, it eliminate the use of protective shielding glass for eyes.

(4) Flux cover also eliminate weld spatter and arc radiations and thus improves the efficiency of process.

(5) Both AC and DC power sources are used but, DC with ECEP (reverse polarity) is preferred.

(6) The process is limited to the flat or down hand position, because of use of flux.

(7) The submerged arc weld is most suitable for marking butt or fillet joints in low carbon steel. With some pre-heat and post-heat precautions, medium carbon steel, alloy steel, some cast irons, copper alloys and nickel alloy can also be welded. The process is not suitable for high carbon steel, tool steel aluminium, magnesium, titanium, lead or zinc.

8.11 GAS WELDING

Gas welding (also called as oxy-fuel welding) obtains the heat for welding by the combustion of a fuel gases. The process is fusion welding process where the joint is completely melted and no pressure is used. Filler metal may or may not be used. The fuel gas most widely used is acetylene (C_2H_2). When mixed with oxygen acetylene burns to produce temperature of the range of 3100°C.

Principle of Oxy-Acetylene Welding

This is most common of all oxy-fuel gas welding process. As mentioned earlier in this case acetylene (C_2H_2), which is produced by chemical reaction between water and calcium carbide (CaC_2) mixes with pure oxygen to produce welding heat. The combustion of acetylene with pure oxygen takes place in two stages.

In the first stage acetylene and oxygen react to produce carbon mono oxide and hydrogen and releasing intense heat which is present as small inner white cone close to welding torch tip as shown in Fig. 8.29 (*a*)

$$C_2H_2 + O_2 \rightarrow 2CO + H_2 + 448 \text{ KJ/mol.}$$

In second stage carbon mono-oxide and hydrogen produced in first stage further react with atmospheric oxygen and given rise to outer bluish flame.

$$4CO + 2H_2 + 3O_2 \rightarrow 4CO_2 + 2H_2O + 812 \text{ KJ/mol.}$$

Though higher amount of heat is produced in second stage but it is not concentrated and is distributed over a larger area. The temperature achieved is of the order of 1200–2000°C. This blue flame may be used to preheat the metal and at the same time provide shielding from oxidation. The temperature of inner white cone is around 3100°C which is used to melt the steel joint.

8.11.1 Welding Flames

There are three types of flames obtained in gas welding by varying the oxygen/acetylene ratio.

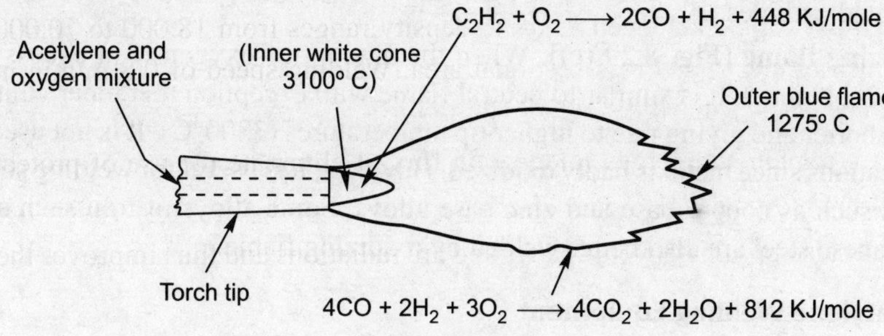

(a) Neutral flame

Neutral flame is with inner white cone and outer blue flame

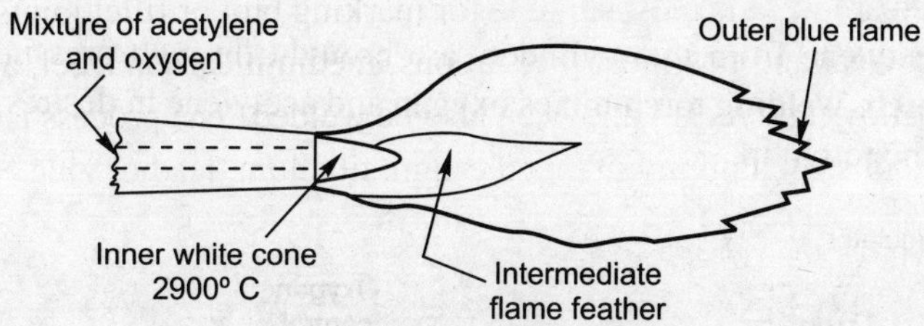

(b) Carburising flame or reducing flame

Reducing flame with intermediate flame feather. The length of feature indicates amount of excess acetylene.

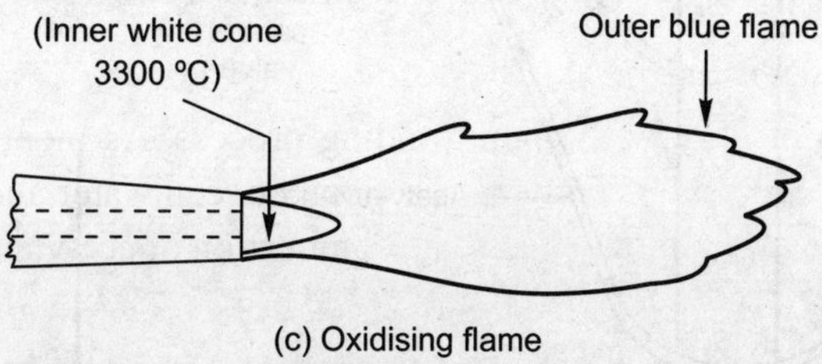

(c) Oxidising flame

In oxidising flame inner white cone is somewhat shorter which gives rise to high tip temperature.

Fig. 8.29 Welding flames.

(*a*) **Neutral Flame [Fig. 8.29 (a)].** Certain amount of oxygen is required for the complete combustion of acetylene (or fuel gases). In neutral flame all the acetylene is completely burned and thus all the available heat in the acetylene is released. The flame has two sharply defined zones. Most of the acetylene welding is done with use of neutral flame.

Neutral flame is used for welding Cast Iron, Aluminium, copper, stainless steel.

(*b*) **Reducing or Carburising Flame [Fig. 8.29 (b)].** When the amount of oxygen is provided less than that theoretically required for complete combustion of fuel then reducing flame is obtained. Reducing flame is similar to neutral flame with the addition of third phase in between inner white cone and outer blue flame. It is known as intermediate flame feather. The length of feather is an indication of excess acetylene present. This unburnt carbon present goes to the weld metal pool and causes the steel to become extremely hard and brittle. The reducing flame is preferable for only a few applications in oxyacetylene welding.

It is used for oxygen free copper alloys, high carbon steels, *cast irons*, high speed steels and cemented carbides.

(*c*) **Oxidizing flame [Fig. 8.29 (c)].** When the amount of oxygen is in excess then oxidizing flame is obtained. It is similar to neutral flame with exception that inner white cone is some what shorter and giving rise to higher tip temperatures (3300°C). It is not used in most of the applications since metal is badly oxidized. However, it is useful for welding some non-ferrous alloys such as copper base and zinc base alloys. Some alloys of iron such as cast iron and manganese steel are also better welded by oxidizing flame.

8.11.2 Oxy-Acetylene Welding Equipment

Oxy-acetylene welding equipment is shown in Fig. 8.30. It contains oxygen and acetylene cylindrical containers, gas pressure regulators and welding torch which mixes two gases before they are ignited.

Oxygen and acetylene from two cylinders are brought through pressure regulators and hose pipes to the welding torch. Welding torch mixes oxygen and acetylene in desires proportions, burns the mixture at the end of the torch tip.

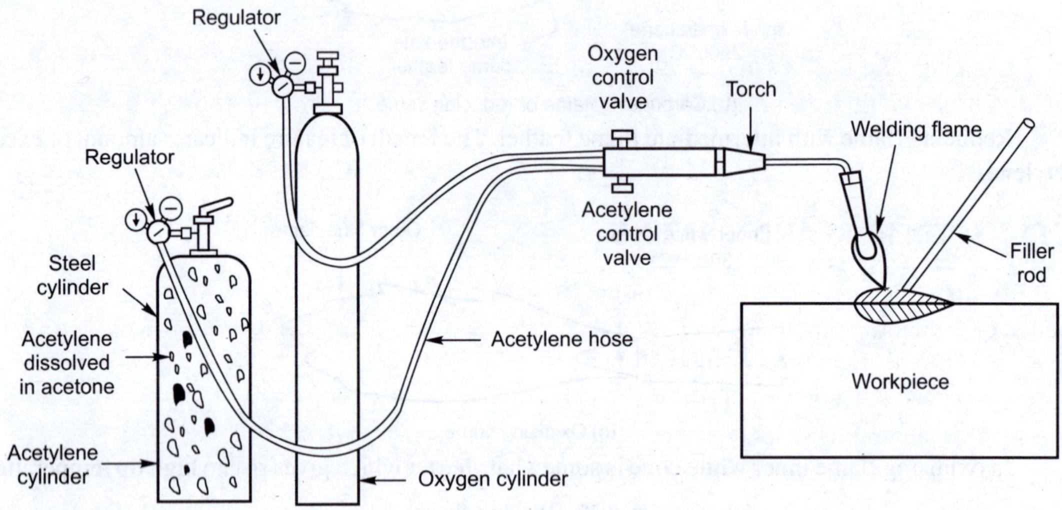

Fig. 8.30 Oxy-acetylene welding equipment.

Oxygen is normally stored in strong steel cylindrical containers at a pressure ranging from 13.8 MPa–18 MPa.

Acetylene is not stored safely at pressures higher than 0.1 MPa. It is usually dissolved in acetone. The storage cylinders are filled with around 80% porous filler such as calcium silicate. Acetone is absorbed into the voids in filler material and serves as medium for dissolving the acetone. These cylinders can hold 9 m^3 of gas at the pressure of 1.75 MPa.

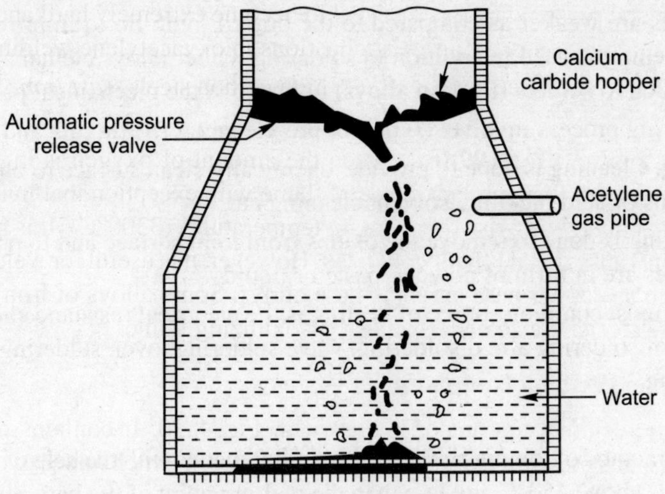

Fig. 8.31 Acetylene generator.

It is also possible to have acetylene gas obtained by dropping the lumps of calcium carbide in water as shown in Fig. 8.31. The reaction between calcium carbide and water is instantaneous and produces acetylene as shown below.

$$CaC_2 + 2H_2O \rightarrow C_2H_2 + Ca(OH)_2$$

Calcium carbide is stored in a hopper near the tip of acetylene generator. A pressure regulated valve controls the flow of CaC_2 into water depending upon the pressure of acetylene in the generator. Acetylene is taken out through a pipe. This may be permitted to produce acetylene upto a safe pressure of 0.1 MPa.

8.12 SOLDERING AND BRAZING

In the previous article we have discussed about the welding processes, where the base metal is melted or brought to the plastic stage and then pressure may or may not be applied. Though welding gives higher joint strength, it is likely to cause metallurgical damages because of higher temperatures. Two distinct advantages of soldering and brazing are:

(*i*) Heating of base metal is negligible to cause any change in their structure or properties.

(*ii*) The process can join two materials which are insoluble in each other.

Soldering and brazing processes have wide commercial applications in joining small assemblies, electrical and electronic parts.

(a) Soldering

Soldering is the process of joining similar or dissimilar metals by means of a filler material (essentially lead-tin alloys), whose melting point is less than 450°C. As this temperature is lower than the melting temperature of base metal, the strength of joint is largely determined by the adhesive qualities of the filler metal. In soldering the gap between the metals to be soldered is filled by filler metal through capillary action.

Soldered joints are weaker as compared to the brazed joint. So to improve the strength of the joint mechanical fastening is used in addition to soldering. Other alloys elements like zinc, silver and cadmium are also added to solders (lead tin alloys) to improve the mechanical properties of the joint.

A good soldering process involve: (*i*) proper pre-cleaning, (*ii*) fluxing and (*iii*) heating.

Pre-Cleaning. Cleaning is done to provide chemically clean surface to obtain proper bond and may be done by means of acid pickling, solvent cleaning etc.

Fluxing. Fluxing is done to remove the oxides from joint surface and to prevent the filler metal from oxidizing. Fluxes are in form of powder, paste or liquid.

Heating. The most common source of heating is the electrical resistance heating with soldering iron. Other methods of soldering are: disoldering, wave soldering, oven soldering, induction soldering and infra red soldering.

(b) Brazing

Brazing is the method of joining similar or dissimilar metals with the help of molten filler material whose melting point is above 450°C and less than the melting point of the base metals. The filler metal is copper alloys (copper zinc and copper silver alloys) and distributed between the surfaces the base metals by capillary action. Joint strength is largely determined by the stength of filler metal used. Almost all the similar and dissimilar metals can be joined by brazing except Al and Mg, which are not easily joined by brazing. Like soldered joints brazed joints are also not suitable for high temperature service.

Brazing process also involves proper joint designing and

(*i*) pre-cleaning

(*ii*) fluxing

(*iii*) heating.

(*i*) **Pre-cleaning.** Joint is properly pre-cleaned using proper solvents. Scales and oxides are removed by acid pickling. Cleaning is essential because the oil, dirt or grease prevent the flow of filler metal.

(*ii*) **Fluxing.** Fluxes are added to the brazed joint to remove any of oxides present and to prevent the formation of oxides in base metal and filler metal. The fluxes generally used are combination of borax, boric acid, chlorides and fluorides. For ferrous metals fluxes used are mixture of borax and boric acid in powder form.

(*iii*) **Heating.** Heat sources used for brazing are electrical resistance heating, molten bath of brazing filler material, oxy-acetylene torch, induction heating, controlled atmosphere furnaces etc.

First joints are pre-cleaned and properly prepared. Then fluxes are applied by spraying, brushing or with the help of pressurised applicator. Now joint is heated to bring it to the melting temperature of filler metal then filler material is applied so that it flows in the joint by capillary action. The filler metal after solidification gives necessary strength.

(c) Braze Welding

Braze welding is similar to ordinary brazing except that the filler metal is not distributed by capillary action. The filler metal is melted and deposited at the point where the weld is to be made.

8.13 COMPARISON AMONG WELDING, BRAZING AND SOLDERING

Welding	Soldering	Brazing
1. In welding process fusion is obtained by heat and/or pressure.	In soldering joint is obtained by means of filler material whose melting point is less than 450°C and less than the melting point of base metal.	In brazing the joint is obtained by means of filler material whose melting point is above the 450°C and less than the melting temperature of base metal.
2. The strength of the joint is highest comparing to soldering and brazing.	The soldered joints are weakest among the soldering, welding and brazing.	The brazed joints are stronger than soldered joint but weaker than welded joints.
3. The joint strength at times may be equal to even greater than the strength of the base metals.	The joint strength depends upon the adhesive qualities of filler material and can never reach the strength of base metal.	The joint strength depends upon the adhesive qualities of filler material and can never reach the strength of base metal.
4. The filler material may or may not be used.	The filler metal is essentially used because joint is made by filler material only.	The filler material is essentially used because joint is made by filler material only.
5. The composition of filler metal is normally same as that of base metal (the metal to be welded)	Essentially filler metals are alloys of lead and tin.	The filler metals are alloys of copper (e.g. copper zinc and copper silver alloys).
6. Welded joints can withstand high temperature.	Soldered joints are not suitable for high temperature service because of lower melting point of the filler metal.	Brazed joints are also not suitable for high temperature service because of lower melting point of filler metal.
7. The process cannot joint dissimilar metals which are insoluble in other.	The process can join dissimilar metals which are insoluble in each other.	The process can join dissimilar metals which are insoluble in each other.
8. The parent metals are heated to the melting temperature, so welding is likely to cause the metallurgical damage and change in the properties.	Heating of parent metal is negligible to cause any change in their structure and properties.	Heating of parent metal is negligible to cause any change in their structure and properties.
9. Welding has wide applications in construction and maintenance of boilers, pressure vessels, tanks bridges, building constructions, cutting tools dies, and furnaces.	Soldering is widely used in joining small assemblies, electric and electronics parts.	Brazing is used for joining electrical parts, joining carbide tips to tools, heat exchangers, radiators etc.

8.14 WELDING DEFECTS

During welding process if proper care is not taken the following defects may arise. They may be because of the faulty technique used by operator or sometimes they are inherent in the process:

1. Incomplete fusion
2. Slag inclusion or oxides
3. Poor penetration
4. Porosity
5. Cracking
   (*a*) Hot cracking
   (*b*) Cold cracking

(1) Incomplete Fusion

Incomplete fusion results from poor joint preparation improper, cleaning, incorrect welding perimeter like welding current. If welding current is lower than the required all the faces of joint would not be sufficiently heated to promote proper fusion.

(2) Oxides and Slag Inclusion

Oxidation is the number one enemy to welding and they should be removed because they tend to get entrapped in the solidifying metal. Fluxes are after used to remove oxides. Fluxes react with oxides and other impurities and form and slag which being lighter floats on the top of weld pool. However, sometimes because of stirring action slag goes into the weld pool and there is not enough time for it to float. It may get solidified inside the fusion zone and ends up as slag inclusion.

(3) Poor Penetration

Poor penetration results from too little heat input. Because welding on cool base metal requires low travel speed and as the base metal absolute heat and temperature rises, speed may be increased. It may be also caused because of the incorrect edge preparation and may lead to cracking.

(4) Porosity

It results from oily, wet or dirty base metal and insufficient gas shielding. It is also caused by the presence of gases such as hydrogen, oxygen and nitrogen which get entrapped during solidification. Hydrogen is the main cause of porosity in weld. Hydrogen is absorbed by weld pool if an incorrect welding technique is applied.

(5) Cracking

Cracking is one of the most serious defect and may occur due to incorrect welding technique or using a filler material having different rate of contraction compared to that of parent metal. It may occur at two stages during the cooling of weld metal.

(*a*) **Hot cracking.** Hot cracking is influenced by the sulphur and carbon contents of mild steel weld metals. The hot cracking can be avoided by using sufficient manganese. It can also be prevented by preheating the base metal and increasing the cross-sectional area of root weld bead.

(*b*) **Cold cracking.** Cold cracking occurs at room temperature after the weld is completely cooled and can be seen in heat effected zone. Cold cracking can be reduced by pre and post heating of the weldments and reducing the cooling rates.

The defects mentioned above do not significantly affect the static strength of welded joint. However, these defects have a serious consequence if the joint is subjected to fatigue loading or is prone to the brittle fracture.

The welding defects are identified by several methods like visual inspection, ultrasonic testing, eddy current testing, magnetic particle inspection etc.

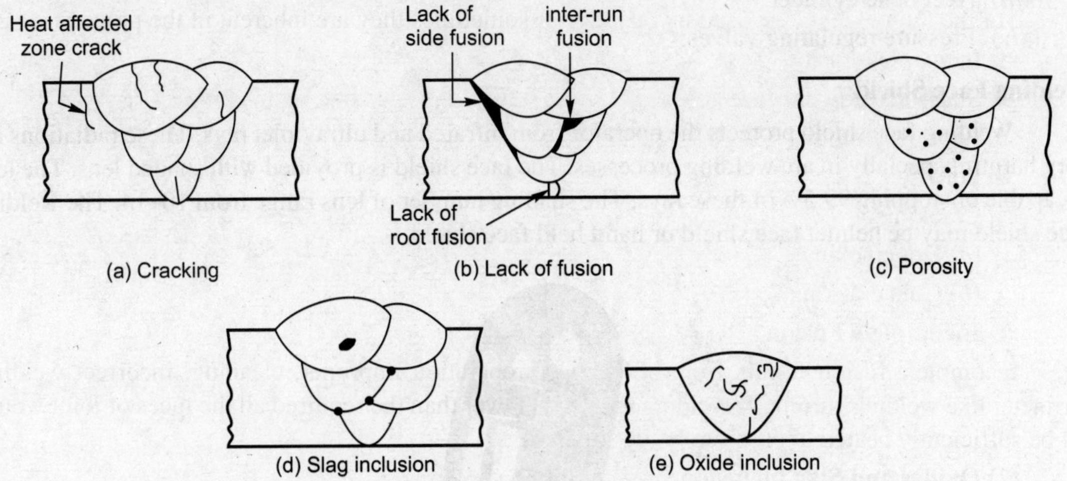

Fig. 8.32 Welding defects.

8.15 WELDING TOOLS AND EQUIPMENTS

Welding equipments can be broadly classified in two categories

1. Personal Protective Equipments
2. Welding Tools Equipments

1. Personal Protective Equipments. Welding is the process in which radiations and toxic gases are released during process. These are very harmful for human body. Personal protective equipments play a very significant role in protecting operator during welding. Some of personal protective equipments are:

(*i*) Welding helmet
(*ii*) Safety goggles
(*iii*) Safety appron
(*iv*) Hand gloves.

2. Welding tools and equipments: Welding tools and equipments play a vital role in making good quality weld. Some of the welding tools and equipments are:

(*i*) Insulated electrode holders
(*ii*) Wire cables
(*iii*) Cable lugs
(*iv*) Ranting clamp
(*v*) Chipping hammer
(*vi*) Tongs
(*vii*) Wire brush
(*viii*) Welding transformer (AC and DC)
(*ix*) Spark igniter (in case of gas welding)
(*x*) Welding torch (in case of gas welding)

(*xi*) Acetylene generator
(*xii*) Oxygen cylinder
(*xiii*) Acetylene cylinder
(*xiv*) Pressure regulating valves.

Welding Face Shield

Welding face shield protects the operator from infrared and ultraviolet rays. These radiations are very harmful specially in arc welding processes. The face shield is provided with shaded lens. The lens is capable of stopping 99.5% of these rays. The shading number of lens range from 10–14. The welding face shield may be helmet face shield or hand held face shield.

Fig. 8.33 Face shield.

Safety Goggle

Safety goggles protect the eyes from harmful radiation

Safety Apron

It protects the operator body from harmful radiations and from minor burns. Safety apron or jackets may be worn according to the safety standards of particular welding facility.

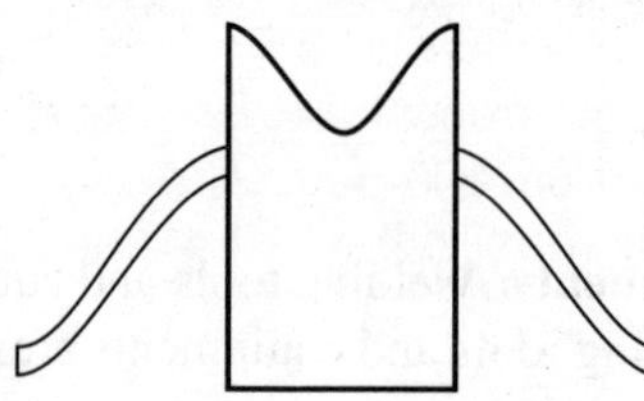

Fig. 8.34 Safety apron.

Hand Gloves

Long hand gloves are considered as a necessary protection from burns during welding. Gloves are usually made from leather, asbestos or non-flammable cloth.

Chipping Hammer

Chipping hammer are used from removing the slag from weld bead. It is chisel shaped and pointed on one end for removal of slag.

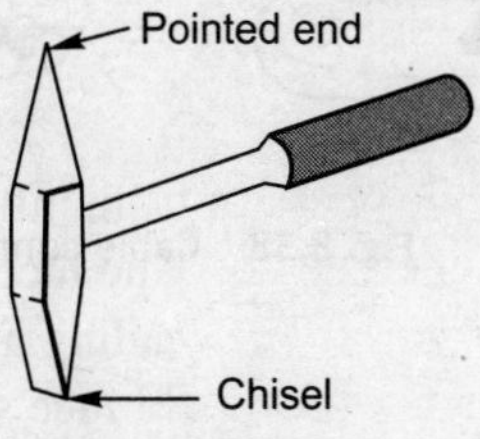

Fig. 8.35 Chipping hammer.

Wire Brush

Wire brush removes small particles of slag. It is made of stiff steel wire embedded in wood.

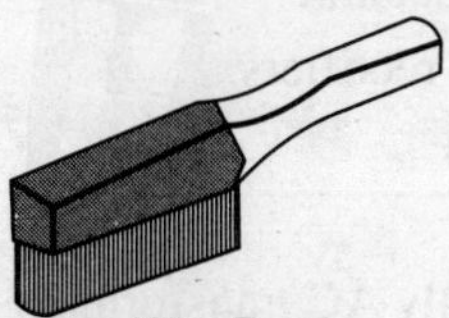

Fig. 8.36 Wire brush.

Tongs

Tongs are used to hold and transfer hot work pieces from welding work table to other place. Tongs are discussed in detail in forging and smithy shop (please refer 5.4).

Electrode Holder

Electrode holders are meant for holding the electrodes. The grooves are cut in the jaws of holders which enables the electrode to hold at various angles. A handle with spring is provided at electrode holder for inserting or removing the electrode. Electrode holders come in various sizes according to amperage output of welding (ranging from 100–500 amperes).

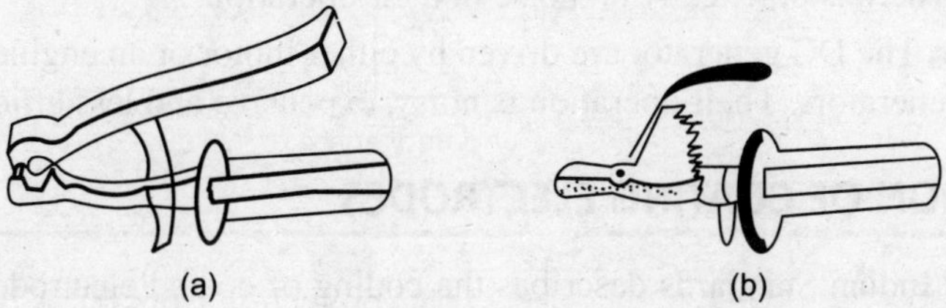

Fig. 8.37 Electrode holder.

Wire Cables

These are the cables which carry the current from welding transformer to workpiece. These are very flexible and are made of aluminium or copper. Those are insulated by two heavy rubber coatings.

Cable Connectors

The mechanical connectors used to connect wire cables to welding machine or to welding work tables.

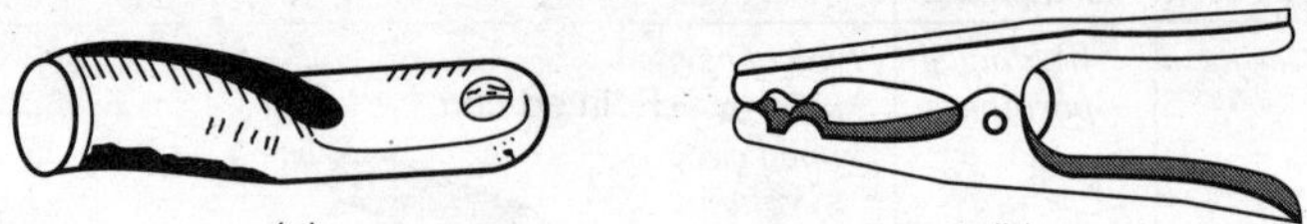

(a) (b)

Fig. 8.38 Cable connector.

Arc Welding Machines

The power source of arc welding may be either AC or DC. So welding machine are of two types:

(1) AC (Alternating current) welding machine

(*i*) AC transformers

(*ii*) Motor or engine driven alternator

(2) DC (Direct current) welding machine

(*i*) AC transformers with DC rectifiers.

(*ii*) Motor or engine driven DC

(1) AC welding Machine

For AC welding ,machines normally AC transformers are used because they does not have any moving part (except cooling fan) and as a result they operate with less maintenance cost. It has low operational cost and there is practically no noise in the operation of AC welding transformer.

AC transformer consists of primary and secondary coils. The input power is given to primary coil. The transformer steps down the voltage. The basic circuit using AC transformer is shown in Fig. 8.40.

(2) DC welding Machines

In DC arc welding machines AC transformer with rectifier or DC generator can be used for supplying required Direct Current (DC).

AC Transformer with DC Rectifiers. In this case power AC power supply is first stepped down by means of transformers then SCR (Silicon Control Rectifiers) are used to convert AC to DC. These machines are also not having any moving part (except cooling fan) and have low maintenance and operational cost. There is practically no noise in their operation.

DC Generator. The DC generator are driven by either motor or an engine. There are obviously moving parts in DC generators. Their operation is noisy, expensive and less efficient.

8.16 DESIGNATION OF COATING ELECTRODES

The Bureau of Indian Standards describes the coding of coated electrodes to be used for mild steel and low alloy steel as follows.

Method of manufacturing	*Performance characteristics*			*Mechanical properties*			*Special characteristics*
Prefix	*1st digit*	*2nd digit*	*3rd digit*	*4th, 5th and 6th digit*			*Suffix*
	Type of covering	*Welding position*	*Type of current polarity and open circuit voltage*	*Tensile and yield strength* N/mm^2	*Min. elongation mm*	*Temperature*	
E → solid extruded R → extruded with reinforcement	1. → High contents of cellulose 2. → High contents of titanium with ionizing, producing a fairly viscous slag 3. → High amount of titanium producing a fluid slag 4. → High contents of iron and manganese oxides or silicates or combination of both producing a inflated slag 5. → High contents of iron oxides or iron silicates or combination of both producing a solid slag. 6. → High contents of calcium carbonate and fluoride. 7. → Any other type of material not mentioned above.	O → F, H, V D, O 1 → F, H, V, O 2 → F, H 3 → F 4 → F, H (fillet) 9 → any other position not mentioned above F: Flat H: Horizontal V: Vertical up D: Vertical down O: Overhead	O → D+ 1 → D+, A90 2 → D–, A70 3 → D–, A50 4 → D+, A70 5 → D±, A70 6 → D±, A70 7 → D±, A50 9 → others D+: DC with electrode positive (reverse polarity) D–: DC (direct current) with electrode negative (straight polarity) DI: DC with electrode positive or negative A90: AC (Alternating current with open circuit voltage not less than 90. A70: AC with open circuit voltage not less than 70 A50: AC with open circuit voltage not less than 50				H: Hydrogen controlled electrode J: Iron powder covering providing metal recovery of 110–130% K: Iron powder covering providing metal recovery of 130–150% L: Iron powder covering providing metal recovery more than 150% P: Deep penetration

Fig. 8.39 The coding of coated electrodes as per bureau of Indian Standards.

The electrodes are marked with 6 digits along with a prefix and a suffix.

1. **Prefix.** It represents the method of manufacturing. It may be *E* or *R*.

2. **Performance Characteristics**

1st digit. It provides the type of flux covering. It may be 1, 2, 3, 4, 5, 6, 9.

2nd digit. It provides the type of welding position for which the given electrode is most suitable. It may be 0, 1, 2, 3, 4, 9.

3rd digit. It provides the type of current polarity and open circuit voltage, it may be 0, 1, 2, 3, 4, 6, 7, 9.

3. **Mechanical Properties of Deposited Weld Metal**

4th, 5th and 6th digits represents the tensile strength and yield strength, minimum elongation percentage, and temperature for minimum impact value.

4. **Suffix.** It represents the special characteristics of coated electrodes.

8.17 WELDING EXERCISE

Aim. To make butt joint by performing shielded metal arc welding (SWAM) using coated consumable stick electrode.

(1) Cut the required size mild steel piece from iron section.

(2) Prepare the edges of pieces to be welded. (file them to form single V as shown in Fig. 8.39 a.)

(3) Set the welding parameter like welding current, voltage etc.

(4) Wear protective apron, goggles and hand glove.

(5) Keep the work pieces in butt joint position (1 mm away from each other) at the welding table. Ensure that circuit is properly completed and earthing is done.

(6) Insert a coated stick electrode in electrode holder and produce arc by striking the electrode with metallic surface of the welding table.

(7) Perform the tacking operation on both ends.

(8) Perform the butt welding operation.

(9) Two major parameters during SMAW are the welding speed and feed of electrode forward weld pool. These parameter should be properly controlled to ensure a defect free weld.

(10) After welding, hold the workpiece with tong and keep it on floor.

(11) Remove the slag with help of chipping hammer.

(12) Clean the work piece with help of wire brush.

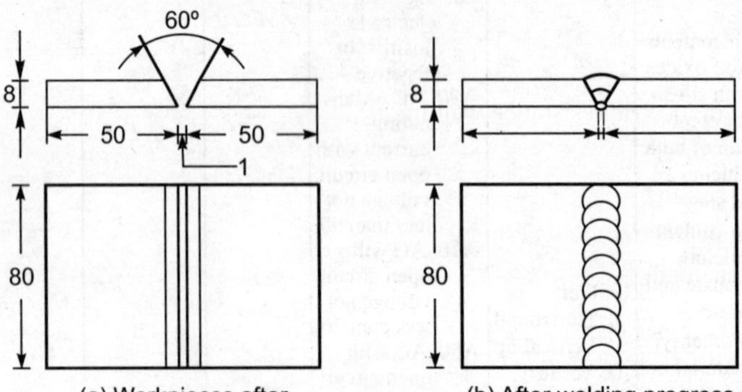

(a) Workpieces after edge preparation (b) After welding progress

Fig. 8.40 Single V bult joint.

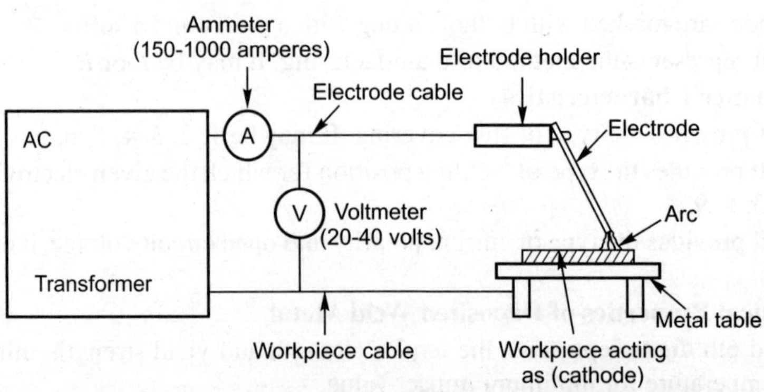

Fig. 8.41 Basic arc welding circuit using AC as power source.

8.18 SAFETY PRECAUTIONS WHILE WORKING IN WELDING SHOP

1. Before starting welding machine verify that welding parameter like current, voltage and electrode specifications are correct.
2. Make sure that circuit is properly completed and earthed.
3. Always wear protective apron, goggles and gloves during welding process.
4. During and after welding always use the tong to hold the workpiece.
5. Preparation of edges before welding ensures good quality of weld.
6. The work pieces should be free from dust and oil etc. before weld. Properly cleaned work pieces contribute in achiving a good quality weld.
7. Never touch the work piece or welding table with bare hands during welding operation. Because your body may become the part of circuit and there may be electric shock.
8. Electrode holder should never be left on the welding table in contact with metallic surface.
9. Welding cables should be completely insulated.
10. In welding section opaque shields or tinted curtain of polyvinyl chloride plastic sheet should be used to protect others from exposure to radiation.

OBJECTIVE TYPE QUESTIONS

1. Which of the following joining method is widely used in electronic industries

(*a*) resistance welding (*b*) arc weld
(*c*) gas welding (*d*) soldering

2. The heat source in electric resistance welding is

(*a*) electric arc (*b*) resistance heating
(*c*) inductance heating (*d*) eddy currents

3. The heat source in electric arc welding is

(*a*) inductance heating (*b*) eddy currents
(*c*) electric arc (*d*) none of above

4. Which of following is not a fusion welding process

(*a*) electric arc welding (*b*) forge welding
(*c*) gas welding (*d*) electroslag welding

5. Spot welding is

(*a*) electric arc welding (*b*) electric resistance welding
(*c*) forge welding (*d*) none of above

6. Seam welding is

(*a*) arc welding process (*b*) same as spot welding
(*c*) continuous spot welding (*d*) same as projection welding

7. The resistance welding process used for obtaining leak proof joint is

(*a*) spot welding (*b*) seam welding
(*c*) projection welding (*d*) all of above

8. In arc welding process electric arc may created between

(*a*) work piece and electrode (*b*) between work pieces to be joined
(*c*) between two electrodes (*d*) all of above

9. Spot welding is extensively used for
   (*a*) joining thin sheets in lap joints
   (*b*) joining thick sheets in lap joints
   (*c*) joining thin sheets in butt joint
   (*d*) joining thin sheets in butt joint
10. Projection welding is
   (*a*) same as spot welding
   (*b*) multispot welding
   (*c*) arc weld
   (*d*) continous spot welding
11. The gas welding (oxy-acetylene welding) flame produces the temperature or
   (*a*) 2000°C
   (*b*) 3100°C
   (*c*) 4000°C
   (*d*) 3500°C
12. During gas welding neutral flame is obtained when
   (*a*) when complete combustion of acelytene takes place
   (*b*) when unburnt acetylene is left because of less supply of oxygen
   (*c*) when oxygen is provided more than required for complete combustion of acetylene
   (*d*) none of above
13. The electrodes used in seam welding are
   (*a*) disk electrode
   (*b*) sphierical electrodes
   (*c*) cylinderical electrode
   (*d*) none of above
14. Consumable electrodes are used in
   (*a*) TIG welding
   (*b*) MIG welding
   (*c*) resistance welding
   (*d*) gas welding
15. TIG welding, MIG and SMAW are types of
   (*a*) electric arc welding
   (*b*) electric resistance welding
   (*c*) gas welding
   (*d*) thermit welding
16. Consumable coated stick electrodes are used in
   (*a*) TIG welding
   (*b*) MIG welding
   (*c*) SMAW (shielded metal arc welding)
   (*d*) electric resistance welding
17. In arc welding process the temperature of arc is
   (*a*) 1100 – 1800°C
   (*b*) 2500 – 3000°C
   (*c*) 2000 – 2500°C
   (*d*) above 4000°C
18. The commonly used gas in gas welding process is
   (*a*) accetylene
   (*b*) hydrogen
   (*c*) methane
   (*d*) CNG
19. Welding is
   (*a*) metal joining process
   (*b*) metal forming process
   (*c*) mechanical joining process
   (*d*) machining process
20. Which of the following joining process has highest strength
   (*a*) welding
   (*b*) brazing
   (*c*) soldering
   (*d*) all have same strength
21. The function of flux in welding processes is to
   (*a*) prevent oxidation
   (*b*) prevent dirt
   (*c*) prevent distortion of parts
   (*d*) join the parts
22. In arc welding process heat is developed between electrode and work piece due to
   (*a*) high current
   (*b*) contact resistance
   (*c*) time of flow of current
   (*d*) high voltage

23. The filler metal used in soldering are

(*a*) lead tin alloys (*b*) copper silver alloys

(*c*) copper zinc alloys (*d*) none of above

24. The filler materials in brazing are

(*a*) lead tin alloys (*b*) copper zinc and copper silver alloys

(*c*) aluminium alloys (*d*) magnesium alloys

25. The filler materials whose melting point is less than 450°C are used in

(*a*) brazing (*b*) soldering

(*c*) braze welding (*d*) arc welding

26. The filler materials whose melting point is more than 450° and less than melting temperature of base metals are

(*a*) soldering (*b*) braze welding

(*c*) brazing (*d*) gas welding

27. The resistance welding process uses

(*a*) low voltage low current (*b*) high voltage low current

(*c*) low voltage, high current (*d*) high voltage, high current

28. The perameters needs to controlled on resistance welding are

(*a*) pressure (*b*) current

(*c*) voltage (*d*) all of above

29. The perameter needs to controlled in electric are welding are

(*a*) air gap to maintain arc (*b*) voltage

(*c*) current (*d*) all of above

30. Which of following joining methods are based on capilary action

(*a*) resistance welding (*b*) soldering

(*c*) thermit welding (*d*) adhesive bonding

VIVA VOCE QUESTIONS

1. What is the difference between fusion welding and solid state welding?
2. What are various type of joint used in welding shop?
3. Name any five important welding process.
4. Explain the significance of edge preparation for butt joints.
5. What is the function of flux in welding processes?
6. What are the various welding position used during welding?
7. Explain the difference between fore hand and back hand welding.
8. What is the difference between resistance spot welding and resistance seam welding?
9. What is meant by polarity in arc welding?
10. Explain the difference between straight polarity and reverse polarity.
11. What is the difference between consumable and nonconsumable electrode?
12. What is the function of coating in stick electrodes?
13. Name various flames used in gas welding process.
14. Name the equipments used in (SMAW) arc welding setup.
15. Name the equipments used in oxy-acetylene welding setup.
16. What is the difference among welding, soldering and brazing?
17. Explain the function of wire brush in the welding shop.

Chapter

9 Metal Casting Processes (Foundry Shop)

INSIDE THIS CHAPTER

9.1 INTRODUCTION TO FOUNDRY (CASTING PROCESS)

Metal casting process consist of making patterns and moulds, preparing and melting the metal, pouring liquid metal into the moulds and then allowing it to solidify. When solidified, the metal takes the shape of cavity. It is then taken out from mould either by breaking the mould or taking the mould apart. The solidified object is called casting. The process is also known as foundry process.

History

The art of foundry is ancient, even in prehistoric time as for back as 5000 BC, metallic object in form of knives, coins, arrows, and household articles were in use, as observed from the excavation of Mohanjodaro and Harappa. One of man's first operation with metal was melting the ore and pouring it into suitable moulds. Copper and bronze castings were commonly used for ornaments, weapons, tools and utensils.

"Most of our obstacles would melt away if instead of cowering before them we should make up our minds to walk boldly through them."

It appears that iron casting technology in India has been in use from the times of invasion of Alexander the great, around 300 B.C. The famous iron pillar presently located near the Qutub Minar in Delhi is an example of metallurgical skills of ancient Indians. It is 7.2 m long and made of pure malleable iron. It is assumed to be of period of Chandergupta II (375–412 AD). The rate of rusting of this pillar which stands outside is practically zero and even the buried portion is rusting at extreme slow rate.

9.2 ADVANTAGES AND DISADVANTAGES OF CASTING PROCESSES

Advantages

1. The most intricate shapes both external and internal may be cast and as a result many other operations such as machining, forging and welding may be minimised or eliminated.
2. It is possible to cast practically any material be it ferrous or non-ferrous.
3. There are certain metals and alloys which can only be processed by casting and not by any other process like forging because of the metallurgical consideration.
4. Necessary tools required for casting moulds are very simple and inexpensive.
5. As the most complicated shape can be cast, it provides designer a lot of freedom to place the amount of material where exactly required. As a result weight reduction can be achieved.
6. Extremely large, heavy metal objects can be made with casting process which are difficult or economically impossible to produce by any other process. The examples are large pump housing, valve, hydroelectric plant parts. Casting of even 200 tonne can be made.
7. Metal casting process is highly suitable for small lot as well as mass production. For example large number of part required for automotive industry, agriculture field, home appliances are all produced by casting processes.
8. Casting process produces the metal with more favourable engineering properties e.g. machinability and vibration damp capacity in cast iron.
9. Castings are generally cooled uniformly from all sides and therefore, they are expected to have no directional properties.

Limitations

1. The dimension accuracy and surface finish achieved by normal sand casting process is not adequate for final applications in many cases.
2. With some materials it is often difficult to remove defects arising out of moisture present in sand castings.

Applications of Metal Casting Processes

The applications include cylinder block, liners, machine tool beds, pistons, piston rings, mill rolls, wheels, large pump housings, valves, hydroelectric plant parts, turbine blades, water supply pipes, and bell, agricultural parts, aircraft engine blades etc.

Materials for Casting Processes

The metals most frequently cast are iron, steel, aluminium, brass, bronze, magnesium, certain zinc alloys, and nickel based super alloys. Of these cast iron is the dominant casting material, primarily because of its low cost, good fluidity, low shrinkage, ease of control and wide range of properties including useful strength and rigidity.

9.3 CASTING TERMS (Fig. 9.1 to Fig. 9.5)

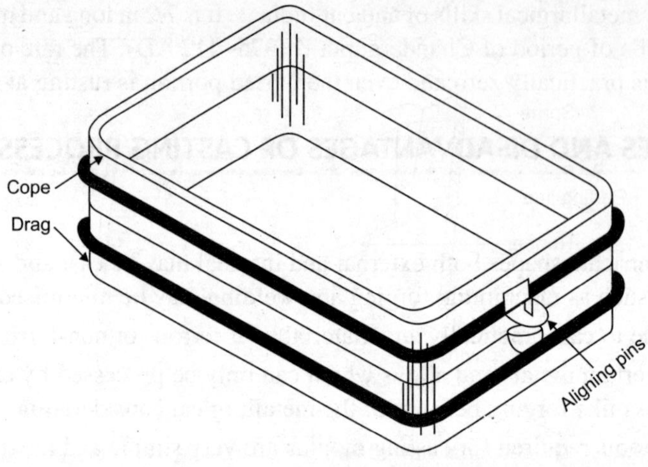

Fig. 9.1 Two part mould flask (or moulding box). The moulding flask is without top and bottom base aligning pins are provided in drag with corresponding hole in cope to ensure proper alignment of cope and drag during moulding.

(*a*) **Moulding Flask.** The box like frame without top and bottom base into which sand is rammed is called "flask". It holds the sand mould intact. It is made up of wood for temporary applications and more generally of metal for long term use. The flask is shown in Fig. 9.1. It is generally made of two or more parts.

Drag. Lower or bottom moulding flask is known as drag.

Cope. Upper moulding flask is termed as cope.

Cheek. It is an intermediate moulding flask in three piece moulding.

Drag is provided with the pins and cope is provided with lugs with hole which fit together and ensure proper alignment during the placement of cope upon drag.

(*b*) **Parting Line or Parting Surface.** It is the surface or line that separates the cope and drag halves of the mould. In split piece pattern it is also the dividing line between two halfs of the pattern (Fig. 9.3).

(*c*) **Bottom Board [Fig. 9.3 (a)].** This is the board normally made of wood which is used at starting of mould making. The pattern and drag is first kept on the bottom board, sand is sprinkled and ramming is done in the drag.

(*d*) **Pattern.** Pattern is the duplicate of final casting to be made with some modifications. It is used for making the mould cavity [Fig. 9.3 (*c*)].

(*e*) **Core [Fig. 9.3 (d)].** Core is placed in moulding cavity to make hole (or hollow cavities) through the castings. Core is made up of core sand.

Basic difference between core and pattern. Although both core and pattern are used to give desired shape to the castings. The removable pattern never comes in contact with molten metal because its purpose is to form mould cavity and it is always removed before pouring the molten metal. Whereas core is surrounded by molten metal and is always subjected to severe thermal stresses as core is present in the mould while pouring the molten metal. The materials used for pattern are wood, plastic and may be metal when large number of casting are required. The material used for making core is core sand.

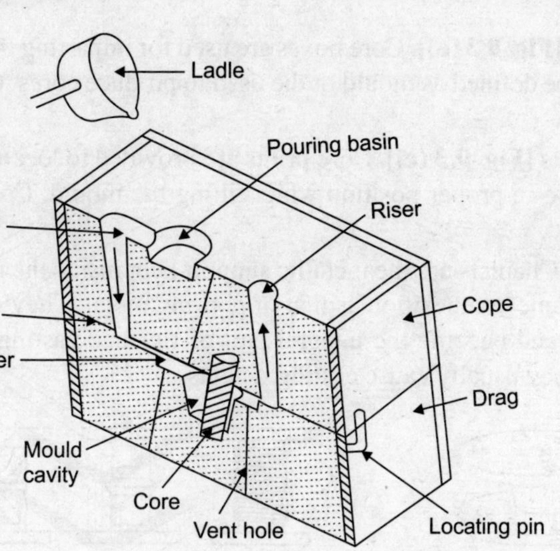

Fig. 9.2 Cross-sectional view of two box mould and its various components.

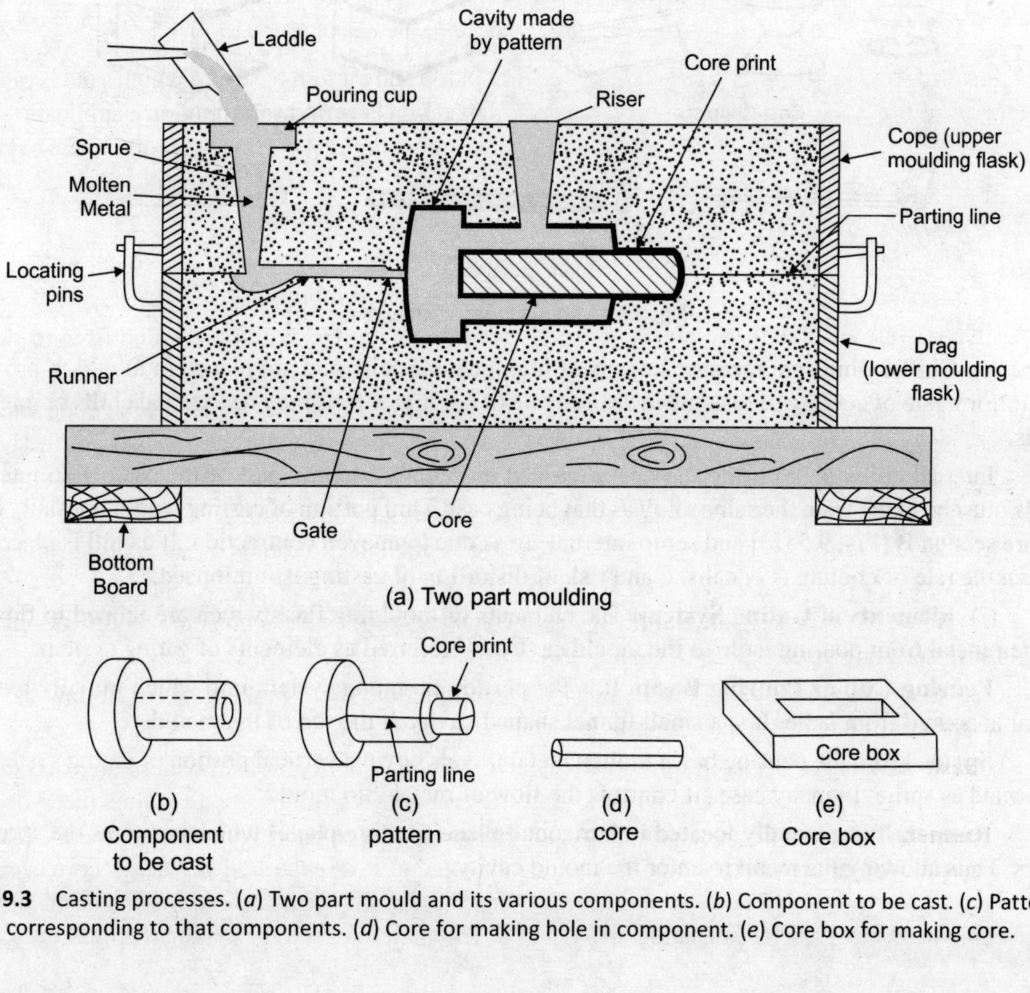

Fig. 9.3 Casting processes. (*a*) Two part mould and its various components. (*b*) Component to be cast. (*c*) Pattern corresponding to that components. (*d*) Core for making hole in component. (*e*) Core box for making core.

(*f*) **Core Box [Fig. 9.3 (e)].** Core boxes are used for imparting the desired shape to the core sand. Core boxes can also be defined as mould or die used to produce cores. Core boxes are made up of wood or metal.

(*g*) **Core Prints [Fig. 9.3 (c)].** Core prints are provided to locate and support the core within the moulds and hold them in proper position while filling the mould. Core print is a region added to the pattern, core or mould.

(*h*) **Chaplets.** Chaplets are the metallic support kept inside the mould cavity to support the core. These should be of same composition as that of pouring metal. They melt and fuse with molten metal during solidification and become the integral part of finished casting. The use of chaplets should be minimised because they usually cause casting defects.

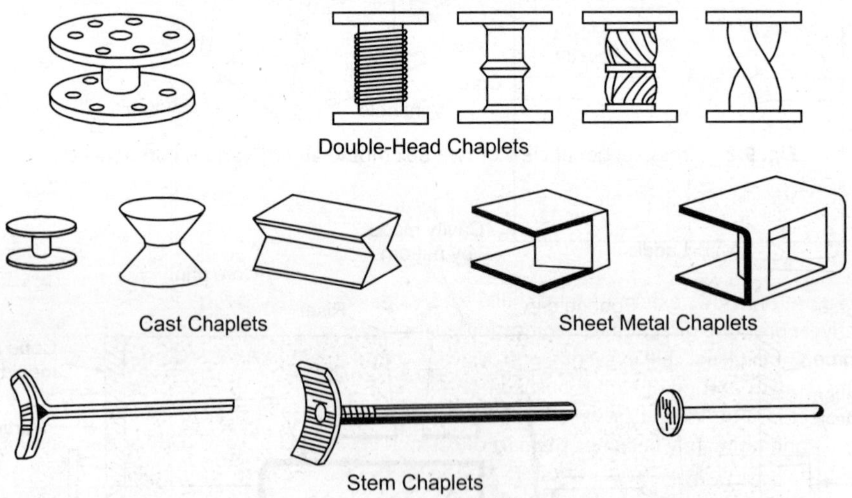

Fig. 9.4 Various types of chaplets.

(*i*) **Chills.** Chills are metallic objects of high heat capacity and are placed in mould to provide the uniform rate of cooling or to achieve desired rate of cooling. Chills may be external chills or internal chills.

Internal chills melt during the operation and ultimately become part of the casting so internal chills must be made from the same alloy as that being cast. Thin portion of casting would normally cool before section B [Fig. 9.5 (*b*)] and setup internal stress due to uneven contraction. If a chill is placed as shown the rate of cooling is equalised and risk of distortion of casting is minimised.

(*j*) **Elements of Gating System.** The elements of moulding flask which are related to flow of molten metal from pouring ladle to the mould cavity are referred as elements of gating system.

Pouring Cup or Pouring Basin. It is the portion of gating system into which initially molten metal is poured from ladle. It is a small funnel shaped cavity at the top of the mould.

Sprue. From the pouring basin molten metal travels down a vertical portion of gating system. It is termed as sprue. In many cases it controls the flow of metal into mould.

Runner. It is generally located in horizontal plane (parting plane) which connects the sprue to gates. Thus allowing the metal to enter the mould cavity.

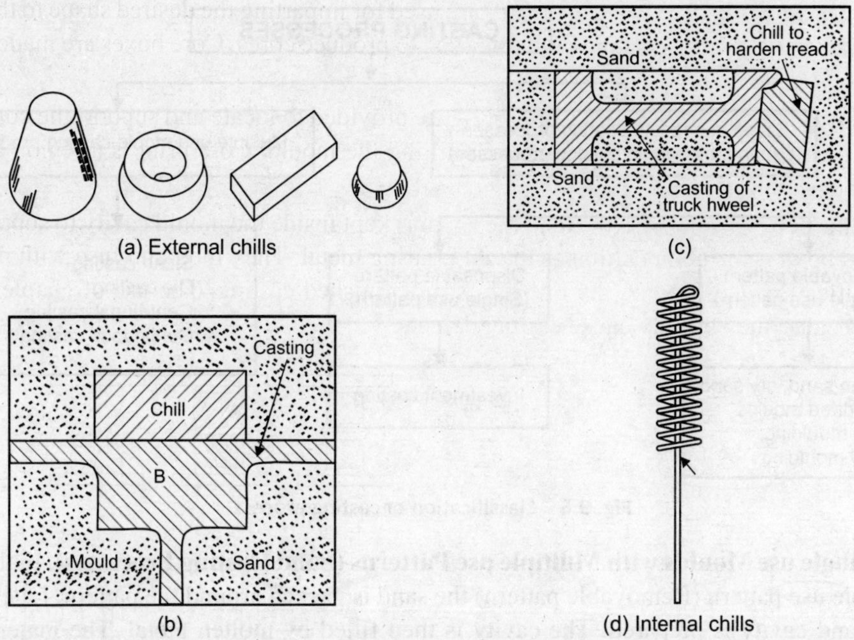

Fig. 9.5 Figure (*a*) shows external metal chills. These are used to accelerate the cooling of various regions. Chills effectively promote directional solidification. Figure (*b*) shows the use of external chill. Figure (*c*) shows application of external chill in casting of truck wheel. Figure (*d*) shows common type of internal chill. The straight portion of wire is pushed into the moulding sand allowing the coiled portion to project into the mould cavity. Internal chill absorbs the heat and fuses together with metal and ultimately becomes the part of casting. Explain ABCD before the abcd itself.

Gates. This is the opening through which molten metal enters the mould cavity. Gate may be also defined as the actual entry point through which molten metal enters the mould cavity. Gates and risers are removed from the casting while performing cleaning and finishing operation.

Riser. Almost all the metals shrink during solidification. Riser is the reservoir of molten metal provided in the casting to compensate for the liquid shrinkage. To perform this function molten metal in riser must solidify after the casting solidification.

9.4 CLASSIFICATION OF CASTING PROCESSES

As there are number of casting processes used to manufacture item, it is helpful to have some form of process classification. Depending upon the type of mould and pattern used, casting process may be broadly classified as:

1. Single use moulds (Expandable mould casting process)
2. Multiple use moulds.

1. **Single Use Moulds (Expandable Mould Casting Processes).** In single use moulds after solidification the mould is broken and finished casting is removed. Because mould is destroyed, a new mould must be made for each casting. In single use mould removal of casting is not a serious problem. Single use moulds may be further classified as:

(*a*) Single use mould with Multiple use pattern (Fig. 9.3)

(*b*) Single use mould with Disposable pattern (Fig. 9.7)

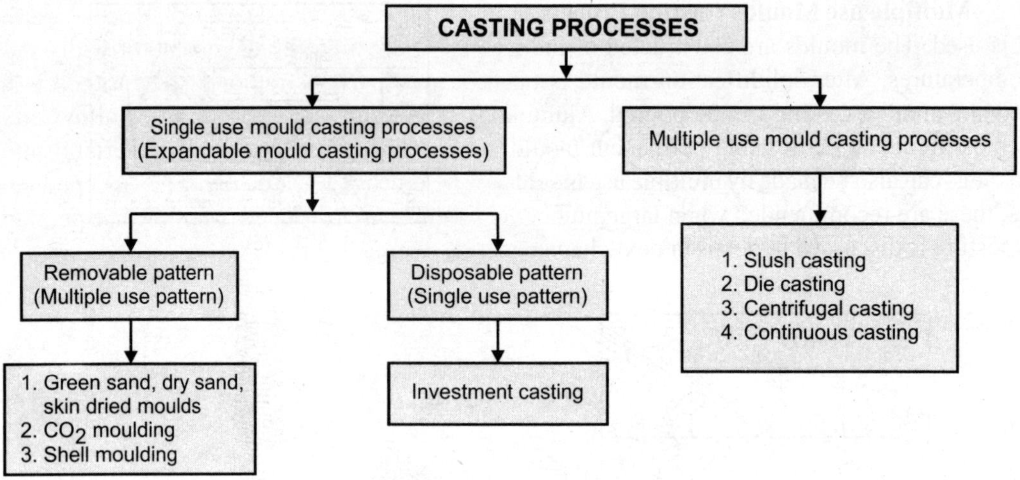

Fig. 9.6 Classification of casting process.

(*a*) **Single use Moulds with Multiple use Patterns (Sand Casting Processes).** In the processes using multiple use pattern (Removable pattern) the sand is packed around the pattern. Later the pattern is removed and cavity is prepared. The cavity is then filled by molten metal. The material used for multiple use pattern is metal, wood or plastic. The material is basically determined by number of casting produced, size and shape of casting and moulding process. Throughout this chapter we will be studying about single used sand mould casting processes with multiple use patterns (Removable patterns) *i.e.* sand casting process.

(*b*) **Single use Moulds with Disposable Pattern (Investment Casting).** Single use pattern (Disposable patterns) are made from polystyrene. Each pattern can be used once only. These are not removed from the mould and are vaporised when the molten metal is poured into the mould. The examples of single use pattern are investment casting and lost foam casting. These are discussed in detail in next chapter.

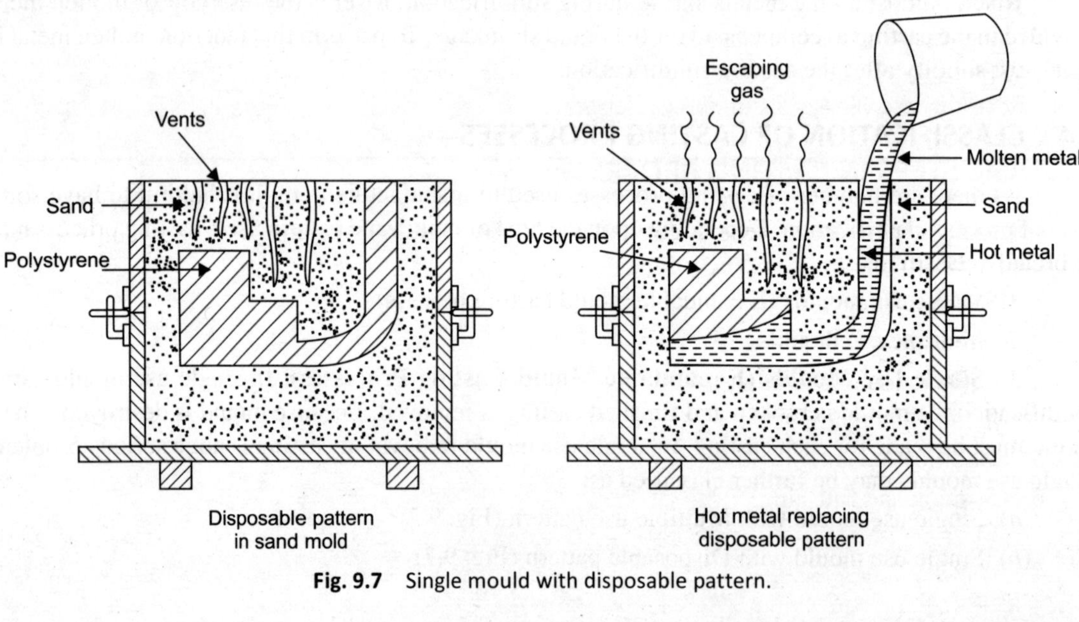

Fig. 9.7 Single mould with disposable pattern.

2. **Multiple use Moulds Casting Processes.** In multiple use mould casting processes reusable mould is used. The moulds are constructed of the metals. These moulds are capable of withstanding high temperatures. After solidification mould is opened and casting is removed. The mould is then reclosed and another casting can be poured. Aluminium, magnesium, and copper based alloys are the metals most frequently cast using permanent moulds. If graphite is used as mould material, iron and steel castings can also be made by multiple use mould casting process. Because of high cost of permanent moulds, these are recommended when large number of castings are produced. Multiple use permanent mould casting is discussed in detail in next chapter.

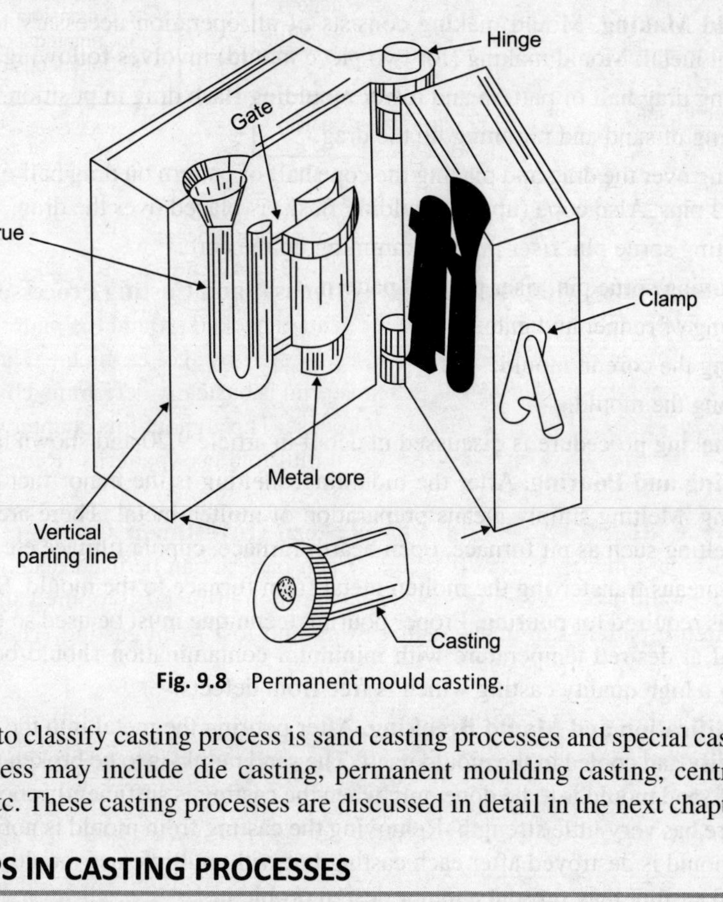

Fig. 9.8 Permanent mould casting.

Another way to classify casting process is sand casting processes and special casting processes. Special casting process may include die casting, permanent moulding casting, centrifugal casting, investment casting etc. These casting processes are discussed in detail in the next chapter.

9.5 BASIC STEPS IN CASTING PROCESSES

Following are the basic steps in the making of sand castings. Any detailed operation may be categorized as belonging to one of these fundamental steps:

1. Pattern making
2. Core making
3. Mould making (Moulding)
4. Melting and pouring
5. Solidification and mould breaking
6. Cleaning and finishing.

1. **Pattern Making.** The casting process starts with construction of a pattern. Pattern is the duplicate of the final casting with some modifications. (These modifications are discussed in pattern allowances). Patterns are required to make moulds. The mould is made by packing the moulding sand around it. When the pattern is withdrawn its impression provides the mould cavity which is ultimately filled with molten metal to make casting.

2. **Core Making.** Cores are made of core sand. Basically core is a sand shape that is kept inside the mould cavity to produce holes or hollow cavities in the castings. With the core any complicated shape can be easily obtained which can not be normally produced by pattern alone.

3. **Mould Making.** Mould making consists of all operation necessary to prepare mould for receiving molten metal. Mould making (for two piece mould) involves following operations:

(*i*) Placing drag half of pattern and lower moulding flash drag in position.

(*ii*) Pouring of sand and ramming up the drag.

(*iii*) Rolling over the drag and placing the cope half of pattern on drag half of pattern with help of dowel pins. Also cope (upper moulding flask) is placed over the drag.

(*iv*) Locating sprue pin, riser pin and ramming up the core.

(*v*) Removing sprue pin, riser pin and pattern.

(*vi*) Making of runner and gate.

(*vii*) Setting the core in mould.

(*viii*) Closing the mould.

Mould making procedure is discussed in detail in article 9.20 and shown in Fig. 9.46.

4. **Melting and Pouring.** After the moulding, melting is the major factor which controls the quality of casting. Melting simply means preparation of molten metal. There are number of methods available for melting such as pit furnace, open hearth furnace, cupola furnace etc.

Pouring means transferring the molten metal from furnace to the mould. Some type of pouring device or ladle is required for pouring. Proper pouring technique must be used so that adequate amount of molten metal at desired temperature with minimum contamination should be supplied to mould. Thus producing a high quality casting which is free from defects.

5. **Solidification and Mould Breaking.** After pouring the metal into the mould, the casting is allowed to solidify and cooled in the mould itself. The sand mould is to be broken to extract the casting. The breaking of sand mould is to be done only when the casting is sufficiently cooled since the metal at high temperature has very little strength. Removing the casting from mould is not very difficult in sand casting as the mould is destroyed after each casting but with multiple use mould, however, removal of complex shaped casting may present a major design problem.

6. **Cleaning and Finishing.** After the castings are removed from the mould various cleaning and finishing operations are required to remove of sand, scale and excess metal from the casting surface. Excess metal can be cut off with some cutting devices. For removing the sand particles sticking to metal surface sand blasting is normally used. Tumbling is also used for cleaning the casting surface.

Each of these steps are considered in detail as we go through the chapter.

9.6 PATTERNS TYPES AND ALLOWANCES

We had discussed earlier that pattern is the duplicate of the final casting with some modifications. The main modifications are:

1. The pattern allowances
2. Core prints
3. Elimination of fine details which are obtained by further processing.

Patterns are used to make cavity in the mould. Pattern making is the first step of casting process. Lets study about pattern allowances, pattern types and pattern materials.

9.6.1 Pattern Allowances

Pattern allowances means the dimensions of the pattern are usually slightly larger than from the final dimensions of the casting. This difference is required because of various reasons which are as follows:

(*a*) **Shrinkage Allowance.** All metals shrink while cooling (except bismuth). To compensate for shrinkage the pattern is slightly larger in every dimension.

➢ The rate of shrinkage (or contraction) with temperature is different for different materials. For example, steel contracts to a higher degree as compared to aluminium (Table 9.1).

Table 9.1 Shrinkage allowances for various metals

Material (Non-ferrous)	Shrinkage allowance (mm/mm)
Aluminium	0.0130
Aluminium bronze	0.0200 to 0.0230
Brass	0.0155
Copper	0.0100 to 0.0160
Lead	0.0260
Monel	0.0200
Magnesium	0.0130
Magnesium alloys	0.0160
White metal	0.0060
Zinc	0.0100 to 0.0150
(Ferrous)	
Grey cast iron	0.0105
White cast iron	0.0160 to 0.0230
Plain carbon steel	0.0210
Chromium steel	0.0200
Manganese steel	0.0250 to 0.0380

It may be noted that there are two types of shrinkages in casting—liquid shrinkage and solid shrinkage.

➢ Liquid shrinkage means reduction in volume when the metal changes from liquid to solid state. To account for this shrinkage risers are provided.

➢ Solid shrinkage is the reduction of volume caused when solid metal loses its temperature (from high temperature to room temperature). In solid state the shrinkage allowance are provided to take care of this reduction. Figure 9.9 shows the dimensions of all casting to be made and corresponding dimensions of pattern.

- To compensate shrinkage allowance all the dimensions of casting are altered uniformly unless they are constrained in some way. For example, dry sand core at the centre of casting may constrain the casting from contraction but edges are not constrained so it may be desirable to provide a higher shrinkage allowance for outer dimensions as compared to those which are constrained.

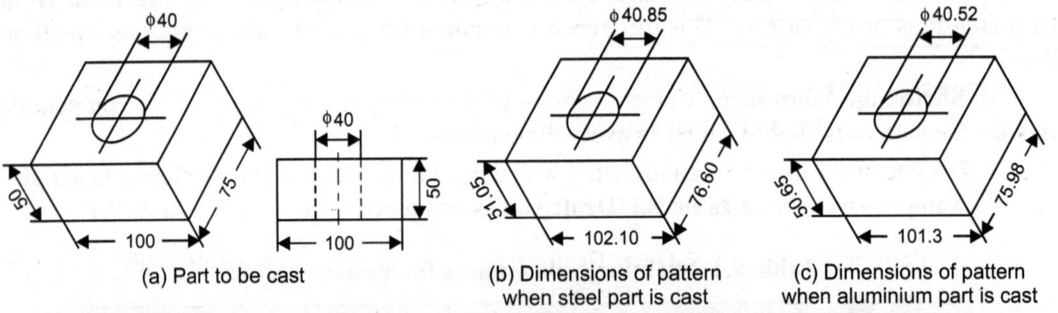

Fig. 9.9 Shrinkage allowance. (*a*) The dimensions of a part to be cast. (*b*) Shows the dimensions of pattern considering shrinkage allowance when steel part is cast. (*c*) The dimensions of pattern considering shrinkage allowance when aluminium part is cast. As steel shrinks more than aluminium, more shrinkage allowance is provided in pattern for steel castings.

(*b*) **Draft Allowance.** While withdrawing the pattern from sand mould the vertical faces of the pattern may damage the mould cavity. To reduce the chances of this happening vertical faces of pattern are always tapered from parting line. Thus the pattern horizontal section is largest at the parting line and becomes smaller the deeper it goes into the sand. This is called draft.

(*i*) Draft allowance varies with the complexity of the job. For external surfaces it is normally about 1° and for inner surfaces, holes and pocket upto 3° may be provided.

(*ii*) One thing is to be noted that draft is added to the original dimension of casting thus making pattern slightly larger.

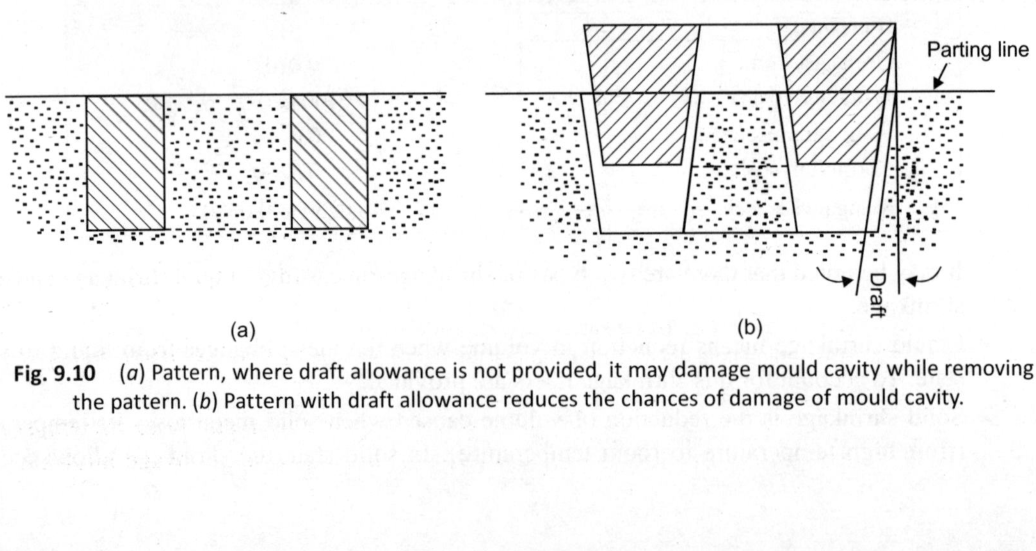

Fig. 9.10 (*a*) Pattern, where draft allowance is not provided, it may damage mould cavity while removing the pattern. (*b*) Pattern with draft allowance reduces the chances of damage of mould cavity.

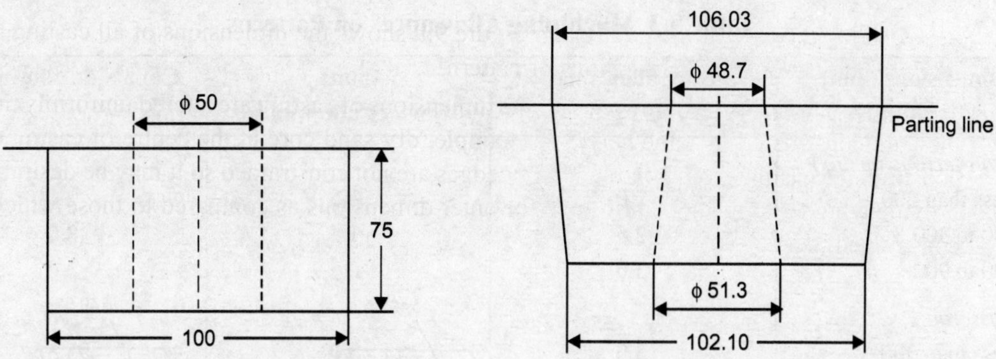

(a) Dimensions of required casting with hole

(b) Dimensions of pattern with draft allowance

Fig. 9.11 Draft allowance provided for a hole in a pattern.

Table 9.2 Draft value for patterns

Pattern	Outer Draft Angle (in degrees)	Inner Draft Angle
Plastic	0.25 to 1.00	0.35 to 2.25
Wood	0.25 to 3.00	0.50 to 3.00
Metal	0.35 to 1.50	0.50 to 3.00

(*c*) **Finish or Machining Allowance.** The finish and dimensional accuracy obtained in sand casting are generally poor and when dimensional accurate part is to be produced, it is achieved by subsequent machining operation.

(*i*) Also ferrous material would have scale on surface which is to be removed. Hence extra metal is provided which is removed by machining or cleaning process. The amount of extra metal depends on the dimensions, casting material and surface finish required.

(*ii*) Machining allowance of some materials are shown in table.

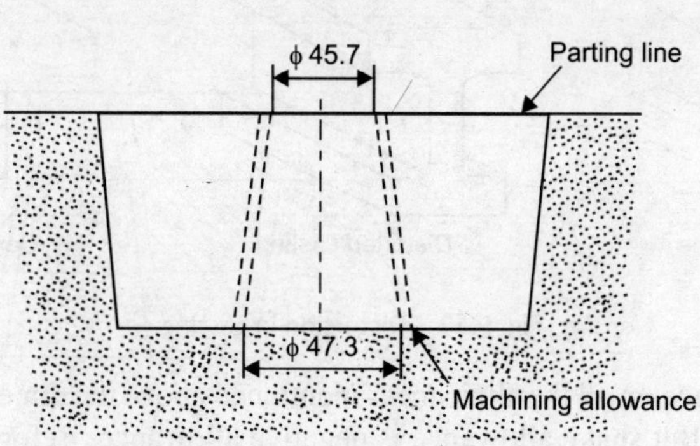

Fig. 9.12 Pattern with finish allowance
(Assuming hole dia = 50 mm and machining allowance of 1.5 mm).

Table 9.3 Machining Allowances on Patterns

Dimensions (mm)	Bore allowance	(mm) Surface allowance	Cope Side allowance
(Non-ferrous metals)			
Less than 200	2.0	1.5	2.0
200 to 300	2.5	1.5	3.0
300 to 900	3.0	2.5	3.0
Cast iron			
Less than 300	3.0	3.0	5.5
300 to 500	5.0	4.0	6.0
500 to 900	6.0	5.0	6.0
Cast steel			
less than 150	3.0	3.0	6.0
150 to 500	6.0	5.5	7.0
500 to 900	7.0	6.0	9.0

(*d*) **Distortion Allowance.** Certain objects such as long flat plate, U-shaped casting, V-section or complicated casting which have thin and long section are distorted in the process of cooling because of metal shrinkage. So extra metal is provided to reduce the distortion of such object. Figure 9.13 shows distortion of I section.

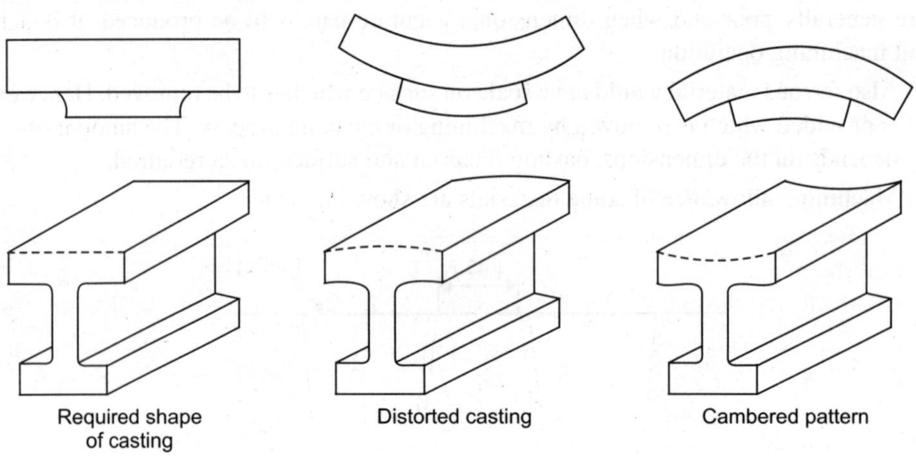

Fig. 9.13 Distortion in casting.

(*e*) **Shake Allowance.** In all the allowances mentioned above the dimensions of pattern are larger than the cast component. But shake allowance is negative allowance. Before the pattern is withdrawn from sand mould, it is rapped to enlarge mould cavity slightly which make the removal of pattern easy. In average size casting this increase in size can be ignored but in large size casting shake allowance is considered by making the pattern slightly smaller. One way of reducing this allowance is to increase the draft which can be removed during subsequent machining.

9.6.2 Pattern Types

There are many types of pattern used in foundry industry. Selection of pattern is usually based on complexity of the part, number of castings required and moulding procedure adopted.

Following are the pattern types used in foundry shop.

(*a*) **Single Piece Pattern.** As the name suggest single piece pattern are the simplest and often least expensive type to make.

(a) Casting (b) Pattern with draft allowance

Fig. 9.14 Single piece pattern.

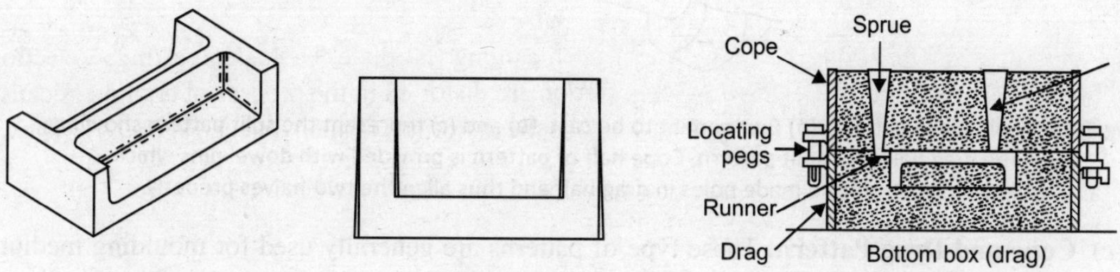

(a) Casting (b) Pattern (c) Mould ready for pouring

Fig. 9.15 (*a*) Component to be cast. (*b*) Pattern corresponding the casting with draft allowance. (*c*) Mould ready for ponring. It may be noted that pattern is entirely in drag.

(*i*) These are used when the shape of casting is simple and does not create problems while withdrawing the pattern from mould. The Fig. 9.14 represent casting and single piece pattern corresponding to the casting. As discussed earlier various allowances are provided while manufacturing the pattern. Draft allowance is shown in pattern.

(*ii*) These are generally used when number of casting required is rather small.

(*iii*) The pattern is expected to be entirely in drag as shown in Fig. 9.15. If the shape of pattern is complex and contains no flat surface then moulding may become complicated with the necessity of follow board. This is a time consuming process and requires a skilled worker.

(*b*) **Split Piece Pattern.** Many pattern can not be made in single piece because they cannot be removed from mould. To eliminate this problem some pattern are made in two parts. Lower half of pattern rests in drag (lower moulding flask) while upper half of pattern is in cope (upper moulding flask).

The split surface of pattern is same as the parting plane of the mould. The two halves of pattern should be aligned properly by making use of dowel pins which are fitted to the (upper half) cope half of pattern. These dowel pin match with the precisely made hole in the lower half (Drag half) of the pattern.

These patterns are also used when the depth of casting is too high or when moderate quantities of casting are to be made.

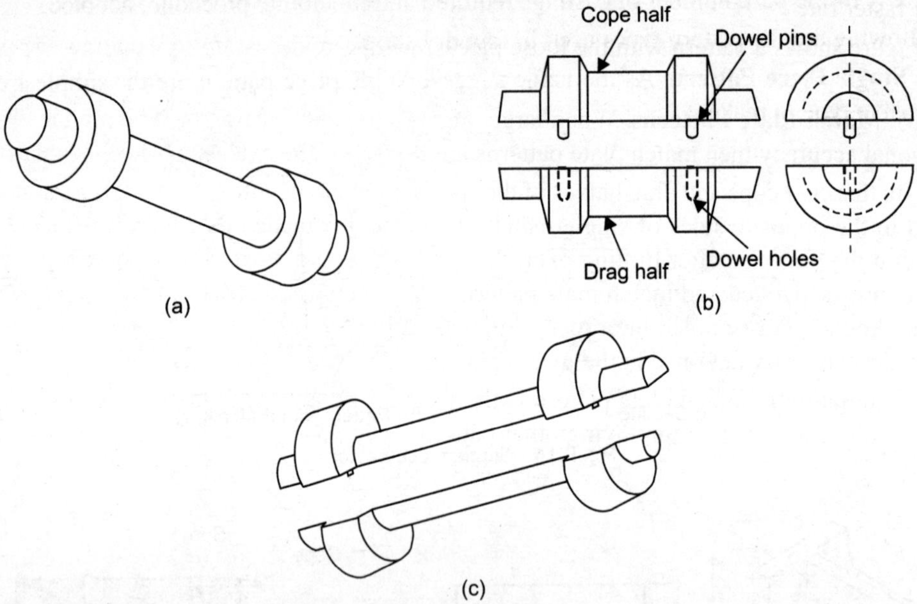

Fig. 9.16 Split piece pattern. (*a*) Component to be cast. (*b*) and (*c*) represent the split pattern showing cope and drag halves of split pattern. Cope half of pattern is provided with dowel pins which will match with precise made holes in drag half and thus align the two-halves properly.

(*c*) **Cope and Drag Pattern.** These type of patterns are generally used for moulding medium and large castings. However, they can also be used for producing small castings. These are similar to split pattern. Cope and drag half of the pattern along with gating and rising system are attached separately

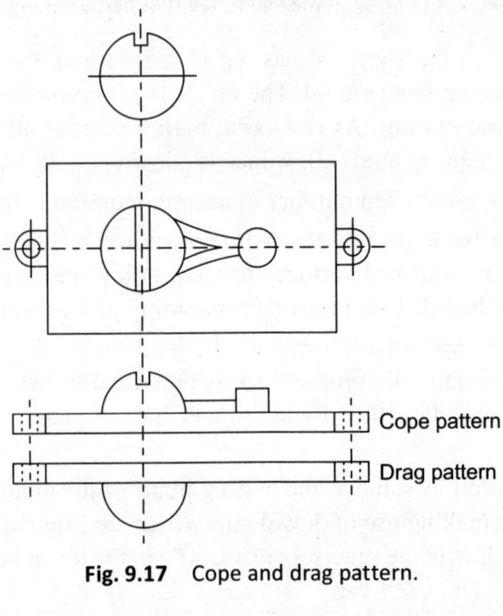

Fig. 9.17 Cope and drag pattern.

to metal or wooden plates along with alignment pins. Thus cope and drag mould can be produced independently on moulding machine using these plates. Then they are assembled to form complete mould. In this way large mould can be handled more easily in separate mould and small moulds can be made at faster rate.

Figure shows a pattern with the loose pieces, which are necessary to withdraw the pattern from mould.

(*d*) **Match Plate Patterns.** When large quantity of small casting is to be produced with higher dimensional accuracy then match plate patterns are used. It is the extension of cope and drag patterns.

In this case cope and drag halves of the split pattern along with gates and risers are permanently attached to the opposite sides of single match plate which is made up of wood or metal. One side is moulded in the drag and after turning over, the cope side is moulded. The upper and lower moulding boxes are now separated and match plate pattern is removed. Using lugs with hole and alignment pin, cope and drag are reassembled in proper alignment. Fig. 9.18 shows match plate with patterns. The 8 components can be cast at once in the match place with pattern as shown in Fig. 9.18.

A limitation in match plate pattern is the weight of mould and flask. Heavier work is done on larger moulding equipment, employing other pattern equipments.

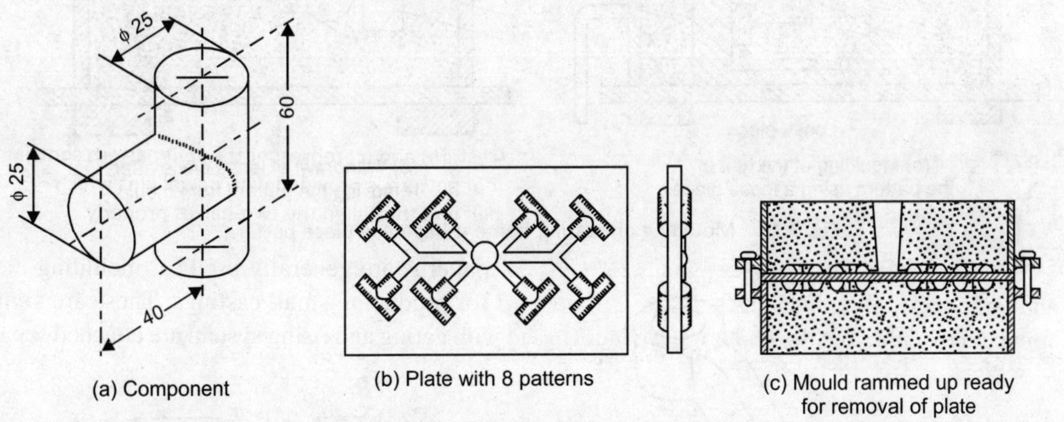

Fig. 9.18 Match plate patterns.

(*e*) **Loose Piece Pattern (Fig. 9.19)**

When the shape of part is such that one piece or split pattern could not be removed from moulding sand, loose piece pattern may be used. During the moulding obstructing part of the pattern is held as loose piece by a wire. After the moulding is over first the main pattern is removed then the loose pieces are removed through the cavity generated by removal of main pattern. One disadvantage of loose piece is that some shifting is possible during mould making. Loose piece pattern are expensive and moulding with loose piece is a highly skilled job. However, they enable the sand casting of complex shape which would otherwise require other processes. Whenever, possible design changes should be considered that would eliminate the need for loose piece pattern.

The component may be cast using split pattern but as the overhanging bosses do not allow the removal of pattern from mould, so it is held as loose piece during moulding and is removed after the cavity generated by removal of pattern.

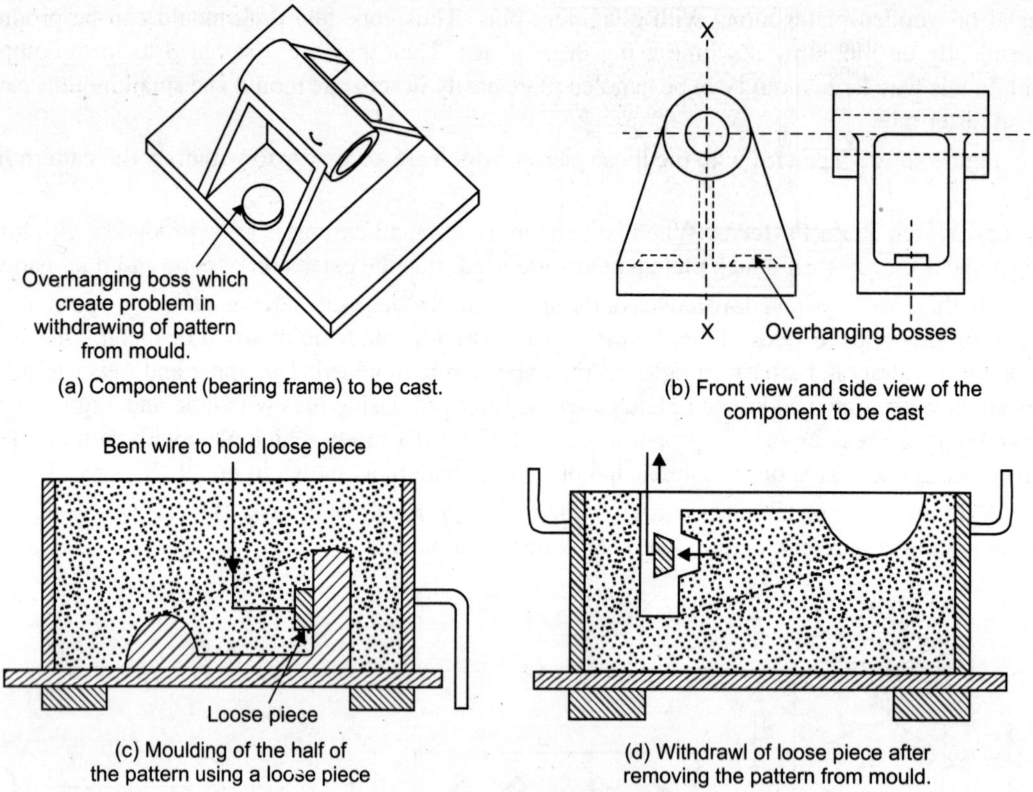

Fig. 9.19 Moulding of bearing frame using loose piece pattern.

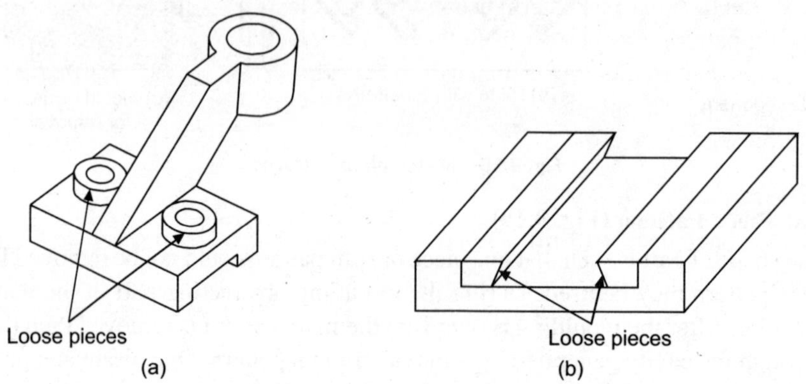

Fig. 9.20 Another example of components where loose piece pattern may be used.

(*f*) **Follow Board Pattern.** This type of patterns are used where some portion of pattern are structurally weak and if not supported properly are likely to break under the force of ramming. Hence the bottom board is modified as follow board. Fellow board is provided with projection or cavity corresponding to the shape of pattern. Actually fellow board acts as a seat for the patterns. During the preparation of core no fellow board is necessary because the sand which is rammed in drag will support the weak portion of pattern.

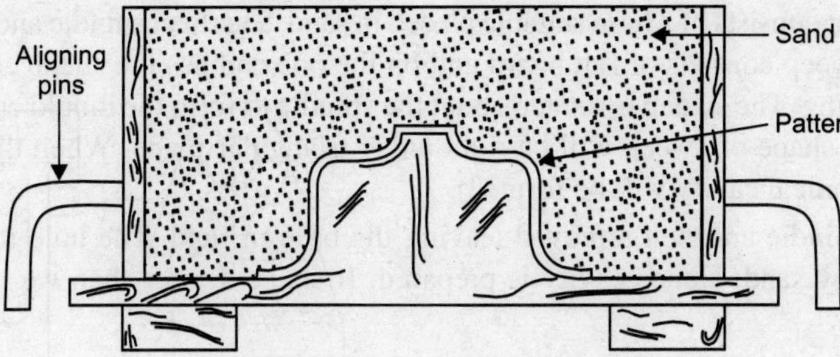

(a) Bottom board is modified as fellow board

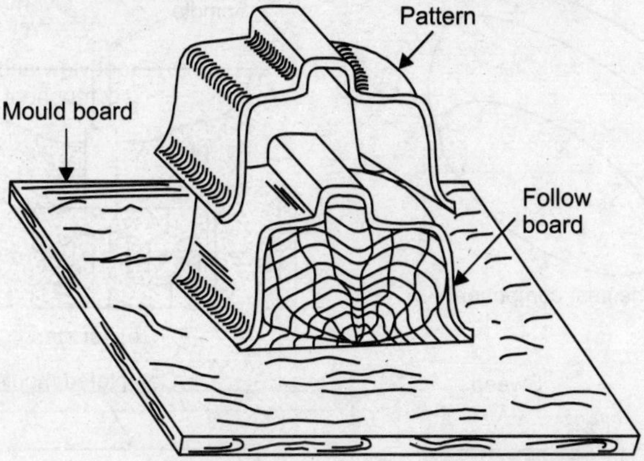

(b)

Fig. 9.21 Fellow board pattern. Here bottom board is modified as follow board to closely fit the contour of weak pattern and thus supports it.

(*g*) **Master Pattern.** A master pattern, usually made of wood, is used as an original for casting metal patterns. Several pattern may be cast from the master patterns and mounted on the pattern plate after they have been finished to the proper dimensions. Master pattern may be the first step to obtain match plates. Most metal pattern are cast in sand mould from a wooden master pattern provided with double shrinkage allowance.

9.6.3 Special Patterns

The patterns we have discussed till now are the duplicate of the cast component except some modification. But special patterns are not exact three-dimensional duplicate of the casting. These patterns may be just a plane figure, a part of pattern or skeleton of pattern. These are employed when only a few large castings are to be produced. In that case preparation of regular solid pattern is too expensive and cost of regular pattern is considerable portion of the cost of casting. Following are the special patterns:

1. **Sweep patterns.** Used for making axis symmetrical shapes such as wheels, rims, bell shapes.
2. **Segmental patterns.** Used for producing large ring shaped castings.
3. **Skeleton pattern.** Used for casting turbine casings, water pipes, chutes, cylinders.

(1) **Sweep pattern (Fig. 9.22).** These patterns are used for producing large casting, particularly which are circular in cross-section and are axis symmetrical like bell shape castings or large kettles of cast iron. The main advantage of this pattern is that it eliminates the cost of manufacturing in a three-dimensional pattern.

The equipment consists of a base suitable placed in sand, a vertical spindle and a wooden template called sweep. The sweep consists of flat board and on the one end has the shape corresponding to the shape of desired casting. The sweep is rotated about the spindle to form the mould cavity (like if conical cavity is desired the shape of sweep will be right angle triangular plane. When this plane will rotate about the edge the conical cavity will be formed).

Sweep and spindle are then removed leaving the base in sand. The hole made by removal of spindle is filled up by sand. Hence cavity is prepared. If shape desires then we may place core and chaplets in the cavity.

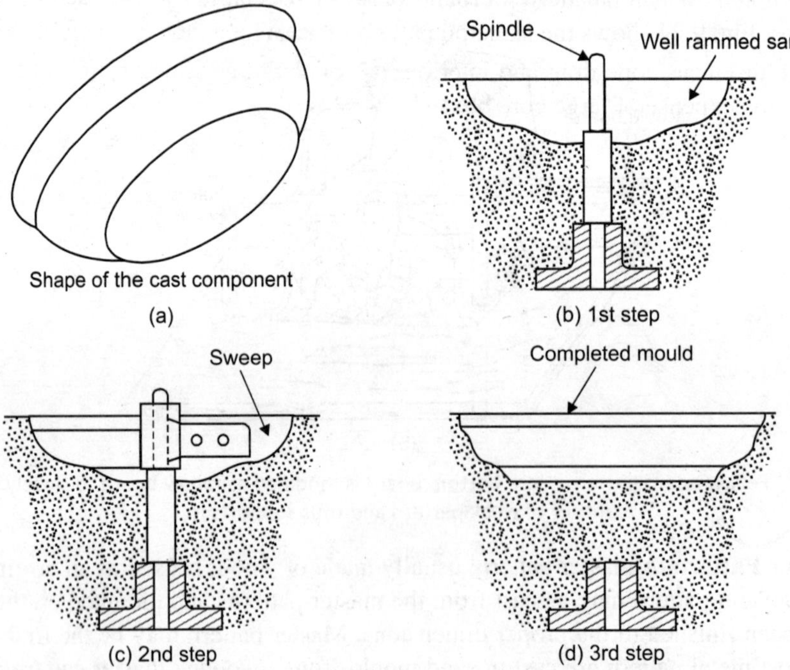

Fig. 9.22 Sands moulding process using sweep pattern. Here cavity is generated by a plane sweep which greatly reduces the cost of three dimensional pattern.

(2) **Segmental Pattern (Fig. 9.23).** These pattern are used for producing large circular castings avoiding the use of solid pattern of exact size. Segmental pattern differs from sweep pattern in the sense that sweep is given a continuous revolving motion to generate desired shape, whereas a segmental pattern is a portion of solid pattern itself and mould is prepared in parts by it. In the use of this pattern

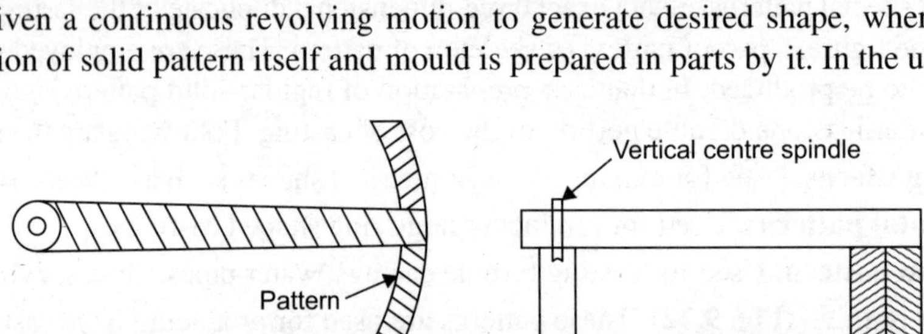

Fig. 9.23 Segmental pattern for making a ring shaped casting. A vertical centre spindle is firmly fixed near the centre of moulding flask. After ramming the sand, segmental pattern is fastened to the spindle and cavity is made in segments *i.e.* steps.

a vertical central spindle is first firmly fixed near the centre of drag and after preparing the part mould in one position the segment is moved to the next position. The process is continued until a complete circular mould cavity has been made. A large cast gear can be moulded with a segmental pattern with only three to four teeth.

(3) **Skeleton Pattern**

These patterns are ideal when preparation of a regular solid pattern would he too expensive and the shapes to be cast permit their use. A skeleton pattern is merely a wooden frame. The space between wooden frame is filled with moulding sand. The exact overall shape is obtained by firmly pressing the sand and strickling it. Skeleton pattern for round shape such as pipes and cylinders are prepared in cope and drag halves. Fig. 9.24 shows the skeleton pattern for casting a large diameter flanged pipe.

In some instances cope drag and interior core may all be formed from same skeleton pattern; thereby saving the expense of large core box.

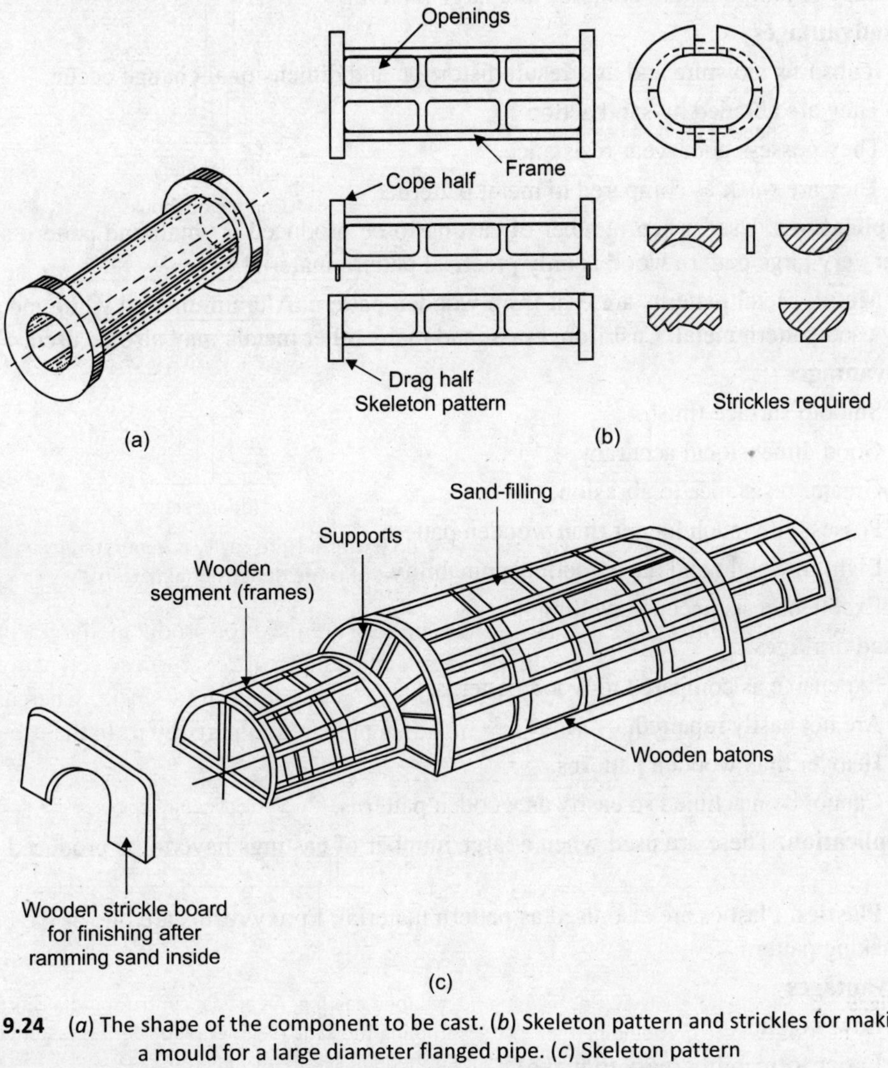

Fig. 9.24 (*a*) The shape of the component to be cast. (*b*) Skeleton pattern and strickles for making a mould for a large diameter flanged pipe. (*c*) Skeleton pattern

9.6.4 Pattern Material

The pattern material is determined primarily by number of castings to be made. It is also influenced by size and shape of the casting, desired dimensional accuracy, and the moulding process. The usual pattern materials are wood, metal and plastics.

(*a*) **Wood.** The most common material for making patterns of sand castings

Advantages

1. Easily available in large quantities.
2. Low weight.
3. Can be easily machined.
4. Relatively cheap.
5. Easy to obtain good surface finish.
6. Easy to join to obtain complex and large pattern.

Disadvantages

1. It absorbs moisture and as a result distortion and dimensional change occurs.
2. They are abraded by sand action.
3. They possess poor wear resistance.
4. They are weak as compared to metal pattern.

Applications. Used when number of casting to be produced is small and pattern size is large because for very large pattern wood is only practical pattern material.

(*b*) **Metal.** Metal patterns are cast from wooden pattern. Aluminium and white metal are most commonly used pattern metal. Cast iron, brass, and many other metals may also be used.

Advantages

1. Smooth surface finish.
2. Good dimensional accuracy.
3. Greater resistance to abrasion.
4. Possess life much longer than wooden pattern.
5. Light in weight and have good machinability.
6. Excellent resistance to swelling.

Disadvantages

1. Expensive as compared to wood patterns.
2. Are not easily repaired.
3. Heavier than wooden patterns.
4. Cannot be machined so easily as wooden patterns.

Application. These are used when a large number of castings have to be produced from same pattern.

(*c*) **Plastics.** Plastics are also used as pattern material. Epoxy resins are one of the best plastics used for making pattern.

Advantages

1. Low weight.
2. Easier formability (easy to make).

3. Smooth surface.
4. Durable.
5. Can be easily cleaned.
6. Moisture resistant.
7. Can be easily withdrawn.
8. Low solid shrinkage.

Limitations

1. Plastic patterns are weak, thus this section may need metal reinforcements.
2. Plastic patterns may not respond well when subjected to condition of severe shock as in machine moulding.

Applications. Used for making small, intricate patterns and core boxes.

Expanded polystyrene and wax is the common material used for moulding processes with disposable (or single use) patterns.

9.7 COMPOSITION OF MOULDING SAND

The main ingredients of moulding sand are:

1. The silica grains (SiO_2)
2. Clay (which acts as binder)
3. Water (moisture activates the clay and provides plasticity).

Some other binder and additives may also be added to the sand to achieve the specific properties.

1. **Silica Grains.** Silica grains form major portion of moulding sand (90–95%), the rest being the other oxides such as alumina and magnesium oxide. This is river sand which is used with or without washing. The grain size may vary in size from a few micrometer to a few millimeters.

The shape of silica grains may be angular, sub-angular, round or compounded.

Effect of silica grains on moulding sand properties

The silica grains are most important constituent because size and shape of silica grain greatly affect the properties of the moulding sand.

- For small and complicated casting, requiring smooth surface a fine-grained sand is desirable to allow the details of mould to be brought out sharply. However, fine grains lower the permeability. Fine grain has lower refractoriness and have more green strength as compared to coarse silica grains.
- For large castings, which give up more heat to the mould coarse grain sand is desirable as higher grain size permits the escape of gases easily and has more refractoriness. Sharp, irregular shaped grains are usually preferred because they interlock and add strength to the mould.

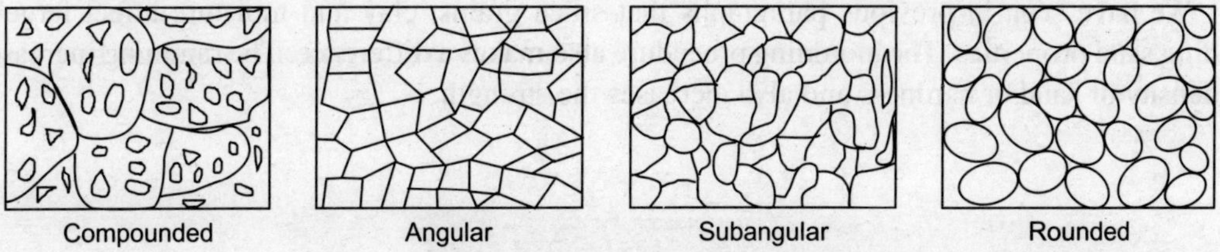

Fig. 9.25 Types of sand grains.

Following are the different shapes of silica grains:

(*i*) **Rounded grain.** Rounded grains have least contact with one another in a rammed structure. These grains have higher permeability, but lack in strength.

(*ii*) **Subangular grains.** Subangular grains have lower permeability and greater strength than the rounded ones.

(*iii*) **Angular grains.** Angular grains have defined edges and surfaces are nearly flat. They produce high strength and lower permeability in the mould than the subangular grains.

(*iv*) **Compounded grains.** In some cases the grains are cemented together such that they fail to separate when screened. They may consist of rounded subangular or angular grain or combination of the three. These are called compound grains and least desirable due to their tendency to breakdown at high temperatures. In practice sand grains contain mixed grain shapes, depending on origin. A subangular to rounded grain mixture would be the best combination.

2. **Clay.** Clays are generally used binding agent because of their low cost and wider utility. Clays impart the necessary bonding strength to the moulding sand so that after ramming mould does not loose its shape. The most popular clay types used are:

Bentonite (Al_2O_3. $ASiO_2$, H_2O, nH_2O)

and

Kaolinite or fire clay (Al_2O_3, $2SiO_2$, $2H_2O$).

Bentonite clays produce the strongest bonds for example. It requires approximately 12% by weight of fireclay to obtain the same green strength as 4 percent of bentonite. For this reason bentonite clays are usually used in synthetic moulding sands.

Effect of clay on properties of moulding sand. Besides sand grains, clay has large influence on the major properties of moulding sand. Clay imparts the bonding strength to the moulding sand. However, it is necessary to control the amount of the clay added, since too much clay will unnecessarily fill the spaces between the grains and thus lower the permeability.

It is also important that bonding clay should be refractory. Pure clays have high refractoriness. Some impurities like lime, alkalies, and other oxides tend to reduce their refractoriness. That is why metal with higher pouring temperature ordinarily require synthetic mould sands.

3. **Water.** Moisture is an important element of moulding sand. Clay acquires bonding action only in presence of required amount of moisture. The amount of water used normally is 2 to 8%.

Effect of water on moulding sand properties. The amount of water used should be properly controlled. A part of water absorbed by clay helps in developing bonding strength. Additional water, however, can act as a lubricant and make sand more plastic and more moldable. There should be an optimum amount of water to be used for given clay contents to obtain the maximum green compression strength. As the amount of water increases beyond optimum amount, green compression strength decreases and plasticity increases.

We have seen in previous paragraphs that silica grains, clay and moisture affect largely the moulding sand properties. The moulding procedure also makes a difference. The ramming increases the bulk density of sand or hardness and also increases the strength.

9.8 SAND PREPARATION

One of the important requirement for preparation of sand is thorough (perfect) mixing of its ingredients. The properly mixed sand meets all the required characteristics more effectively.

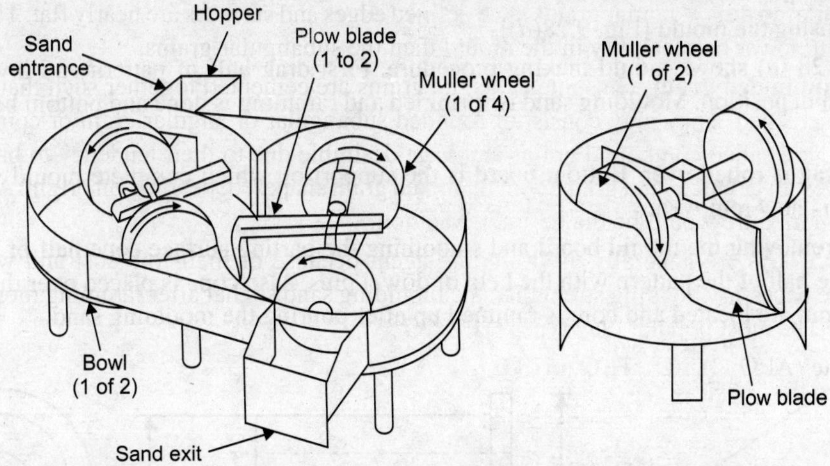

Fig. 9.26 Types of sand mullers.
Fig. 9.26 shows continuous sand muller and batch type sand muller.
Plow blade moves the sand and the muller wheels mix the components.

As we know that moisture content of a moulding sand is very important and controls practically all other properties. Moisture content is the most changeable property because moisture continuously evaporates from damp moulding sand. The purpose of sand preparation is:

(*i*) To add sufficient water to bring the moisture content within desired limits.

(*ii*) To mix the water thoroughly.

(*iii*) To coat the grains of sand uniformly with additive agent.

The mixing is done by putting the sand ingredients in a mixer known as 'muller'. The muller has a vertical cylindrical shell with a closed bottom. It houses one or two muller wheels and equal no. of plough blades. The combined action of blades and wheels knead, roll and stir the sand. Water is added shortly after the start of mixing. The mixing is done for about 3 minutes. After mixing sand may be treated by aerator. Aerator is a device which breaks up and separates the sand into small particles by beating and screening action. Aerator fluffs the sand so that it flows readily around the pattern and mould and easier to handle and use.

9.9 MOULD MAKING

The process of making a sand mould for split pattern (Fig. 9.27) is explained in following step (Fig. 9.28).

1. Placing pattern and lower moulding flask (drag) in position [Fig. 9.28(*a*)].
2. Pouring of sand and ramming up the drag [Fig. 9.28(*a*)].

3. Rolling over the drag and placing the cope in position [Fig. 9.28(*b*)].
4. Locating the sprue pin, riser pin and ramming up the cope [Fig. 9.28(*c*)].
5. Removing sprue pin, riser pin and pattern.
6. Making of runner and Gate.
7. Setting the core in mould.
8. Closing the mould [Fig. 9.28(*d*)].

Fig. 9.28 (*a*) shows mould making procedure. First drag half of pattern and lower moulding flask is placed in position. Moulding sand is sprinkled and ramming is done and bottom board is placed on top.

The drag is rolled over. Bottom board is the support on which complete mould will rest. The mould board is now removed.

After removing the mould board and smoothing the parting surface cope half of the pattern is placed on drag half of the pattern with the help of dowel pins. Also cope is placed over the drag. Sprue pin and riser pin are located and core is rammed up after pouring the moulding sand.

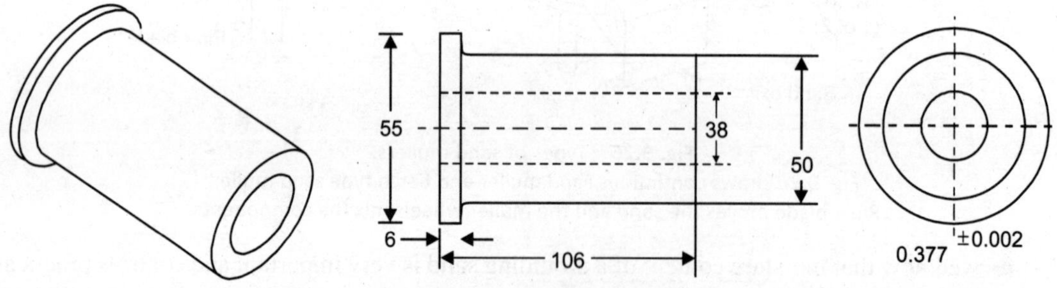

(a) Final shape of cast component

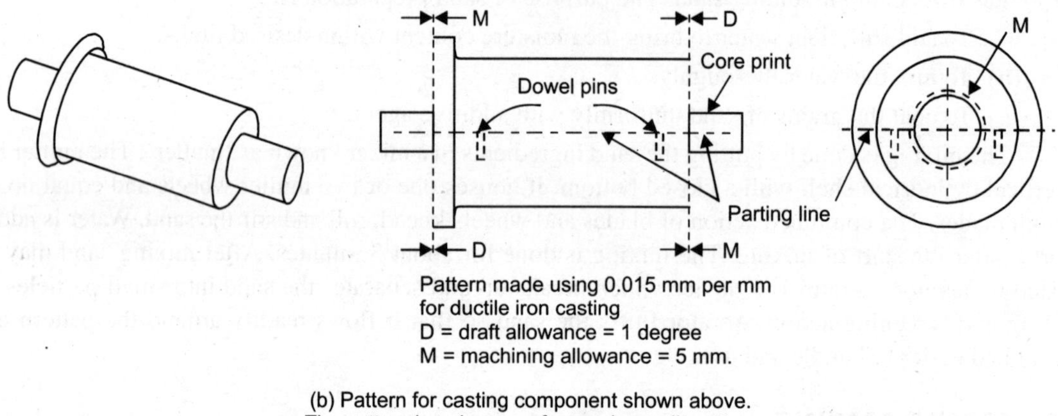

(b) Pattern for casting component shown above.
The pattern is to be manufactured as split pattern

Fig. 9.27 Component to be cast and pattern along with pattern allowances.

After ramming up the cope sprue pin, riser pin and pattern are removed. Pouring basin, runner and gate are cut in mould and after setting the core mould is closed. The Fig. 9.28 (*d*) shows mould ready for pouring. Let us discuss all these steps in detail.

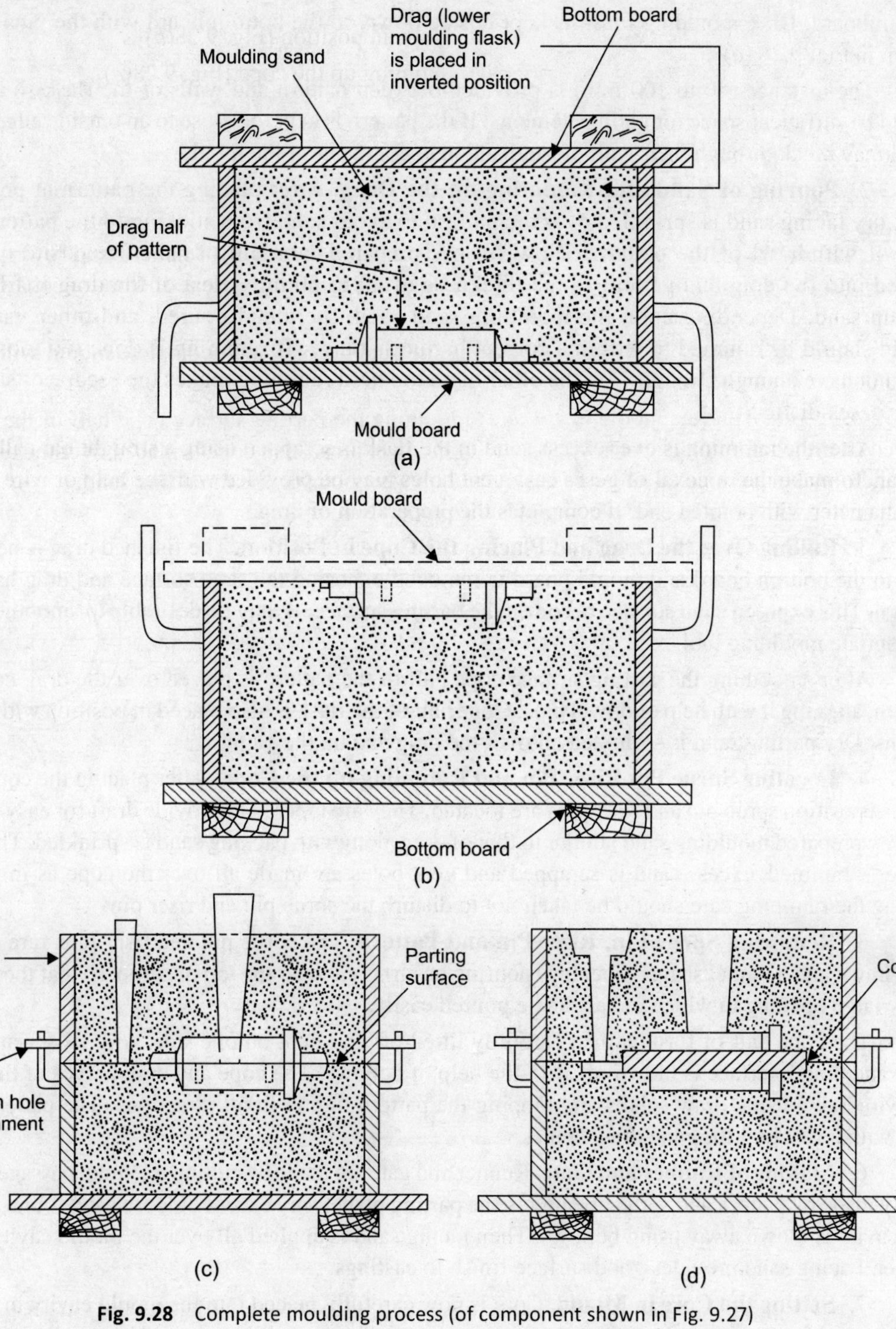

Fig. 9.28 Complete moulding process (of component shown in Fig. 9.27)

1. **Placing Pattern and Moulding Flask in Position.** The first step in making a mould is to prepare the pattern and moulding sand in sand muller. The drag half of the pattern is placed on the

bottom board. Drag moulding Flask is kept upside down on the bottom board with the pins down as shown in Fig. 9.28 (*a*).

The distance (50 to 100 mm) is provided between pattern and walls of the flask so that their should be sufficient space for gating elements. If the pattern is kept too close to an outside edge, molten metal may break through their wall of sand and may come out of the mould.

2. **Pouring of Sand and Ramming up the Drag.** After placing the pattern at position in drag, dry facing sand is sprinkled over the pattern to avoid sand from sticking to the pattern (at the time of withdrawl of the pattern). Now freshly prepared moulding sand of requisite quality is poured into the drag and pattern to the thickness of 40 to 50 mm. Rest of the drag is filled with backup sand. Depending upon the moulding sand used, the metal poured, and other variables a mould should be rammed to within a correct hardness range. If ramming is done too loose mould does not have enough strength. On the other hand too hard ramming makes the escape of steam and other gases difficult.

After the ramming is over excess sand in the flask is scrapped using a straight bar called strike off bar. To make the removal of gases easy vent holes may be provided with the help of wire of 1 to 2 mm diameter with pointed end. It completes the preparation of drag.

3. **Rolling Over the Drag and Placing the Cope in Position.** The finished drag is now rolled over to the bottom board and mould board is removed exposing the sand surface and drag half of the pattern. This exposed sand surface becomes the parting surface. It may be desirable to smoothen it with appropriate moulding tool.

After smoothing the parting surface cope half of the pattern is placed over the drag half of the pattern, aligning it with help of dowel pins. Cope half of the flask is also placed in position with the help of pins. Dry parting sand is sprinkled all over the drag and on the pattern.

4. **Locating Sprue Pin Riser Pin and Ramming up the Cope.** After placing the cope half of flask in position sprue pin and riser pins are located. They are tapered to provide draft for easy removal. Freshly prepared moulding sand similar to that of drag alongwith backing sand is sprinkled. The sand is properly rammed, excess sand is scrapped and vent holes are made all over the cope as in the drag. During the ramming care should be taken not to disturb the sprue pin and riser pin.

5. **Removing Sprue Pin, Riser Pin and Pattern.** The sprue pin and riser pins rare carefully withdrawn and a funnel shaped opening (pouring basin) is cut near the top of sprue, so that there will be fairly large opening in which metal can be poured easily.

The cope half of flask is then carefully lifted off and kept on one side. Any loose sand on the cope and drag interface is blown off with the help of bellows. The cope and drag halves of the pattern are withdrawn using draw spikes and rapping the pattern lightly in all directions. The pattern can be then withdrawn by lifting the draw spikes.

6. **Making of Runner and Gate.** Runner and gate are made by cutting the small passage between cavity made by pattern and sprue opening in the parting plane. Any excess of loose sand found in runner and cavity is blown away using bellows. Then facing sand is applied all over the mould cavity and the runner. Facing sand provides good surface finish to castings.

7. **Setting the Core in Mould.** Core is now carefully placed into the mould cavity in the drag. Core is used for making hole in the casting. It is prepared by ramming core sand in core box. Core rests in the core prints which locate and support it.

8. **Closing the Mould.** The cope is now lifted and placed on the drag moulding flask taking care of alignment pins. Mould is now ready for pouring and similar to that shown in Fig. 9.28(*d*) and Fig. 9.3.

9.10 MELTING

Till now we have discussed all about making the mould. The next step is melting the metal in a furnace. Melting is the major factor which controls the quality of the castings. Metals may be melted in various types of furnaces such as open hearth furnace, electric arc furnace, rotary furnace, cupola furnace etc. The choice of furnace depends upon the kind of metal, quality of metal to be melted, fuel costs, and other factors. All furnaces normally have a refractory wall structure or boundary which contains the high temperature region. Furnaces are constructed so that raw material and fuel and air if needed may enter and product may leave. There should be some means for maintaining temperature and chemistry control of the metal. In the following article we will see some details of cupola furnace.

CUPOLA (Fig. 9.29)

Cupola is a furnace which is most widely used to melt grey cast iron, nodular cast iron and some malleable cast iron. Cupola is simple in construction, economical to operate and melts iron continuously with a minimum of maintenance.

Constructional Details. Cupola is essentially a refractory lined vertical cylinder open at both its top and bottom. The inside dia of cupola ranges from few inches to 7 feet. Small sizes have been built for demonstration and experimental purposes. Cupola consists of drop door at the bottom. After closing drop door a proper sand bed can be prepared. This sand bed provides a necessary refractory bottom for the molten metal and the coke. Immediately above the sand bed is the metal tapping hole which is initially closed with clay till molten metal is ready for tapping. Above the metal tap hole is the slag hole through which slag generated during the melting process is tapped. Above the slag hole is wind box. Air from suitable blower comes through the blast pipe and enters in wind box. The openings for introducing the air to coke bed are called tuyeres. Tuyeres are surrounded by wind box for air supply.

Cupola Operation. While starting cupola a bottom is first prepared by closing the drop door and ramming the moulding sand which slopes gently towards tap hole. Then a coke bed of suitable height is prepared above the sand bottom. While igniting, first the coke bed is ignited through tap hole or any other hole near its bottom. After coke bed is properly ignited, layers of coke, iron charge and lime stone or fluxes are alternatively fed into cupola through charging door maintaining the necessary proportions and rate of charging. Then the air blast is turned on and combustion occurs rapidly within 5 to 10 minutes. Molten metal is collected near the tap hole. When enough molten metal is collected in the well of cupola the slag is drained off through slag hole before opening the tap hole. The molten metal is collected in ladles and transported to the moulds in which it is to be poured.

The charge needed to produce cast iron consists of pig iron, cast iron scrap. Steel scrap is also used when alloy cast iron is needed. The fluxes are added in the charge to remove oxides and other impurities present. The flux most commonly used is limestone ($CaCO_3$). The flux is expected to react with oxides and form compounds which have low melting point and also lighter than molten metal. As a result molten slag tends to float on the metal pool and thus can be very easily separated.

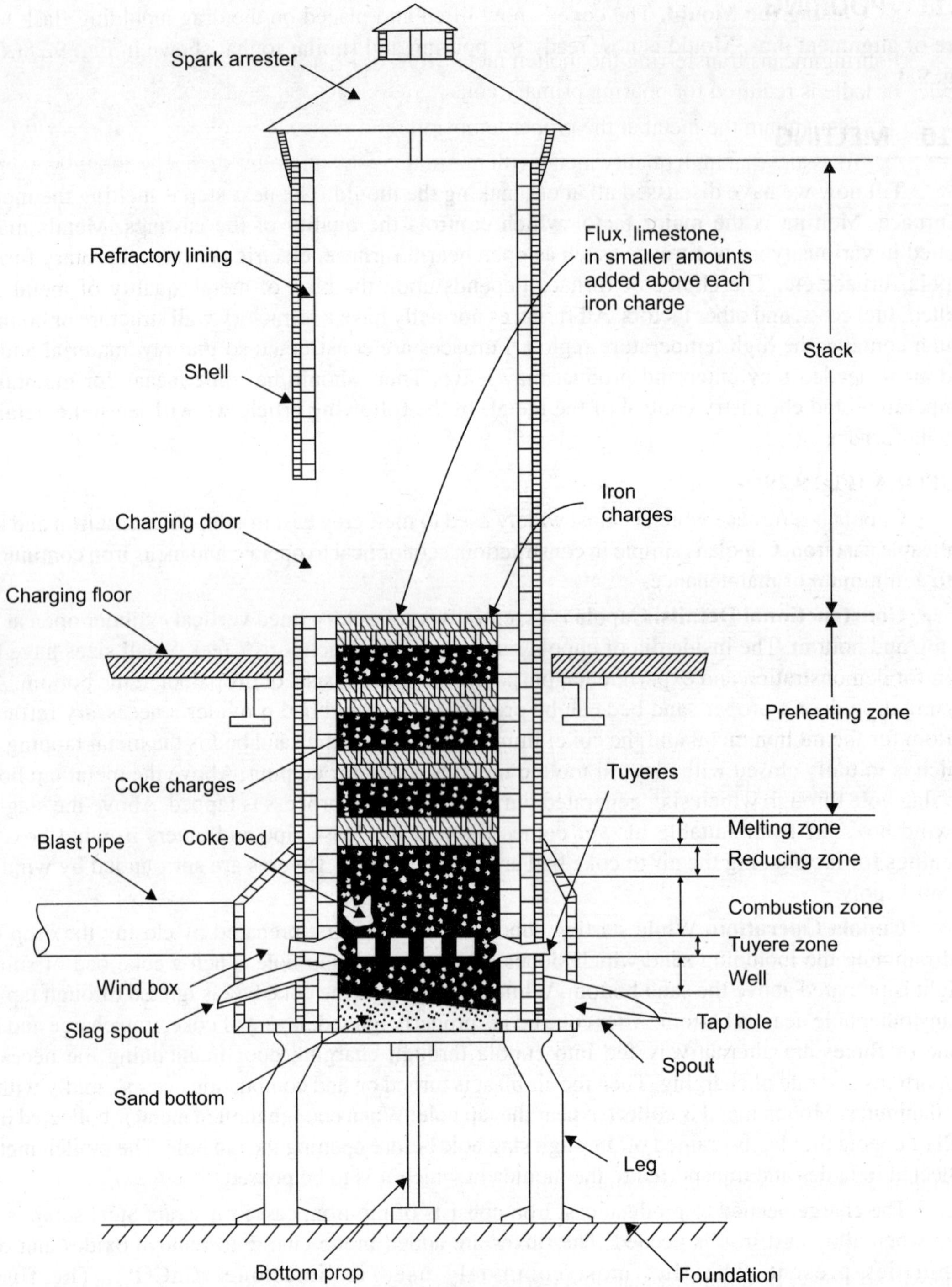

Fig. 9.29 Cupola

9.11 POURING

Pouring means transferring the molten metal from furnace to the mould. Some type of pouring device or ladle is required for pouring primary considerations for this operation are:

1. To maintain the metal at the proper temperature of pouring.
2. To assure that high quality metal with minimum contamination should be supplied to mould.

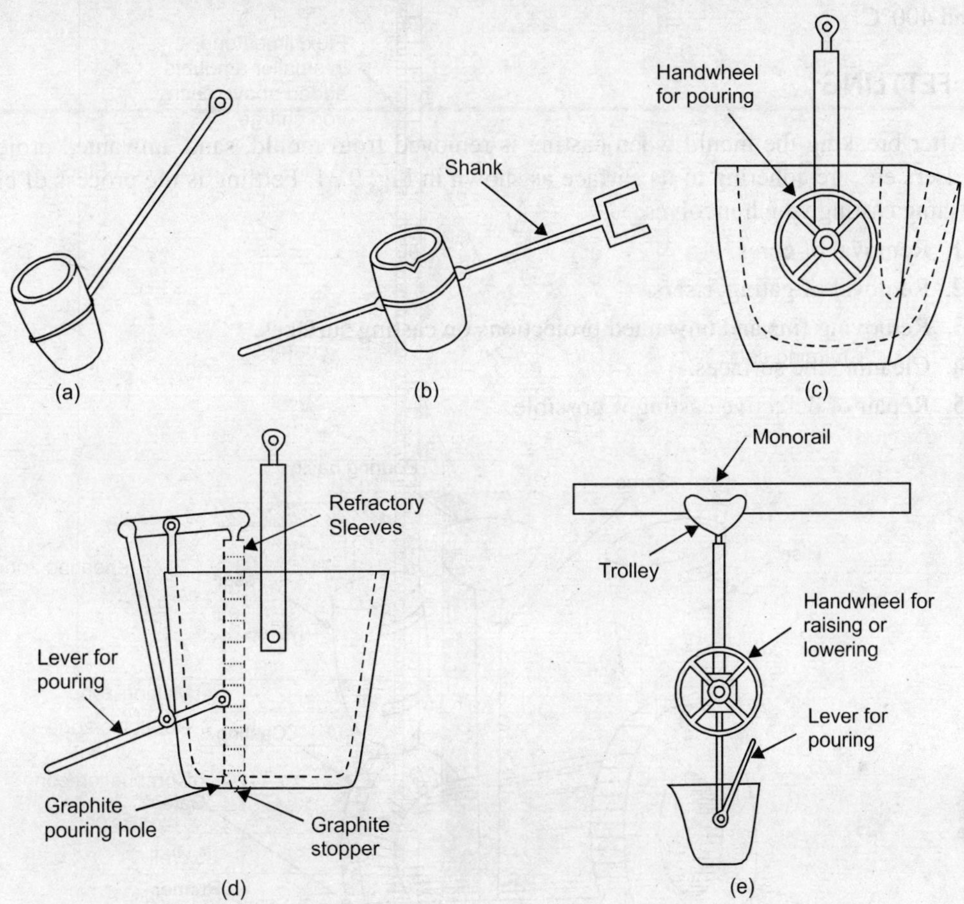

Fig. 9.30 Different types of ladles used for distributing and pouring molten metal into moulds. (*a*) Hand ladle. (*b*) Shank ladle. (*c*) Teapot ladle. (*d*) Bottom pour ladle. (*e*) Ladle handler.

The specific type of pouring ladle is determined largely by the size and numbers of casting to be poured. In low technology foundries hand ladles are used which is handled by one man for pouring metal into moulds. When more metal is required or heavier metal is poured shank ladle is used, which is handled by two men. In larger foundries bottom pour or teapot type ladles are used.

In modern industries engaged in mass production of castings moulds are placed on conveyers and passed slowly by a pouring station. Pouring stations are located permanently next to the furnace or molten metal may be brought to certain points by overhead handling equipment. A programmed amount of molten metal is poured automatically to the corresponding moulds.

9.12 SOLIDIFICATION

After pouring the metal into mould the casting is allowed to solidify and cooled in mould. Then sand moulds are broken to obtain the castings. The sand mould should be broken at proper temperatures when the casting is sufficiently cooled in mould. If the hot castings are exposed to air, there is likely to be faster and uneven cooling which may cause thermal stresses in the castings. For ferrous alloys mould should be broken below 700°C. For thin and delicate casting it should be removed at around 400°C.

9.13 FETTLING

After breaking the mould when casting is removed from mould, sand, unwanted projections, gating, risers etc. are adhering to its surface as shown in Fig. 9.31. Fettling is the process of cleaning and finishing casting which involves:

1. Removal of cores.
2. Removal of gating, risers.
3. Removing fins and unwanted projections on casting surfaces.
4. Cleaning the surfaces.
5. Repair of defective casting if possible.

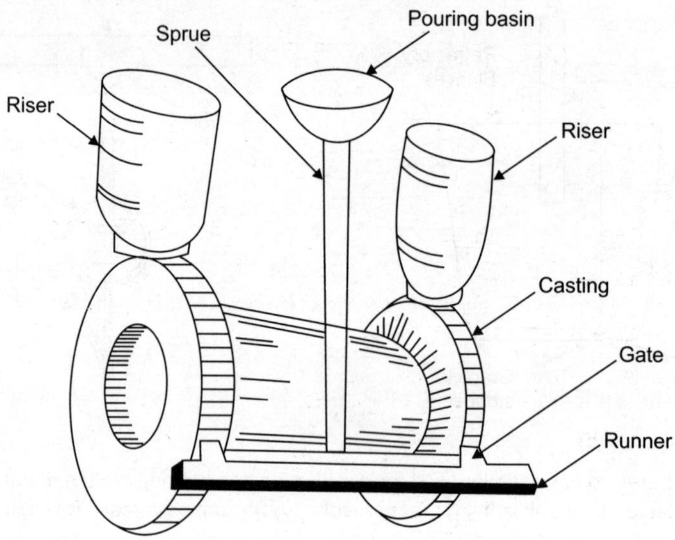

Fig. 9.31 Unfettled casting with sprue, runner, riser and gates.

1. **Removal of Cores.** Cores can be removed simply by knocking off with an iron bar, or with the help of core vibrator. For removal of core from large castings hydroblasting is an effective way.

2. **Removal of Gating System.** Gating and risers can be removed by breaking off, sawing with hacksaw, flame cutting with oxyacetylene torch or by cutting by abrasive cut off wheel. The gates and riser on iron casting may be broken off but to remove them from steel castings cutting torch or cutting of wheel is necessary. The method depends upon the metal, casting size and other conditions.

3. **Removing Fin and Unwanted Projections.** The fins and other unwanted projections may be chipped off with chisel, sawing, flame cutting, and grinding. Another useful method which is commonly used is for removing fins and sand adhering the casting surface is tumbling. Here the casting are kept in a barrel which is completely closed and rotated on about its axis at 30 to 40 rpm. for sufficient time. Other material may be placed in the barrel with castings. Surfaces are cleaned as castings rub against each other and other materials.

4. **Cleaning the Surfaces.** For cleaning the sand particles sticking to the casting surface sand blasting is used. The casting is kept in a closed box and a jet of compressed air with a blast of sand grains or steel grits is directed against the casting surface which removes the scale and cleans the casting. If the operator is in the room directing the abrasive steam he must be properly protected.

9.14 FOUNDRY TOOLS AND EQUIPMENTS

As we know that casting is a multistage process involving large number of tools and equipments at each stage. Depending upon the stages number of tools can be classified as follow:

1. Moulding tools and equipments
2. Metting tools and equipments
3. Pouring equipments
4. Fettling equipments

9.14.1 Moulding Tools and Equipments

1. **Shovel.** It consists of iron pan fitted with a wooden handle. It is used for mixing the moulding sand and pouring it in moulding flask.

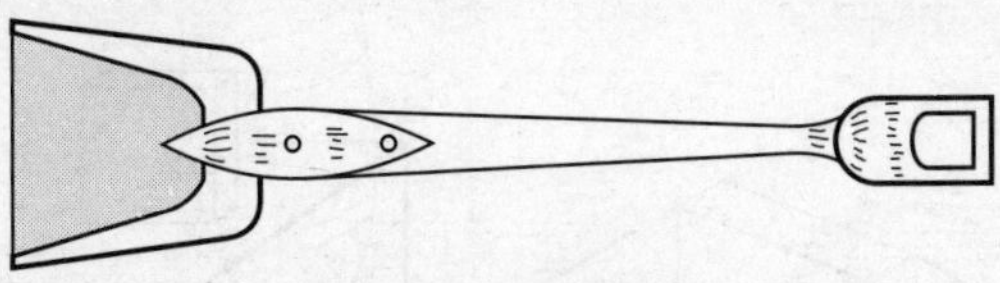

Fig. 9.32 Shovel

2. **Trowels.** It consists of Iron plate with a wooden handle. It is of smaller size as compared to shovel. It is used for mixing the moulding sand. In addition, it is used for finishing flat surfaces and corners inside moulding flask.

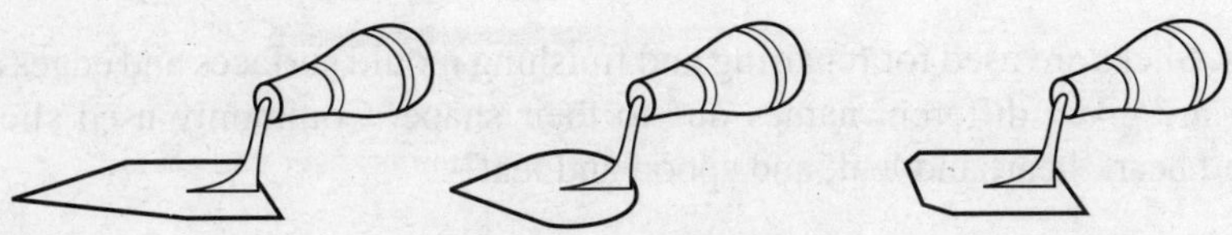

Fig. 9.33 Various types of trowels.

3. **Riddle.** It is also known as sieve. It consists of wooden or metallic frame fitted with a screen of standard wire mesh at the bottom. It is used for riddling of sand to remove foreign particles from sand.

4. **Rammer.** Rammers are used for ramming moulding sand in moulding flask so that it packs uniformly all around the pattern. Common used rammers are:

(*a*) **Hand rammer.** It is smaller in size and usually made of wood. At one end, it carries solid cylindrical shape and other end is wedge type. It is used in bench moulding.

(*b*) **Peen rammer.** It is made up of metallic rod with a wedge shape projection at one end. It is used for packing the sand at the corners of mould. It is used in bench moulding.

(*c*) **Floor rammer.** It is a long metallic bar carrying wedge shape construction at one end and a flat plate at other end. It is used in floor moulding.

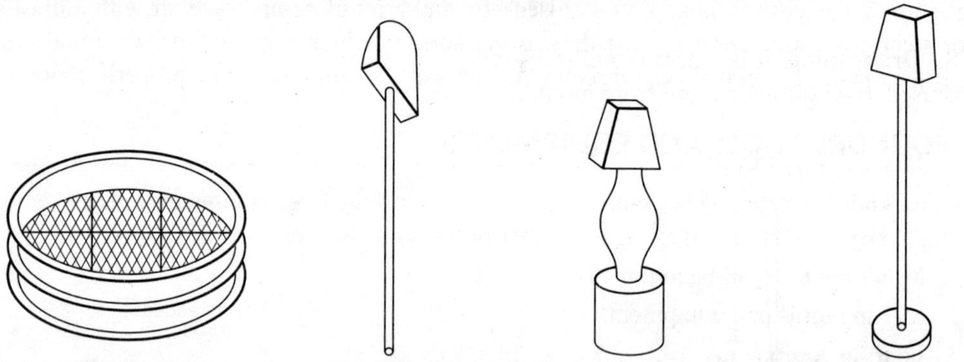

Fig. 9.34 Rammers

5. **Smoothers.** These are finishing tools used for finishing flat or round surfaces, square corners and edges of moulding flask.

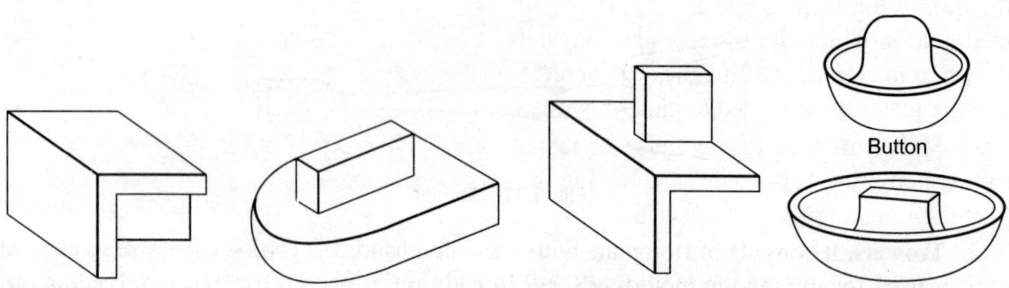

Fig. 9.35 Various types of rammers used in foundry shop.

6. **Slicks.** Slicks are used for repairing and finishing mould surfaces and edges after the removal of pattern. They are given different names due to their shape. Commonly used slicks are heart and spoon, square and heart, heart and leaf, and spoon and leaf.

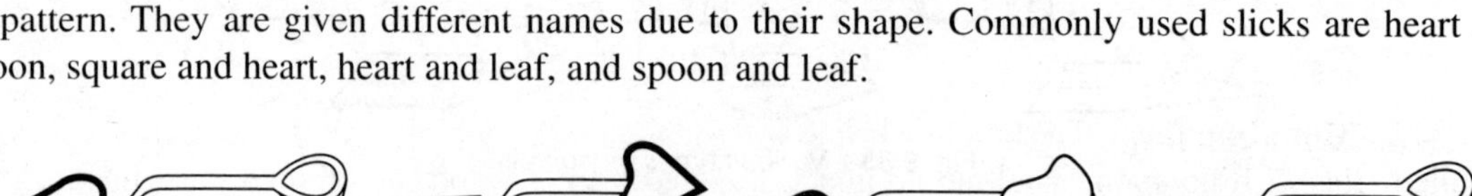

Fig. 9.36 Various types of slicks.

7. **Lifter or cleaners.** Lifters are used for removing the loose moulding sand from mould cavity and thus cleaning and finishing the moulding flask. These are made of metal.

Fig. 9.37 Lifter

8. **Draw spike.** It is tapered steel rod having ring shape at one end and a sharp point at other end. It is used to withdraw pattern from mould.

Fig. 9.38 Draw spike

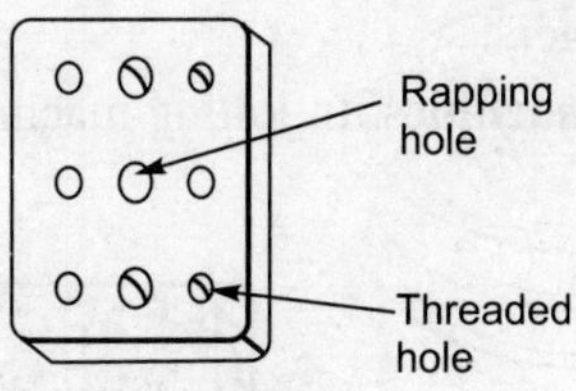

Fig. 9.39 Draw screw and rapping plate

9. **Draw screw.** Draw screw is steel rod carrying loop or ring at one end and machine screw threads at other end. These are used in combination with rapping plate. Rapping plate is provided with threaded hole in which draw screw can move. The pattern may be withdrawn from mould by lifting the draw screw because draw screw and rapping plates are clamped to pattern by means of screws.

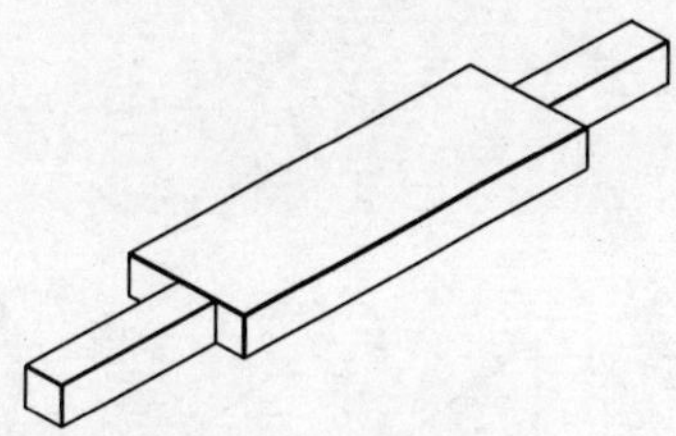

Fig. 9.40 Strike off bar

10. **Strike off bar.** It is wooden or metallic bar having one surface perfectly flat. It is used for removing the surplus sand from mould and for making the mould surface flat.

11. **Vent wire.** Vent wire is mild steel wire or rod of few milimeters diameter. It helps escape of gases from mould cavity. The gases include atmospheric gases and gases produced during pouring of molten metal in mould cavity.

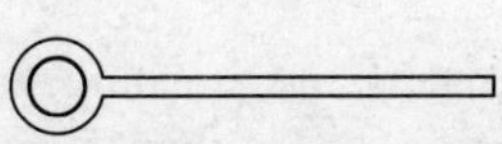

Fig. 9.41 Vent wire

12. **Moulding flask.** The box like frame without top and bottom base into which sand is rammed is called "flask". It holds the sand mould intact. It is made up of wood for temporary applications and more generally of metal for long term use. The flask is shown in Fig. 9.1. It is generally made of two or more parts.

Drag. Lower or bottom moulding flask is known as drag.

Cope. Upper moulding flask is termed as cope.

Cheek. It is an intermediate moulding flask in three piece moulding.

Drag is provided with the pins and cope is provided with lugs with hole which fit together and ensure proper alignment during the placement of cope upon drag.

13. **Patterns.** Pattern is the duplicate of final casting, to be made, with some modifications. It is used for making cavity in the mould. Patterns are discussed in detail in **Article 9.3 and 9.6.**

14. **Moulding machines.** Hand moulding is suitable for producing few castings or for large size castings. For producing the large number of casting moulding machines are used. In moulding machines, sand is rammed by machine instead of manual rammers. The basic machines used for ramming the sand in moulding flask are:

(*a*) jolting machines
(*b*) squeezing machines
(*c*) combination of jolting and squeezing machines
(*d*) sand slingers.

(*a*) **Jolting machines.** In jolting machines sand is filled in the moulding flask usually from an overhead hopper.

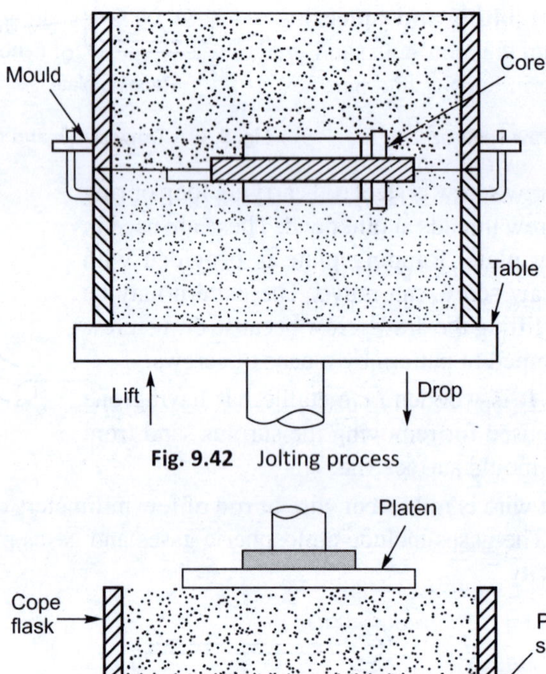

Fig. 9.42 Jolting process

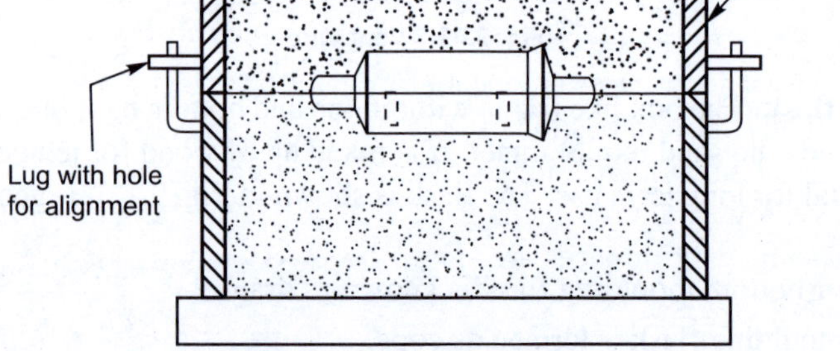

Fig. 9.43 Squeezing process

The work table along with pattern, flask and sand is raised to certain height and allowed to free fall on base of machine. The momentum of falling sand forces the sand to pack around the pattern in moulding flask.

— The sand at the bottom experiences maximum moulding force and thus mould is hardest at pattern surface and parting line. The sand above the bottom is progressively less harder.

— This type of ramming is suitable of horizontal surfaces only.

(*b*) **Squeezing machines**

— In these type of machine sand is filled in flask usually by hopper. Then sand is rammed by a plate (of squeezing machine) slightly smaller than the inside dimensions of moulding flask. The uniform pressure is applied on the plate which results in packing the sand around the pattern in moulding flask. The sand next to plate experiences maximum force and mould is hardest at that place while the sand below is progressively less harder. To achieve the better and uniform mould density and hardness in mould, plate with contour may be used. Diaphragm may also be used to achieve uniform density and hardness. This type of machines are suitable for small casting and shallow flasks.

(*c*) **Combination of jolting and squeezing machines.** These machines utilize a combination of jolting and squeezing to ram the moulding sand. In this case, benefits of both the machines are achieved and thus mould with uniform density and hardness is obtained.

(*d*) **Sand slingers**

— Sand slingers throw the sand in moulding flask with high velocity. The sand is conveyed by belt into slinger head. The slinger head contains a rotor equipment with 4 to 5 blades. These blades pick up the sand and throw it in mould from nozzle.

— The process is vary fast and provides very uniform mould density with hardness. One variant of sand slingers are mould blower which are commonly used for making cores than moulds.

9.14.2 Melting Equipment

After moulding next step in casting is to melt the metal and pour it in mould cavity. For melting of metals various furnaces are available. The processes used in these furnaces are basically remelting processes *i.e.*, ores obtained from mines are not directly used to cast the objects. The selection of particular type of furnace is largely based on the following factors:

1. Type of metal to be melted.
2. Quality of molten metal (*i.e.* capability of melting medium for absorbing impurities).
3. Quantity of molten metal required per hour.
4. Method of pouring.
5. Economic factors, *i.e.* capital investment, maintenance cost and running cost of furnace.

Types of furnaces. The furnaces are meant for melting the metal. Heat required for melting can be created by several means like combustion of fuel, electric arc and electric resistance. All the furnaces have a refractory wall structure which contains high temperature region. The furnaces are designed in such a manner that they allow minimum exchange of heat and gases from outside. On basis of source of heat and type of metal to be melt furnaces are classified as:

1. Blast furnace
2. Cupola
3. Open hearth furnace

4. Air furnace
5. Electric direct arc furnace
6. Crucible furnace
7. Tilting type furnace

1. **Blast furnace.** Blast furnace is explained in detail in Chapter 3 Engineering Materials. Please refer 3.5 extraction of engineering materials.

2. **Cupola.** Cupola is a furnace which is most widely used to melt grey cast iron, nodular cast iron and some malleable cast iron. Cupola is simple in construction, economical to operate and melts iron continuously with a minimum of maintenance. Cupola explained in detail in 9.10 melting.

3. **Open hearth furnace.** Open hearth furnace has a long, shallow basin known as 'hearth'. At the front of open hearth furnace, the number of charging doors are present as shown in figure 9.44. Heat is supplied to hearth by means of combustion of fuel which is usually gaseous (natural gas, producer gas or atomized oil).

—When the hot gases pass over the hearth, because of low roof heat is reflected. That is why it is known as "reverberatory type furnace".

—After the hot gases pass over the hearth. They give up much of remaining heat to the regenerative chambers on the left. After suitable time (say 30 minutes) the flow of gases is reversed and incoming air and gaseous fuel are preheated as they enter through the regeneration chamber on the left. The chambers on left and right are alternatively heated to increase fuel efficiency and temperature.

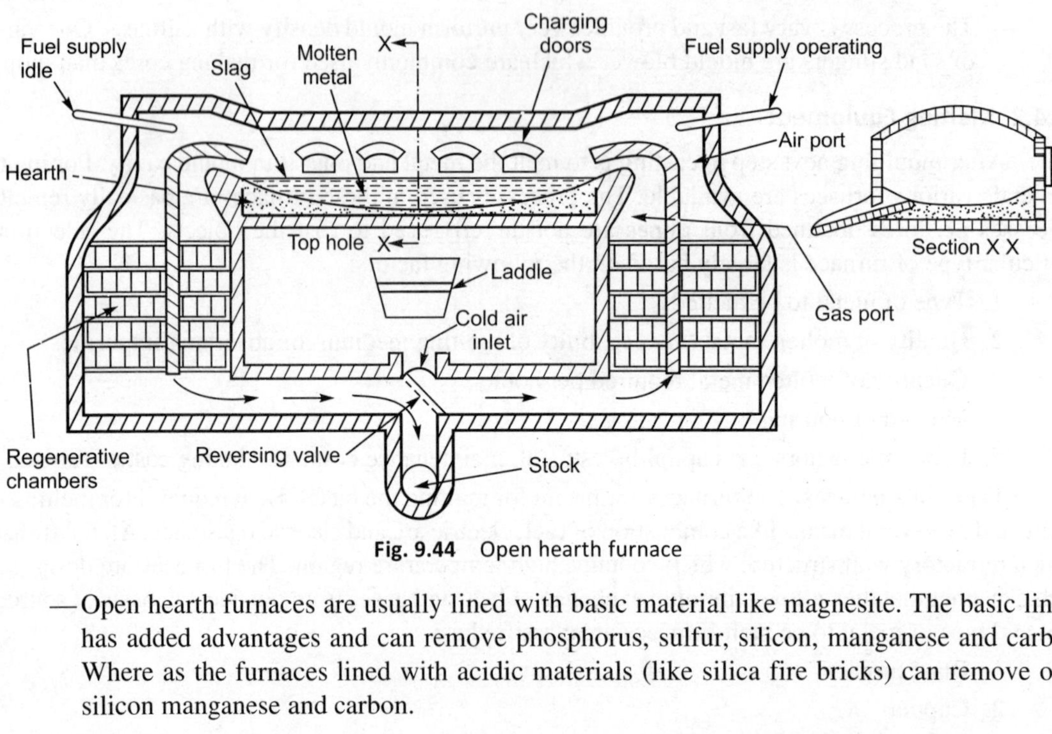

Fig. 9.44 Open hearth furnace

—Open hearth furnaces are usually lined with basic material like magnesite. The basic lining has added advantages and can remove phosphorus, sulfur, silicon, manganese and carbon. Where as the furnaces lined with acidic materials (like silica fire bricks) can remove only silicon manganese and carbon.

4. **Air furnace.** Air furnace is reverberatory type furnace which is primarily used for producing malleable cast-iron. It is similar to open hearth furnace with some important differences:

1. Regenerative chamber are not there for preheating the incoming air.
2. Only one burner is used with the stack at opposite end.
3. Charging is of solid materials is done through removable section of roof known as **'bungs'.**
4. This air furnaces are often used in combination with cupola for supplying molten cast iron.
5. These furnaces are lined with acidic materials. The acidic slag protects the metal from direct exposure to the flame and furnace atmosphere. The major advantage of air furnace over cupola is that chemical composition of liquid metal can be easily controlled.

5. **Electric direct arc furnace.** As the name suggests, in electric arc furnaces heat source used for melting charge (usually steel) is electric arc.

The charge is fed from the top of furnace and placed on the hearth. Three large graphic electrode extend downward through holes in roof. These electrode sit above the charge heap. The arc is generated between any one of electrode and metallic charge and then from charge to next electrode. The temperature produced is around **11000°F** and it quickly melts the charge underneath the electrodes.

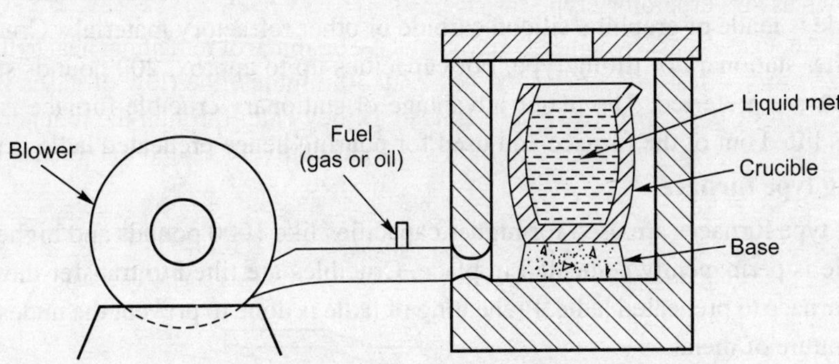

Fig. 9.45 Crucible furnace

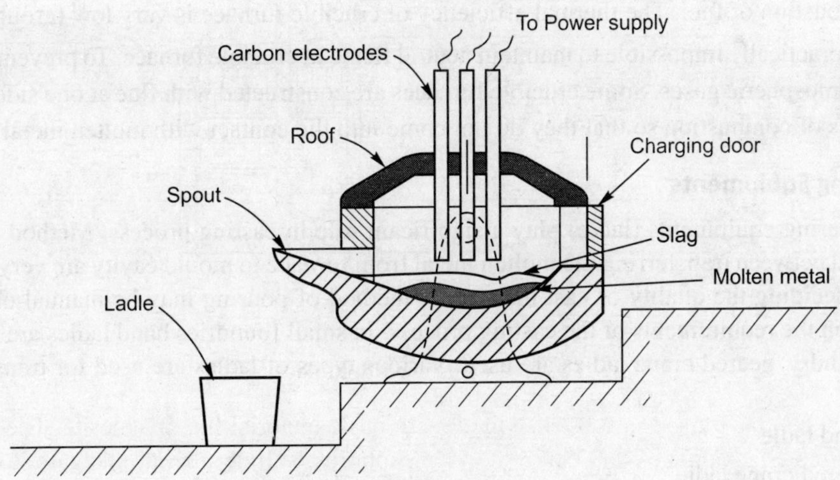

Fig. 9.46 Direct arc electric furnace

—The direct arc furnace may have either acidic or basic lining on the hearth. For acidic lined hearth, silica fire bricks are used and for basic lined hearth magnesite and alumina bricks are used.

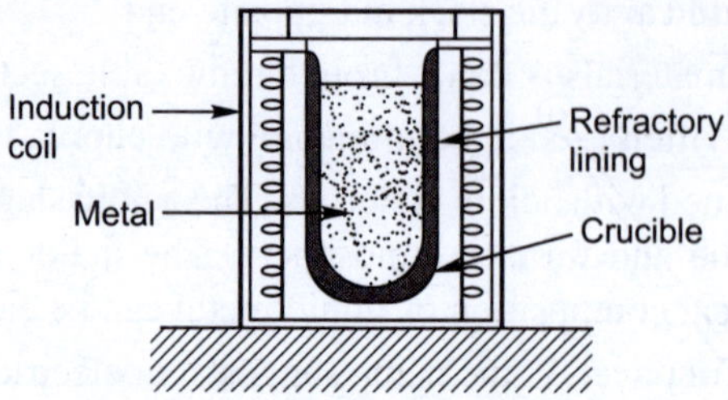

Fig. 9.47 Induction type electric furnace

6. **Crucible furnace.** Crucible furnace are the simplest type of furnaces. Earlier they were used for melting ferrous metals but now a days these are mostly used for melting non-ferrous metals.

—In crucible furnace, metal is placed in refractory cup shaped container known as crucible. Crucible is made of graphite, silicon carbide or other refractory materials. Crucible furnaces are either stationary or tilting type. For capacities up to approx. 200 pounds stationary type furnaces are preferred. The major advantage of stationary crucible furnace is that crucible itself is lifted out of the furnace and used for pouring hence preheated ladle is not required.

7. **Tilting type furnace**

—Tilting type furnaces are used for higher capacities like 1000 pounds and higher. In this case crucible is permanently cemented in place. Crucibles are tilted to transfer the molten metal from furnace to preheated ladle. Preheating of ladle is done to prevent the undesirable drop in temperature of metal.

—The fuel used for crucible furnace is gas, oil or coke. Blower is used to provide the oxygen for combustion of fuel. The thermal efficiency of crucible furnace is very low (around 10%).

—It is practically impossible to maintain neutral flame in crucible furnace. To prevent absorption of atmospheric gases. Some crucible furnaces are constructed with flue at one side to carry of gases of combustion so that they do not come into the contact with molten metal.

9.14.3 Pouring Equipments

The pouring equipments (ladle) play a significant role in casting process. Method of pouring, time consumed between transferring the molten metal from furnace to mould cavity are very significant variables for deciding the quality of cast metal. The method of pouring may be manual or automatic depending upon the requirements of the casting process. In small foundries hand ladles are used where as in large foundry, geared crane ladles are used. Various types of ladles are used for transferring the molten metal.

1. Hand ladle
2. Geared crane ladle
3. Crucible ladle.

1. **Hand ladle.** Hand ladles are used when smaller amount of metal is there. Ladles consist of a steel shell lined with suitable refactory material like fire clay. The ladle can be held with handle for pouring the metal in mould cavity. Two men are required for holding the ladle and transferring the molten metal from ladle to mould cavity.

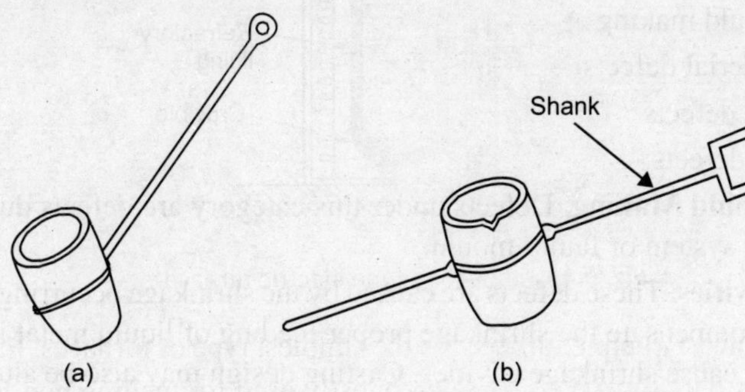

Fig. 9.48

2. **Geared crane ladle.** When greater amount of metal is required to be poured as in case of large casting geared crane ladle may be used. The ladle from furnace moves to mould cavity on overhead crane or monorail. The ladle pours the molten metal by geared mechanism in mould cavity. The overhead crane mechanism is used in combination with geared ladle for pouring of molten metal.

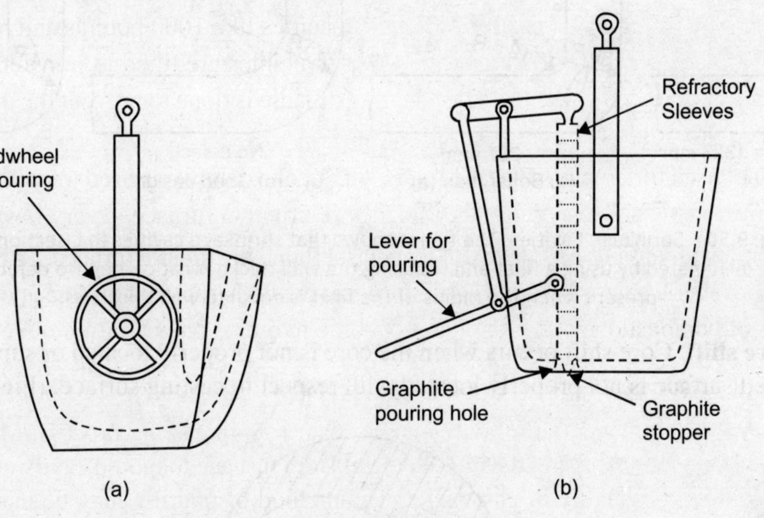

Fig. 9.49

3. **Crucible ladles.** Crucible is used in both melting and pouring purposes. *i.e.* metal is melted in crucible and then crucible is held by handle to transfer the molten metal to mould cavity. Hence crucible acts as both melting and pouring equipments.

9.15 CASTING DEFECTS

There are large number of variables which may cause the casting defects. So exact classification of casting defects is quite difficult.

Casting defects may be classified as:

1. Defects in mould making
2. Moulding material defects
3. Pouring metal defects
4. Metallurgical defects.

1. **Defects in Mould Making.** Defects under this category are defects due to faulty pattern or core boxes, faulty gating system or faulty mould.

(*a*) **Shrinkage cavities.** These defects are caused by the shrinkage occurring during solidification of metal. To compensate the shrinkage proper feeding of liquid metal is required. Improper riser may also cause shrinkage cavities. Casting design may also be altered to eliminate this defect. Fig. 9.50 shows how shrinkage cavities can be eliminated in the casting of an L-section. Fig. 9.51 shows various designs of casting a flanged pipe.

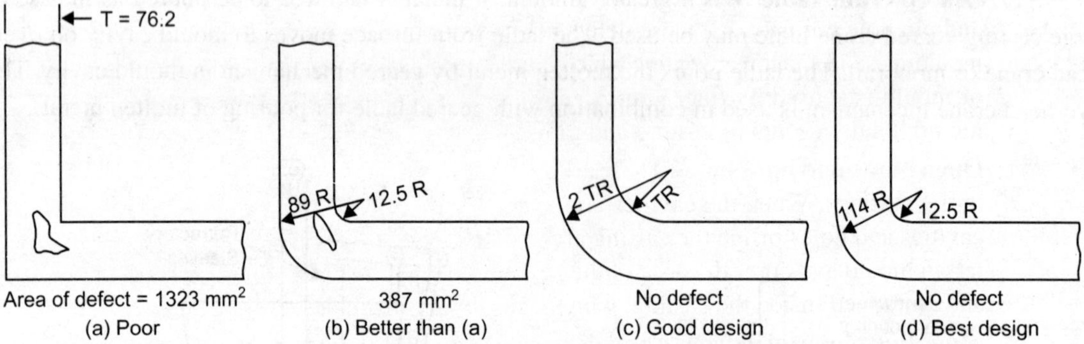

Fig. 9.50 Shrinkage Cavities. The figure shows that shrinkage cavities in L-section can be eliminated by using a fillet and reducing the wall thickness at corner. No defect is present when the radius of the fillet is equal to wall thickness.

(*b*) **Core shift.** Core shift occurs when the core is not properly located or supported. As a result cored surface is not properly located with respect to casting surface. (Fig. 9.51)

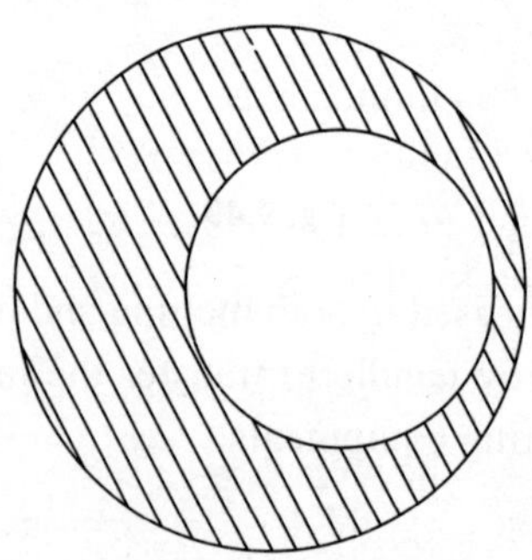

Fig. 9.51 Core shift

(*c*) **Mould shift.** A misalignment between two half of a mould or core may give rise to defective casting. (Fig. 9.52)

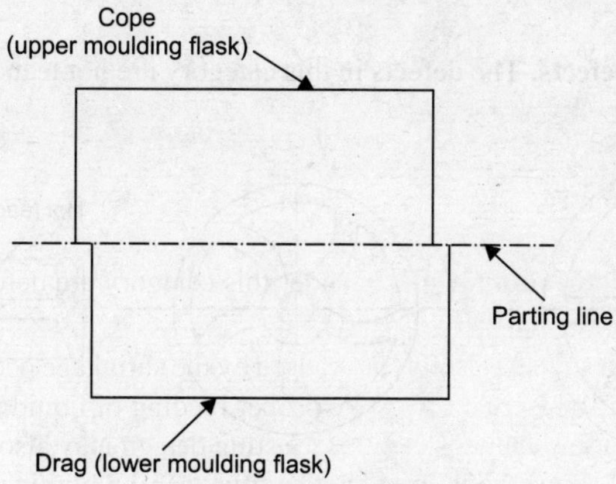

Fig. 9.52 Mould shift

(*d*) **Gas defects.** Defects in this category may be classified as blow, and open blows, air inclusions and pin hole porosity. All these defects are caused largely because of lower permeability of moulding sand and lower venting which does not allow the escape of gases. Lower permeability of mould is caused by finer grain size of sand, higher moisture and excessive ramming of mould. Faulty design of casting and moulding also causes these defects.

Open blows and blow holes. These are smooth walled rounded, flattered or elongated cavities in the castings. When the cavities are on the surface of casting, they are called open blows cavities and holes inside the casting are known as blow holes. These are caused by moisture left in mould and core. Because of the heat of molten metal moisture is converted into steam and entrapped inside the casting, which ends up as blow holes or open blows.

2. **Moulding material defects.** The defect under this category may be because of characteristics of moulding sand like refractoriness, strength of moulding sand, permeability, silica grain size or due to improper ramming.

(*a*) **Metal penetration.** This defect occurs when molten metal enters the gap between sand grains. As a result some sand will tightly adhere to casting surface and will produce a casting with rough surface. The main reason for this defect is that mould has too large sand grains. This could also be caused by too high pouring temperature and insufficient facing sand on mould cavity surfaces.

(*b*) **Run out.** Run out occurs when molten metal leaks out of mould during pouring. The causes are faulty mould making and faulty moulding flask.

(*c*) **Fusion.** It occurs when sand grains are fused with molten metal resulting rough glassy appearance. The main reason for this defect is that moulding sand is of lower refractoriness or pouring temperature is high.

3. **Pouring Metal Defects**

(*a*) **Misrun and cold shuts.** When the molten metal is poured to produce a casting it should first flow into all regions of cavity and then freeze into shape of mould. But when thickness of casting is too small or pouring metal temperature is too low the metal begins to freeze before it has filled the mould completely thus leaving unfilled cavities. This defect is called misrun. Cold shut occurs when two streams of metal which are too cold meet in mould cavity and do not fuse together. Too low fluidity of molten metal is also one of the main reasons to cause this defect.

4. **Metallurgical Defects.** The defects in this category are hot tears and hot spots.

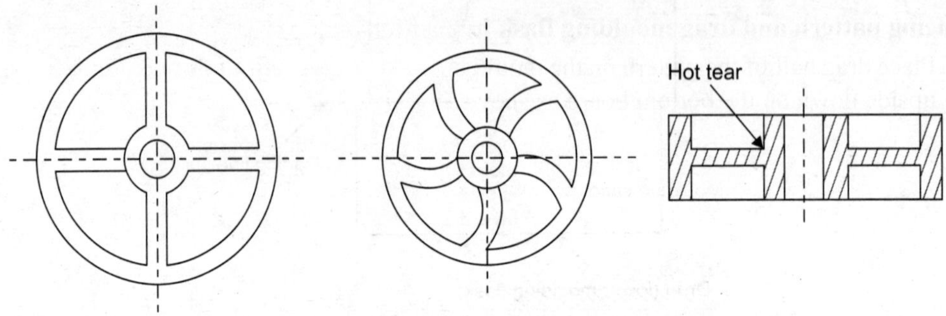

Fig. 9.53 Reducing hot tear by making curved spokes in wheels.

(*a*) **Hot tears.** Immediately after solidification some metals have low strength. Solid shrinkage of casting may cause the sufficiently high stresses at this time and casting may develop fracture. The main reason for this defect is faulty design of casting. The effect may be reduced by changing the design of casting. For example, in the hand wheels, where rims are connected to the hub through spokes hot tears are likely to occur as shown in Fig. 9.54. It is possible to reduce the formation of hot tears by making spokes curved and having them in odd numbers.

(*b*) **Hot Spots.** Hot spots are caused by the chilling effects of moulding sand. It occurs only with certain metals such as grey cast iron with insufficient silicon. Such metal may become hardened by chilling effect of moulding sand. Hot spot will cause difficulties if the castings are to be machined. Proper alloy composition and chilling practices are essential for eliminating the hot spots.

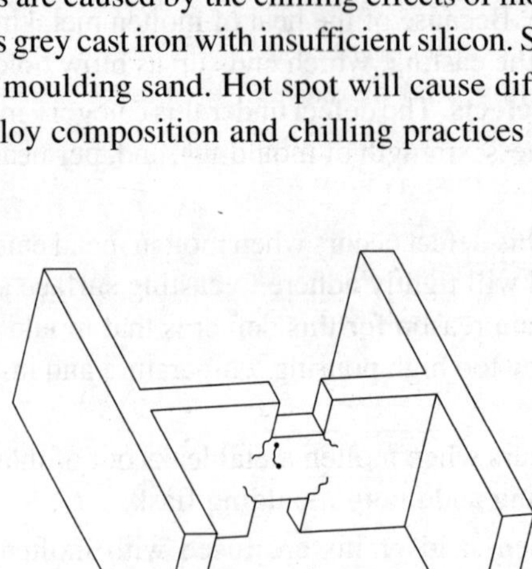

Fig. 9.54 If the reduced section were made the same size as the remaining casting, hot tear might not occur

9.16 FOUNDRY SHOP EXERCISE

Aim. To cast a aluminium wheel shape component.

Equipment used. Riddle, rammer, shovel, smoother, strike off bar, slick, draw screw, vent wire, pattern, moulding flask, crucible furnace, raw aluminium and shank ladle.

Procedure:

Placing pattern and drag moulding flask in position

1. Place drag half of the pattern on the mould board. The drag half of the moulding flask is kept upside down on the bottom board as shown in figure 9.55.

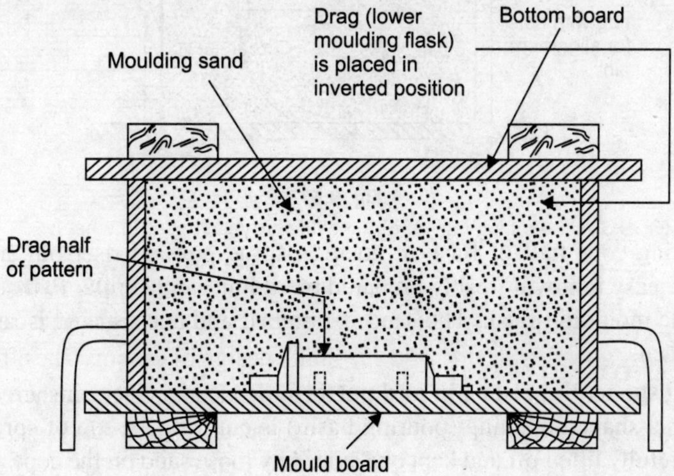

Fig. 9.55

2. Sprinkle dry facing sand on the pattern to avoid sand from sticking to pattern. Now pour freshly prepared moulding sand of required quantity on it and ram the sand with help of rammers. After ramming is over, excess sand in the flask is removed by using strike off bar.
3. The finish drag moulding flask is now rolled over to the bottom board and mould board is removed exposing the sand surface and drag half of the pattern. The exposed surface of sand may be smoothened if required.

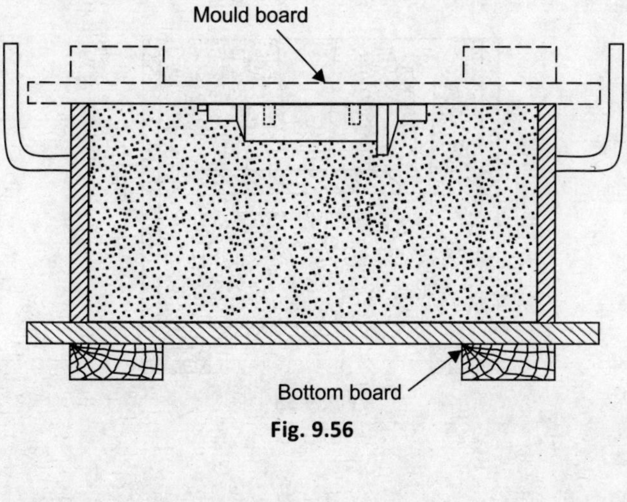

Fig. 9.56

4. Now place cope half of the pattern on the drag half of pattern with help of dowel pins and also place cope half of flask on drag half with help of alignment pin, dry parting sand is sprinkled also over the drag flask.

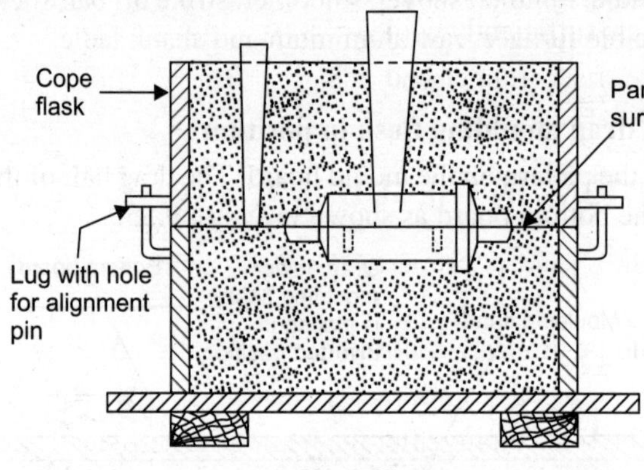

Fig. 9.57

5. After placing cope half of flask in position sprue pin and riser pin are located. They are tapered for easy removal freshly prepared moulding sand similar to that of drag is sprinkled all over the moulding flask. The sand is rammed and excess sand is removed with help of strike off pen.
6. **Removing sprue pin, riser pin and pattern.** Sprue pin, riser pins are carefully withdrawn and a funnel shaped opening (pouring basin) is cut near the top of sprue. The cope half of flask is carefully lifted off and kept one side. Any loose sand on the cope and drag is blown off with help of bellows. The cope half and drag half of the pattern is removal using draw screws.
7. **Making of runner and gate.** Runner and gate are made by cutting small passage between cavity (made by pattern) and sprue opening in the purting plane. Facing sand is applied all over the mould cavity and runner.
8. **Setting the core in mould.** Core is now carefully placed in the mould cavity. Core is prepared by ramming core sand in core box, core rests in core prints.
9. **Closing the mould.** Cope is now lifted and placed on the drag moulding flask taking care of alignment pins. Mould is now ready for pouring as shown in figure.

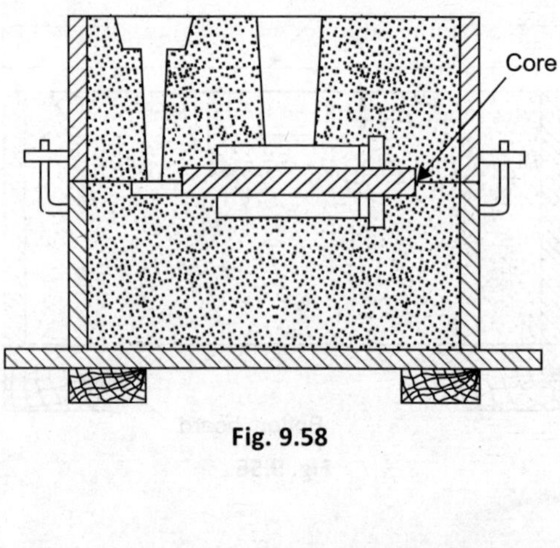

Fig. 9.58

10. **Melting and pouring.** Melt the raw aluminium in crucible furnace and then transfer the metal from furnace to mould cavity with the help of shank ladle handle which is operated by two men.
11. **Solidification and mould breaking.** Allow the metal to solidify in mould. Break the mould only after metal gets sufficiently cooled.
12. **Fettling.** The gate, riser, runner and adhering sand should be properly removed without damaging the casting.

9.17 SAFETY PRECAUTIONS IN THE FOUNDRY SHOP

—Ensure that moulding sand is properly rammed otherwise mould may produce defective casting.

—The moisture should be properly controlled otherwise mould may have less strength.

—The transfer of molten metal from furnace to mould should be done taking utmost care.

—The mould should be broken only after the metal gets sufficiently cooled.

—The gate, riser, runner and adhering sand particles should be properly removed without damaging the casting.

MULTIPLE CHOICE QUESTIONS

1. The pattern is made of
(*a*) wood (*b*) metal (*c*) plastic (*d*) all of above

2. The duplicate of final casting with some modifications is known as
(*a*) pattern (*b*) core (*c*) core print (*d*) chills

3. To make holes (or hollow cavities) in casting, which of following is used
(*a*) core (*b*) chills (*c*) pattern (*d*) chaplets

4. The lower part of moulding flask is
(*a*) cope (*b*) drag (*c*) cheek (*d*) core

5. Upper part of moulding flask is
(*a*) drag (*b*) cope (*c*) cheek (*d*) chaplet

6. Metal casting processes are processes in which
(*a*) removal of metal takes place
(*b*) metal are shaped by forces applied through various dies and tools
(*c*) metals are joined to achieve shape
(*d*) metals are shaped by moulding, melting, pouring and removal of solidified object from mould.

7. The chaplets are used in mould to
(*a*) support the moulding flask (*b*) support the core in moulding flask
(*c*) provide efficient venting (*d*) join cope and drag

8. Chill are used in moulding flask to
(*a*) provide directional solidification (*b*) support the core
(*c*) to provide efficient ventling (*d*) to support the pattern

9. In sand casting processes sand mould can be
(*a*) used once only (*b*) used many times
(*c*) used two to three times (*d*) none of above

10. The function of riser is
(*a*) to compensate for liquid shrinkage of metal
(*b*) to allow the escape of gases during pouring
(*c*) to feed molten metal from pouring basin to gate
(*d*) both (*a*) and (*b*)

11. The function of core print is to
(*a*) make holes in the casting
(*b*) to locate the core in moulding flask
(*c*) to join upper and lower moulding flask
(*d*) none of above

12. The purpose of sprue is to
(*a*) act as reservoir of molten metal
(*b*) feed molten metal from pouring basin to runner
(*c*) achieve directional solidification
(*d*) none of above

13. The element generally located in horizontal plane (parting plane) which connects sprue to gate is
(*a*) riser (*b*) runner (*c*) pouring basic (*d*) sprue

14. The vertical portion of gating system which connects pouring basin to runner is
(*a*) riser (*b*) runner (*c*) sprue (*d*) gate

15. The actual entry point from where liquid metal enters the mould cavity is
(*a*) riser (*b*) runner (*c*) sprue (*d*) gate

16. While considering shrinkage allowance the dimensions of pattern are
(*a*) larger than final dimensions of casting
(*b*) smaller than final dimensions of castings
(*c*) equal to the final dimensions of castings
(*d*) either larger or smaller than final dimensions of casting

17. Green sand means
(*a*) the colour of sand is green
(*b*) the sand contains moisture while pouring liquid metal
(*c*) the sand has green colour additives
(*d*) the sand is used after drying the box

18. The pattern allowance which facilitate the withdrawal of pattern from mould is known as
(*a*) shrinkage allowance
(*b*) machining allowance
(*c*) draft allowance
(*d*) distortion allowance

19. Cores are made up of
(*a*) sand (*b*) wood (*c*) plastic (*d*) rubber

20. Which of following is a element of gating system
(*a*) pouring basin (*b*) sprue (*c*) runner (*d*) all the above

21. Cupola furnace is used for melting
(*a*) cast iron (*b*) steel (*c*) non-ferrous metal (*d*) all the above

22. Permeability of moulding sand refers to
(*a*) ability to allow gases to escape through the mould
(*b*) ability to withstand high temperature of molten metal without fusing
(*c*) ability of sand due to which sand grains bind together
(*d*) the property of sand due to which sand mould automatically collapses after solidification of castings

23. Refractoriness of moulding sand refers to
(*a*) ability to allow gases to escape through the mould
(*b*) ability to withstand high temperature of molten metal without fusing
(*c*) ability of sand due to which sand grains bind together
(*d*) the property of sand due to which sand mould automatically collapses after solidification of casting

24. The maximum amount of water in moulding sand is
(*a*) 2% (*b*) 8% (*c*) 5% (*d*) 12%

25. Fettling process in sand casting is
(*a*) solidification of casting after pouring molten metal
(*b*) method of pouring molten metal
(*c*) process of cleaning and finishing casting after removing from moulding
(*d*) process of removing solidified casting from flask moulding flask

26. When pouring metal temperature is too low and metal begins to solidify before it has filled mould cavity is known as
(*a*) mis run (*b*) hot liars (*c*) cold shut (*d*) hot spot

27. Blow holes and air inclusions are caused due to
(*a*) lack of fluidity of molten metal
(*b*) lack of refractoryness of moulding sand
(*c*) poor permeability of moulding sand
(*d*) hydrostatic pressure on moulding sand

28. Hot tears and hot spots are
(*a*) pouring metal defect (*b*) gas defects
(*c*) metallurgical defect (*d*) none of above

29. Investment castings are used for
(*a*) large size casting
(*b*) very large size castings
(*c*) small size casting
(*d*) small size casting having complicated shapes

30. In centrifugal casting impurities are
(*a*) collected in the outside surface of casting (*b*) collected in the inside of casting
(*c*) collected at the centre of castings (*d*) uniformly distributed in castings

31. The characteristics of permanent mould castings are
(*a*) better dimensional accuracy (*b*) better surface finish
(*c*) complex shapes can be easily produced (*d*) all of above

32. Cold chamber method of permanent die castings is used for
(*a*) metals having low melting point (*b*) metal having high melting point
(*c*) casting of plastics (*d*) casting of ceramics

VIVA VOCE QUESTIONS

1. What are advantages of foundry process over other manufacturing processes?
2. Name five applications of castings in industries.
3. Name five materials suited for casting process.

4. Name basic steps in casting process.
5. What is the basic difference between core and pattern?
6. Name five type of patterns.
7. Name five pattern allowances.
8. Name the materials used for making patterns and cores.
9. What do you mean by elements of gating system?
10. What is the difference between runner and riser?
11. What is the function of riser?
12. Name various types of gates used in casting process.
13. Name one ingredient of moulding sand.
14. Name five desirable characteristics of moulding sand.
15. Name various mould methods.
16. Name various furnaces used for melting ferrous metals.
17. What is meant by fettling of casting?
18. Name various casting defects.
19. Explain the function of core print, chills and chaplets.
20. Name the tools used for making the mould.
21. What is meant by sand preparation.
22. Name the tools and equipment used for pouring.
23. Name the tools and equipment used for fettling of the casting.
24. Explain the function of slicks.
25. Why at all draw screws and draw spikes are used?
26. Explain the function of shovel.
27. Explain the function of strike off bar.

"Success often comes to those who dare and act; it seldom goes to timid who are ever afraid of the consequences."

Chapter **10**

METAL CUTTING PROCESSES (MACHINE SHOP)

INSIDE THIS CHAPTER

10.1 INTRODUCTION TO MACHINE SHOP

Machine shop is a place where material removal processes are carried out. A machine which performs the material removal processes is known as machine tool.

Machine tools are non-portable power driven machines which hold, support and guide the workpiece and cutting tool during the metal cutting process. The relative motion between cutting tool and workpiece causes cutting operation which generates the part of desired shape, size and finish.

Two basic movements which are commonly used in machine tools are circular and straight line. Circular movements are basically rotary motion and, straight line motion are obtained by reciprocating action. The machine tools which use rotary and linear motion are lathe, drilling machines milling machineate. Two basic machine tools which use straight line reciprocating motion to move cutting tool in relation to workpiece are shaper and planner.

"Go confidently in the direction of your dreams. Live the life you have imagined."

10.2 CLASSIFICATION OF MACHINE TOOLS

Machine tools may be classified in number of ways.

1. Based on the production capabilities and application.
2. Based on type of surface generated.

1. **Based on the Production Capability and Application**

(*a*) **Single purpose machine tools.** These are the machine tools designed specifically for performing single operation on the jobs. These are having highest amount of automation and are used in mass production. These machine tools are used specifically for one product and are least flexible. Examples are transfer lines meant for complete machining of a product.

(*b*) **Production machine tools.** In these machines number of functions are automated and operator skill required to produce the components is reduced. A general purpose machine may also be converted in production machine by utilizing jigs and fixtures. Some examples of production machine tools are capstan and turret lathe.

In these machines idle time is reduced and machine utilization is increased. As the cost of tooling design is higher and setting time for a given job is more, production machine tools are justified only for large volume of production.

(*c*) **Special purpose machine tools.** These machine tools are specifically designed for the given application. The setting operation in these machines is practically eliminated and complete automation is achieved. This greatly reduces the manufacturing time and manufacturing cost. These machine tools are expensive and used in mass production. The disadvantage of these machines is that they are restrictive in their application capabilities. Examples of special purpose machine tools are cam shaft grinding machine and piston turning lathe.

(*d*) **General purpose machine tools (Basic Machine tools).** As the name suggests general purpose machine tools are designed to perform a variety of operations on a wide range of components. These machine tools are not suitable for large production as setting time for any given operation is large. Thus machine utilization is poor. Skilled workers are required to run general purpose machine tools. Their utility is in the job production. Examples are lathe, shaper and milling machine tools.

2. **Classification Based on Surfaces Generated**

(*a*) **Machine tools producing flat surfaces.** Shaper, slotter planner, milling machine, broaching machine.

(*b*) **Machine tools producing revolved surfaces**

(*i*) Lathe machine

(*ii*) Drilling machine

(*iii*) Boring machine

(*c*) **Miscellaneous machine tools**

(*i*) Gear cutting machine tools

(*ii*) Thread milling machine tools

(*iii*) Sawing and filing machine tools.

10.3 BASIC CHIP FORMATION PROCESSES

The basic chip formation processes are turning, shaping, planing, milling, drilling, boring, sawing, broaching and grinding. Some of these processes are discussed in detail in coming articles. Let's see the basic movement of cutting tools and workpiece in these processes.

1. **In Lathe Machine.** Cutting movement is provided by rotating workpiece and feed movement is given by tool and carriage.

2. **In shaper.** Cutting movement is provided by reciprocating tool and feed movement is given by work table.

3. **In planner.** Cutting movement is provided by reciprocating work table and feed movement is given by cutting tool.

4. **In horizontal milling machine.** Cutting movement is provided by rotating tool and feed movement is given by work table.

5. **In drilling machine.** Cutting movement is provided by rotating cutting tool and feed movement is also given by cutting tool. Workpiece is stationary and clamped on work table.

6. **In horizontal boring machine.** Tool rotates and feed movement is also given by cutting tool.

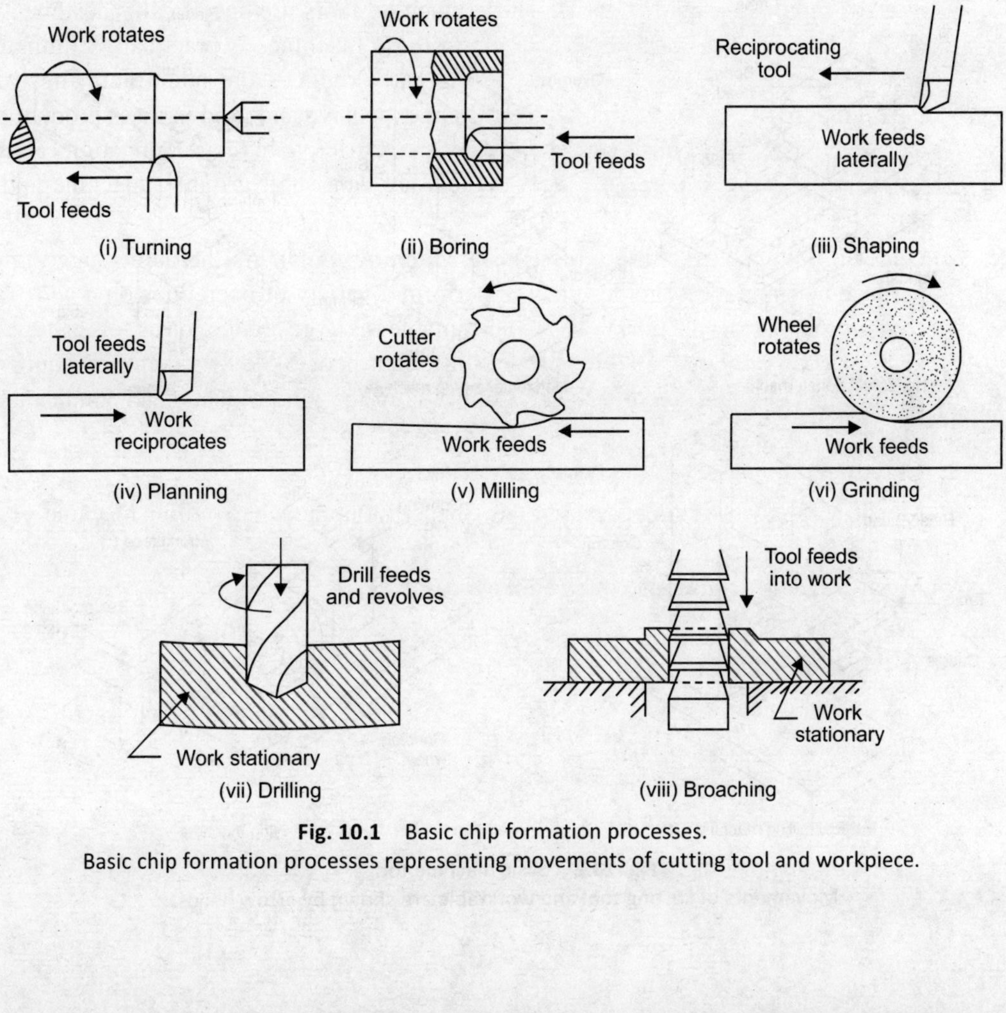

Fig. 10.1 Basic chip formation processes.
Basic chip formation processes representing movements of cutting tool and workpiece.

7. **In cylindrical grinding.** Cutting movement is provided by rotating tool (grinding wheel) and feed movement is given by either tool or table or both may be used.

8. **In broaching machine.** Cutting movement is provided by tool and feed movement is also provided by cutting tool.

9. **In sawing machine.** Cutting movement is provided by cutting tool and feed may be provided by tool or workpiece or both may be used.

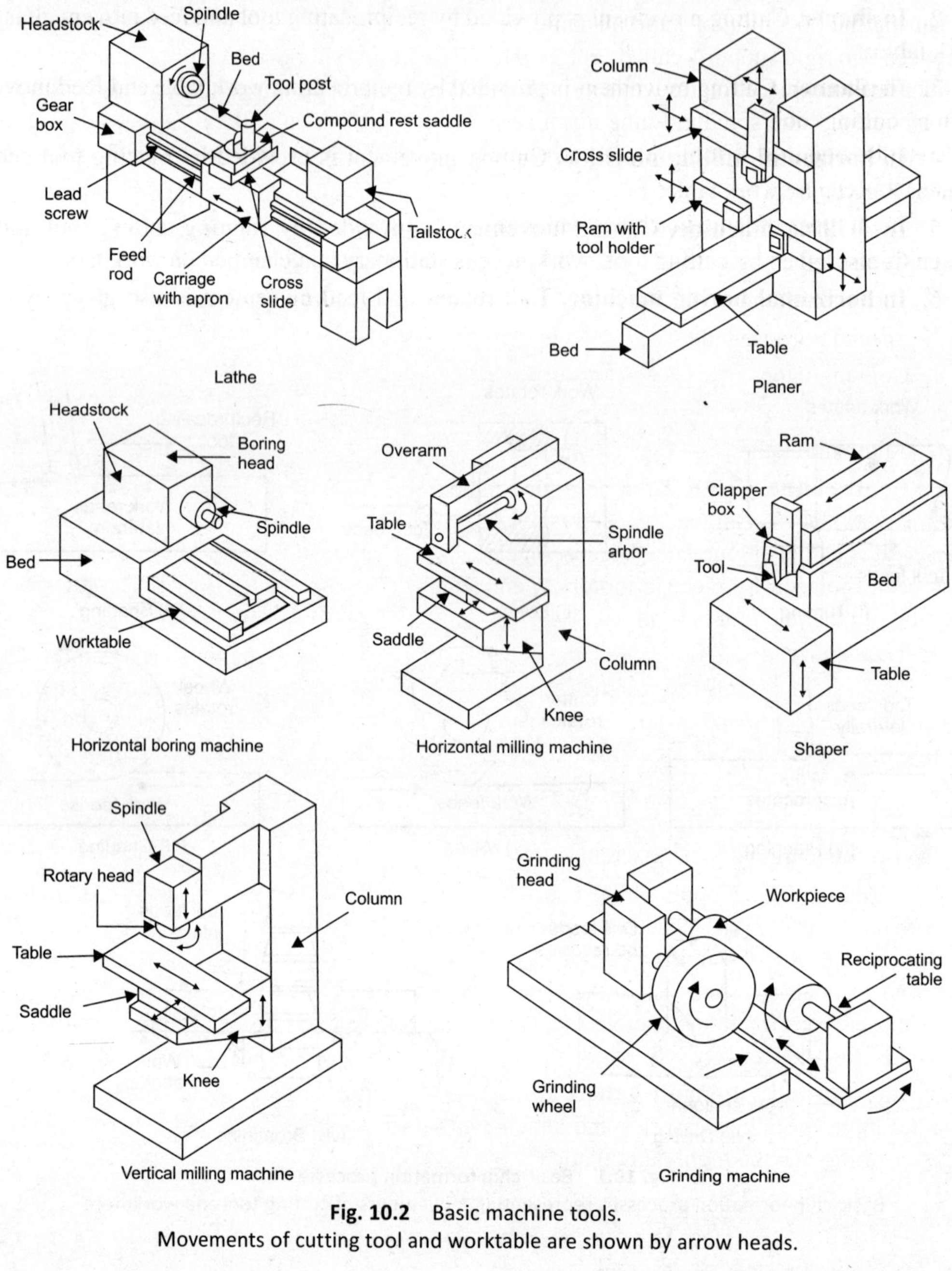

Fig. 10.2 Basic machine tools.
Movements of cutting tool and worktable are shown by arrow heads.

10.4 LATHE MACHINE

Lathe is the oldest machine tool invented. It is most fundamental and versatile machine tool with large number of uses in all manufacturing shops. As number of operations can be performed on a lathe, substitution of lathe machine may require a number of other machine tools to perform these operations on a single workpiece. Therefore, the utility and significance of lathe even in today's scenario can be easily recognised.

During the operation of lathe machine the workpiece are held in lathe chuck, or face plate, or held between two rigid supports called centres. The cutting tool is held in tool post which is mounted m-cross slide. The workpiece rotates while a single point cutting tool advances in a direction parallel to the axis of rotation of lathe spindle. Lathe machine majorly produces cylindrical surfaces. However, flat surfaces can be obtained if tool moves perpendicular to the axis of rotation of spindle. (operation is termed as facing)

Different types of lathe machines which are used in manufacturing shops are:

1. Centre lathe
2. Tool room lathe
3. Special purpose lathe
4. Copying lathe
5. Capstan lathe
6. Automatic lathe.

1. **Centre lathe.** Centre lathe or engine lathe is most common lathe. It is called centre lathe because the workpiece is clamped by centres in a lathe (there are also number of other ways to hold the workpiece). It is called engine lathe because in earlier days lathe were driven by steam engine. Which now has been replaced by electric motors. Centre lathe is used for more general applications and construction of machine tool is rigid.

2. **Tool room lathe.** Tool room lathe is basically meant for making tools, test gauges, dies etc. The accuracy desired in tool room lathe is much higher as compared to centre lathe. The machine handles large range of size and materials. Tool room lathe also has higher range of speed and feed. It has large range of accessories and attachments and is more expensive than comparable sizes of engine lathe.

3. **Special purpose lathe.** Special purpose lathe is used for special applications which can not be handled by conventional lathe machines.

4. **Copying lathe.** Copying lathe is often an centre lathe with copying attachment mounted on it. The copying lathe may be needed to machine complex contours. In copying lathe a template or master piece is used which is similar to the shape of workpiece to be machined. The size of the masterpiece may vary but geometrical shape is similar. The movement of cutting tool is driven by a stylus which traces the master piece for actual contour to be produced. So the shape of the contour of master piece is copied on the workpiece.

5. **Capstan and Automatic lathe.** Capstan and automatic lathes are used in the mass production. These lathes have special accessories and attachments which help them in increasing the production rate. The operation may be highly automatic.

10.4.1 Constructional Details of Lathe Machine

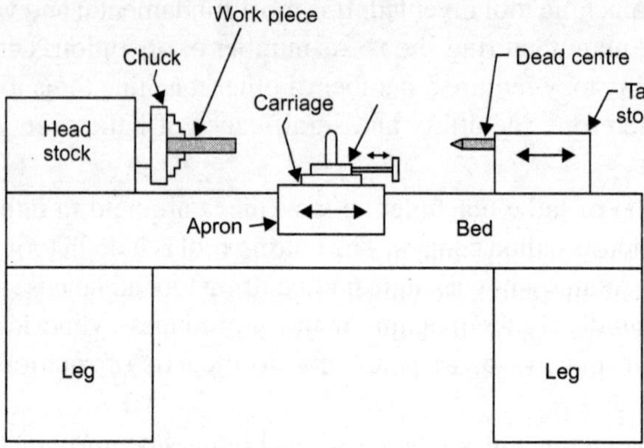

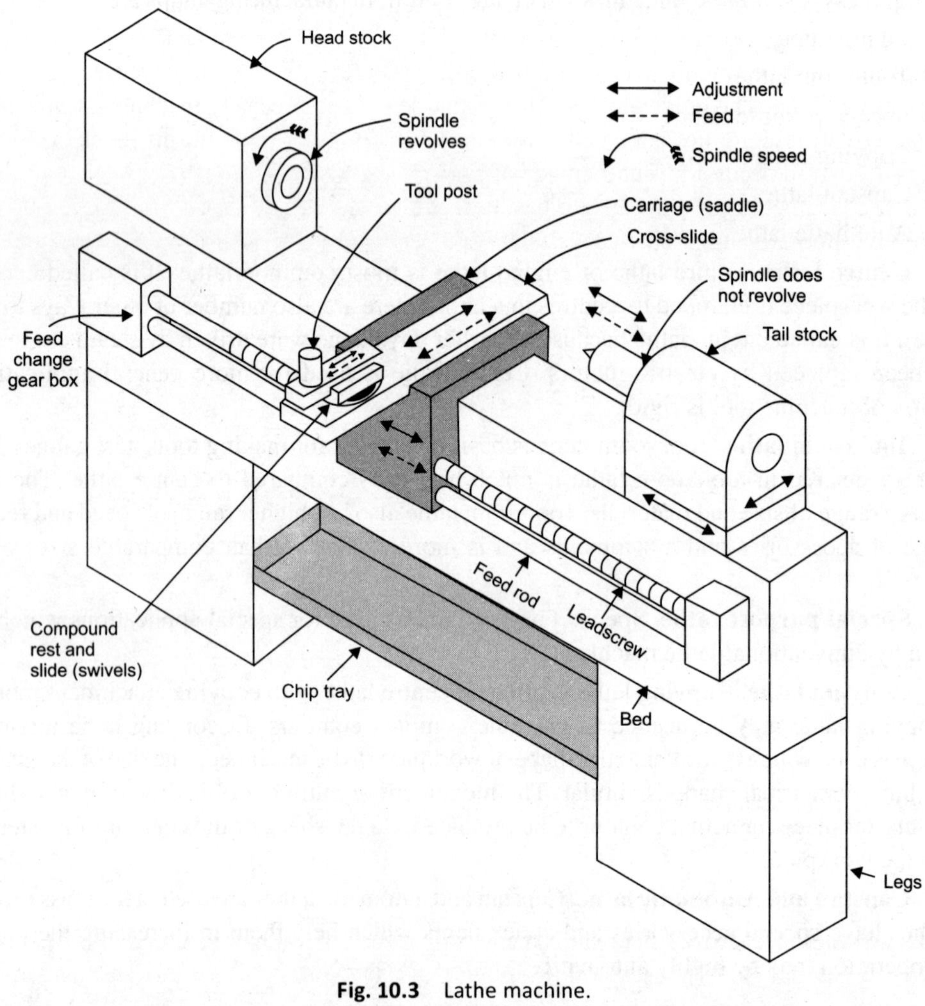

Fig. 10.3 Lathe machine.

The lathe has following main parts:

1. Bed
2. Head stock
3. Tail stock
4. Carriage
5. Feed mechanism
6. Legs.

1. **Bed.** Bed is the base of lathe machine on which different parts of lathe like, head stock, tailstock, carriage assembly are mounted. Bed must be rigid as it has to withstand various cutting forces transmitted to it during the operation of the lathe. It is usually made of single piece casting of grey or modular cast iron. Beds are usually made of gray or duetile cast iron which are well mormalised.

Guideways are provided for a well guided and controlled movement of operating parts like carriage and tail stock. Guideways may be of flat, inverted V or, prismatic shape. Inverted 'V' shape guideways and prismatic ways are preferred over the flat guideways as their shape totally prevents the entry of chips and dust between carriage and bed and thus prevents the contact surfaces from being spoiled due to scratching. The guideways are surface hardened and precisely machined. Care should to taken to assure that these are not damaged. Any inaccuracy in guideways means the accuracy of the entire lathe is destroyed. Beds are usually made of gray or ductile cast iron which are well normalised. The material choosen is cast iron because it has high damping capacity (ability to absorb vibrations), high compressive strength and good wear resistance, which is quite desirable in machine tool beds.

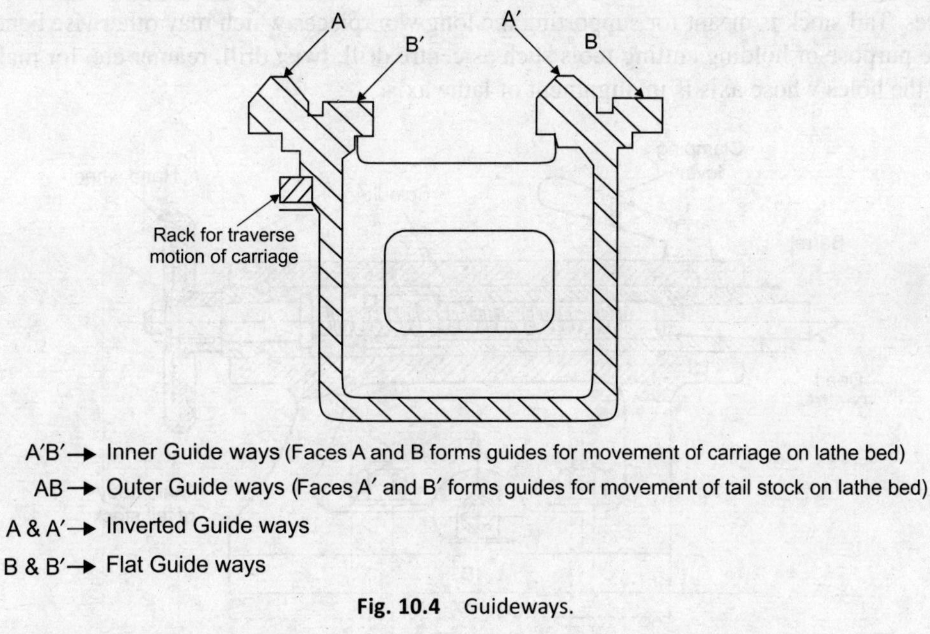

Fig. 10.4 Guideways.

2. **Head stock.** Head stock is towards the left most end of bed. It is mounted in a fixed position on the inner ways. Head stock houses hollow spindle and cone pulley or set of transmission gear or combination of both. Through gear box and/or pulley spindle can be rotated at number of different speeds. Different spindle speeds are required to perform various operations.

- Spindle protudes from gear box and contains means for fastening the workholding devices like chuck, face plate, dog plates, line centre etc. Spindle nose is threaded so that face plate or chuck can be mounted on it to hold and rotate the workpiece. Hollow spindle is tapered at the nose end to receive the live centre having a standard morse taper.
- Spindle is mounted on accurate bearings (which are generally roller or ball bearing) and axis of rotation of spindle is accurately parallel to the bed ways. The spindle is sufficiently rigid to provide accurate motion and maintains perfect alignment with lathe axis.
- Spindle is hollow through out its length through which long bar stock can be fed. The size of spindle hole is an important diameter of lathe machine as it determines the maximum size of bar stock that can be machined when the material must be fed through the spindle.
- Power is supplied to spindle from an electric motor to V belt or silent chain drive.

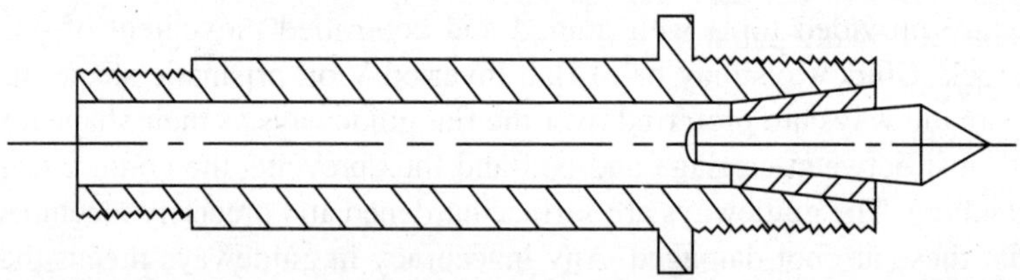

Fig. 10.5 Head stock spindle.

3. **Tail Stock.** Tail stock is towards the right most side of bed. It is a non rotating part which slides on the guide ways and can be clamped to it in any position to accommodate different lengths of workpieces. Tail stock is meant for supporting the long workpieces which may otherwise bend. It also serves the purpose of holding cutting tools such as centre drill, twist drill, reamer etc. for making and finishing the holes whose axis is in alignment of lathe axis.

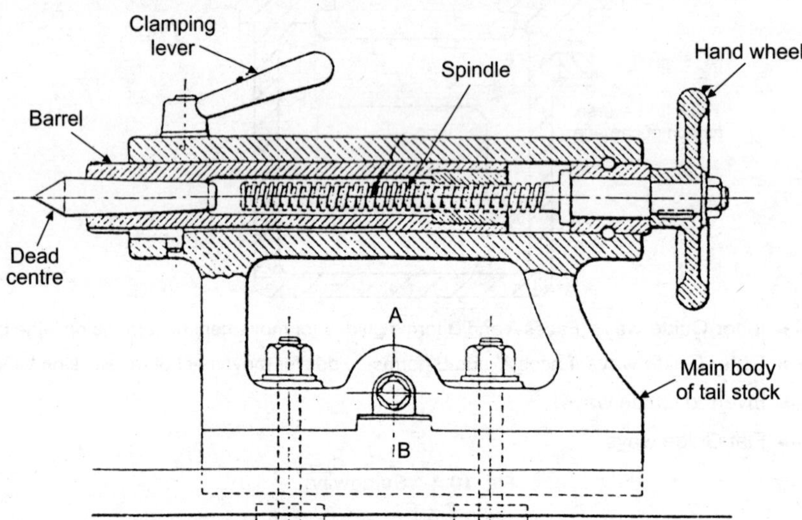

Fig. 10.6 Tailstock.

Constructional details. Tail stock consists of a cast iron body. The base of body is provided with feet which fit in parallel bed ways of lathe.

- The upper cylindrical part of body is hollow and is machined to receive a barrel. Barrel is hollow through out its entire length. One end of hollow barrel is provided with morse taper. Lathe centre (which is strictly known as dead centre) is held in the barrel which supports the workpieces during cutting operations. Cutting tools like drills, reamer are also held in this taper hole.
- Other end of barrel is provided with internal threads in which spindle works. The tail stock barrel is given longitudinal movement with respect to tail stock body with the help of spindle and handwheel.

4. **Carriage.** Carriage like tail stock slides on the ways of bed. It provides necessary motion (which is parallel to the axis of rotation of spindle) to the cutting tool to generate the specific surface. Carriage houses has following parts:

(*a*) Cross slide
(*b*) Compound rest
(*c*) Tool post
(*d*) Apron.

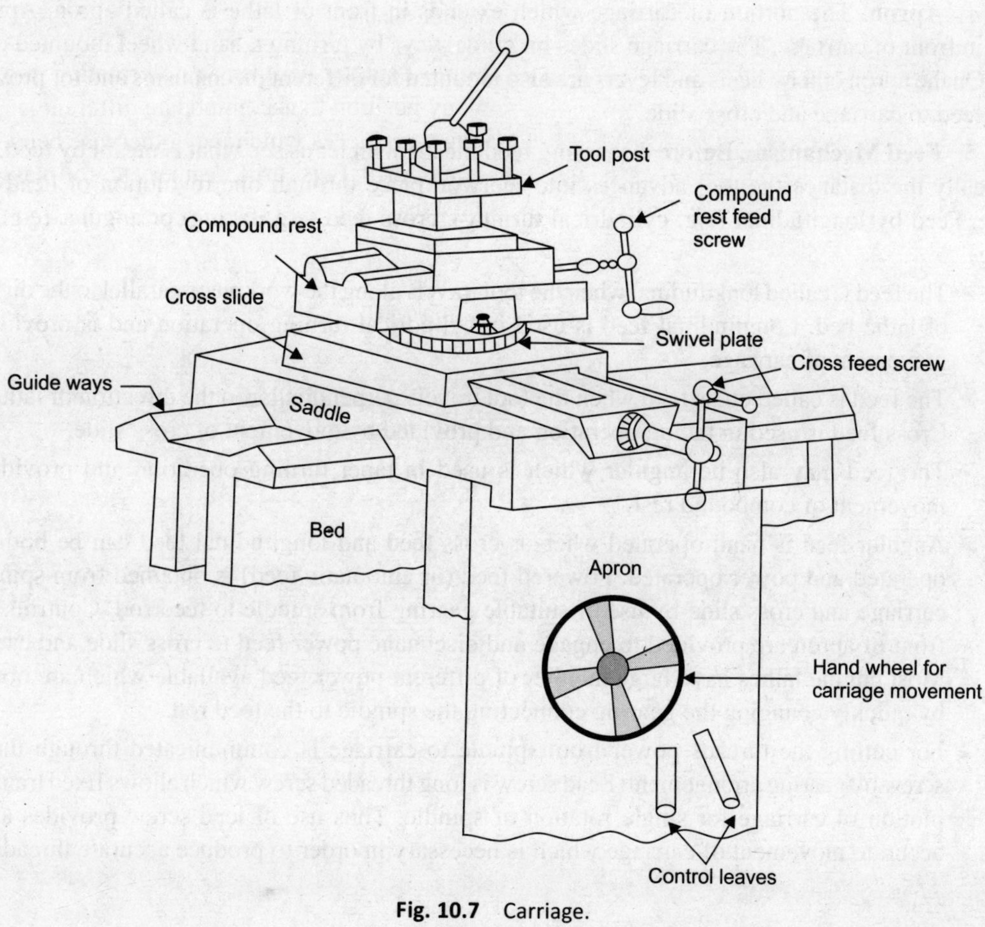

Fig. 10.7 Carriage.

(*a*) **Cross slide.** Cross slide is mounted upon the carriage and its movement is perpendicular to the rotational axis of spindle. This movement is known as cross feed and is used for producing flat surfaces which are perpendicular to the lathe axis. This movement is used in performing facing operations on lathe machines. Cross slide can be moved by means of cross feed screw, which is controlled by a small hand wheel and a graduated dial. Both power feed and manual feed is provided to cross slide.

(*b*) **Compound rest.** Compound rest is another slide which is mounted on cross-slide. Compound rest supports the tool post which holds the cutting tool. Compound rest can rotate and translate with respect to cross slide. Its base is graduated in degrees to indicate angular setting. Compound rest can be clamped to remain at any angular setting which allows the movement of the cutting tool along a path which is not parallel to lathe axis. This movement is used in taper turning operation or in turning of a tapered surface. Compound rest can be moved by means of a screw which is controlled by a small hand wheel and graduated dial. Power feed is not provided to compound rest.

(*c*) **Tool post.** Tool post is mounted above the compound rest. Tool post rigidly clamps the cutting tool in desired position. Fig 10.7 shows 4 way tool post. In a 4 way tool post 4 cutting tool may be mounted at a time. Various bolts are provided to tightly hold the cutting tool. 4 way tool post is useful while machining a number of different components (the desired tool can be brought in action). Every time when any tool is changed, packings may be added or substracted in order to set the tool at a correct height in line with axis of rotation of spindle.

(*d*) **Apron.** The portion of carriage which extends in front of lathe is called apron. Apron is bolted in front of carriage. The carriage slides on guide ways by turning a hand wheel mounted on the apron. On the apron other wheels and levers are also mounted for different mechanisms and for providing power feed to carriage and cross slide.

5. **Feed Mechanism.** Before discussing feed mechanism let us see what is meant by feed. Feed is basically the distance the tool advances into the workpiece through one revolution of head stock spindle. Feed by longitudinal (e.g. cylindrical turning), cross feed (e.g. facing) or angular (e.g. taper turning).

- The feed is called longitudinal when the tool travels along the workpiece parallel to the direction of lathe bed. Longitudinal feed is used in cylindrical turning operation and is provided by movement of carriage.
- The feed is called cross feed when the tool travels perpendicular to the direction of lathe bed. Cross feed is used in facing operation and provided by movement of cross slide.
- The feed may also be angular which is used in taper turning operation and provided by movement of compound rest.
- Angular feed is hand operated whereas cross feed and longitudinal feed can be both hand operated and power operated. Powered feed (or automatic feed) is obtained from spindle to carriage and cross slide by use of suitable gearing from spindle to feed rod. Controls at the front of apron are provided to engage and disengage power feed to cross slide and carriage. Most engine lathes have large number of different power feed available which are obtained by quickly changing the gearing connecting the spindle to the feed rod.
- For cutting the threads power from spindle to carriage is communicated through the lead screw by gearing arrangement. Lead screw is long threaded screw which allows fixed transverse motion of carriage for single rotation of spindle. Thus use of lead screw provides a more accurate movement of carriage which is necessary in order to produce accurate threads.

- Feed screw may also be used for automatic longitudinal feed of cutting tool in the lathe machines when a separate feed shaft is not provided for this function. If the lathe contains both feed shaft and lead screw, always feed rod is used for longitudinal movement of tool because by doing this wear of lead screw is minimised and accuracy is maintained.

6. **Legs.** These are the supports which bear the entire weight of machine. These are the castings which may be integrated with lathe bed. Both the legs are firmly secured to floor by means of foundation bolts.

10.4.2 Specifications of Lathe Machine

Specifications means the important parameters required to completely describe the size of lathe machine and maximum size of workpiece lathe can handle. A number of parameters may be used to completely specify the lathe machine for specific application. However, the major parameters to specify a lathe machine are:

(*i*) Distance between centres (live and dead centre): This would be specifying the maximum length of workpiece that can be turned on a lathe.

(*ii*) Swing over the bed: It specifies the largest diameter of workpiece that can be rotated on the lathe machine without striking the guideways.

(*iii*) Swing over the carriage: It specifies the largest diameter of workpiece that can be turned on a lathe without striking the top of carriage. It is of course less than the swing over the bed.

(*iv*) Maximum diameter of workpiece through spindle: The maximum diameter of hole through the hollow spindle is specified to mention the maximum dia. of workpiece lathe can handle when a workpiece is fed through the spindle hole.

Above mentioned are the basic elements to specify the lathe machine. Following specifications should also be provided to completely specify a lathe machine.

(*v*) Length of bed

(*vi*) Width of bed

(*vii*) Power rating of electric motor

(*viii*) Number of spindle speeds

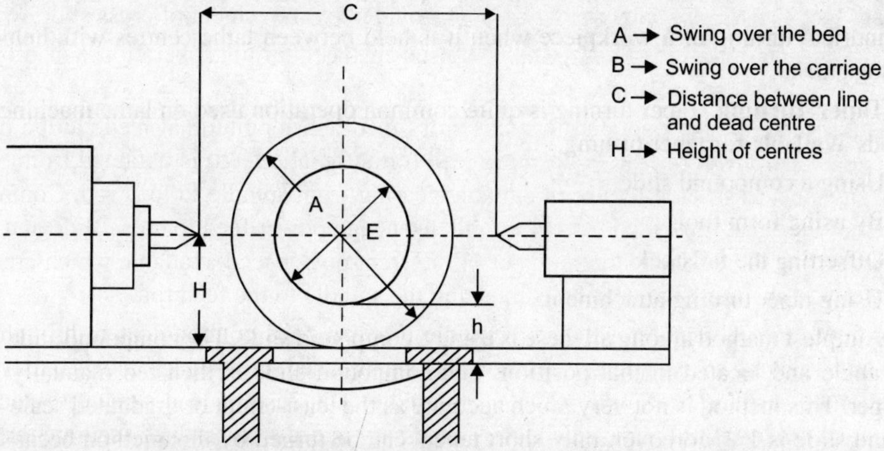

Fig. 10.8 Specifications of lathe machine.

(*ix*) Range of spindle speeds

(*x*) Lead screw diameter and number of threads per cm.

(*xi*) Accuracy achievable.

10.4.3 Operations Carried on Lathe Machine

1. Turning
2. Taper turning
3. Facing
4. Parting and grooving
5. Knurling
6. Drilling
7. Boring
8. Thread cutting.

1. **Turning.** Turning (or cylindrical turning) is most general operation used on lathe machine. In this operation external cylindrical surfaces are generated. The position of cutting tool is first adjusted by providing desired depth of cut using cross slide. Then as the workpiece rotates, tool is made to move in the direction parallel to lathe axis. (This movement of tool is called feed). It can be easily visualised that tool has the helical motion with respect to workpiece and always encounters the untouched layer of workpiece. During turning operation cutting tool continuously removes the material from its circumferencial surface in form of chips and hence turning results in reduced diameter and new cylindrical surface. If good surface finish and accurate size is desired one or more roughing cuts may be provided before finishing cut.

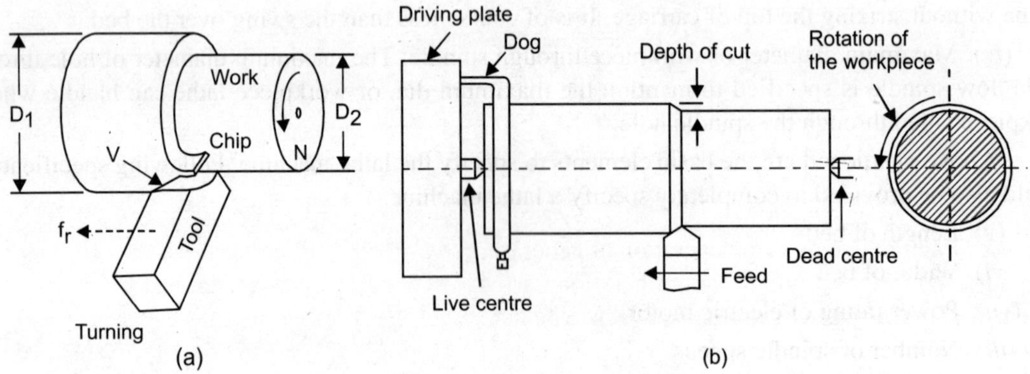

Fig. 10.9 Cylindrical turning operation.

Cylindrical turning of a workpiece when it is held between lathe centres with help of driving plate and dog carrier.

2. **Taper Turning.** Taper turning is quite common operation used on lathe machine. There are four methods available for taper turning.

1. Using a compound slide
2. By using form tools
3. Offsetting the tailstock
4. Using taper turning attachment.

The simplest method among all these is using a compound slide. The compound slide is swiveled to desired angle and located in that position. The compound slide is then fed manually to produce required taper. This method is not very much accurate as the least-count of graduated scale on the base of compound slide is 1°. Moreover, only short tapers can be turned by this method because length of travel of compound slide is quite limited.

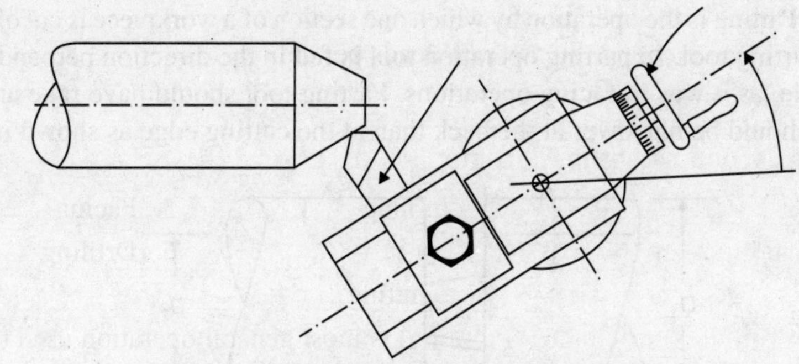

(a) Compound slide method for taper turning

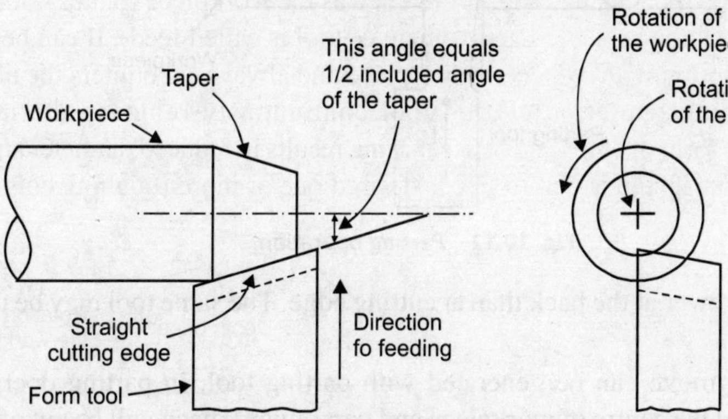

(b) Using form tool for taper turning

Fig. 10.10 Methods for taper turning.

3. **Facing.** Facing is the operation of generating flat surface perpendicular to the rotational axis of spindle. While performing facing operation carriage may be locked to lathe bed to prevent axial movement. Feed is given by cross slide. It is to be noted that in case of turning the depth of cut is given by cross slide whereas in case of facing feed is given by cross slide. Another difference is that in facing operation length of workpiece gets reduced whereas in turning diameter of workpiece gets reduced.

Arrow head represents the straight line motion of tool with respect to circular motion of workpiece. The workpiece is held in lathe chuck.

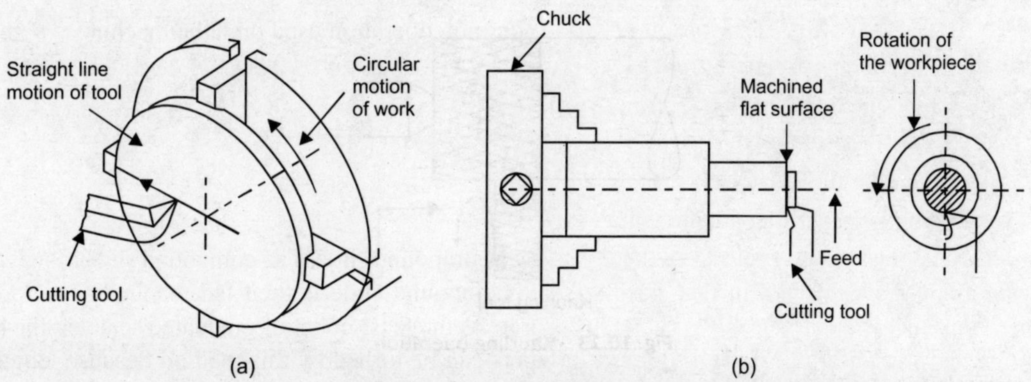

Fig. 10.11 Facing operation.

4. **Parting.** Parting is the operation by which one section of a workpiece is cut off from remaining workpiece with a parting tool. In parting operation tool is fed in the direction perpendicular to the axis of rotation of spindle, as it was in facing operations. Parting tool should have rake angle less than for same material and should be narrower at the back than at the cutting edge as shown in Fig. 10.12.

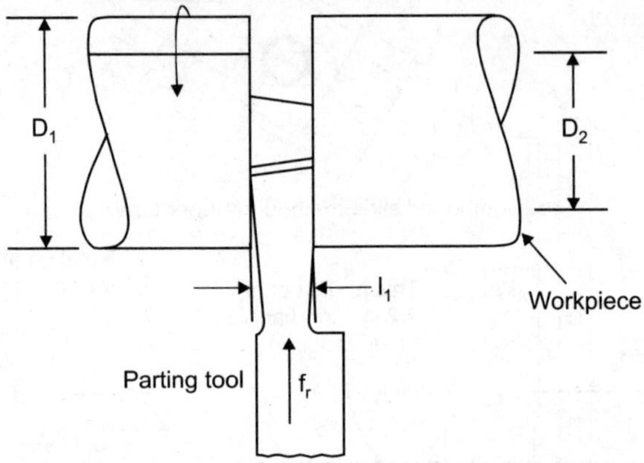

Fig. 10.12 Parting operation.

Parting tool is narrower at the back than at cutting edge. The same tool may be used for grooving operation.

5. **Grooving.** A groove can be generated with parting tool. In parting operation, during the feeding, tool goes beyond the centre of workpiece and part of workpiece will be cut off from remaining material. In case of grooving tool never reaches the centre of work piece and a rectangular groove would be obtained and this operation is known as grooving. It is also feasible to obtain specific groove shape by using form cutting tool.

6. **Knurling.** Knurling is a metal forming operation (cold working process) which is done on lathe. No material is removed in form of chips in this process. Knurling is basically done to make slightly raised diamond shaped-pattern on the cylindrical surface of workpieces. Surfaces are often knurled so that they may be gripped with fingers or by hand.

A knurling tool has two hardened steel rollers with serrations inclined at 45° to the horizontal. Both rollers are mounted on a floating head, which allows the rollers to exert equal pressure as they are pressed against a rotating workpiece.

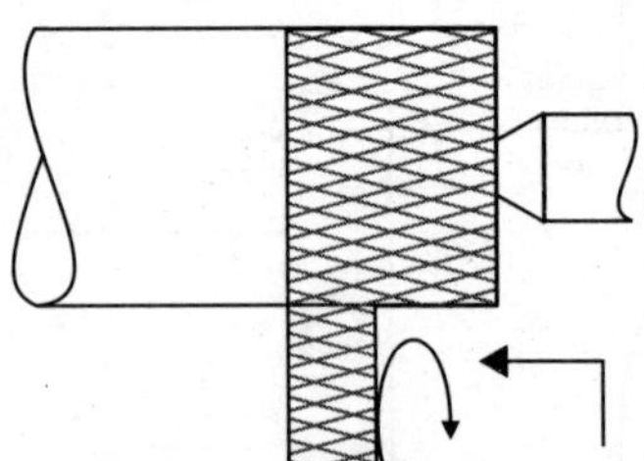

Fig. 10.13 Knurling operation.

7. **Drilling.** Drilling is the operation of making cylindrical holes into the solid material. Drill is mounted in the tapered hole of tailstock barrel. Drill is fed manually against the rotating workpiece by turning the hand wheel of tail stock. The drill should be located concentric with the axis of rotation of spindle. The operation is limited to holes which are through the axis of rotation of workpiece. Reamers, boring tool, counter bores and other cutting tools also may be held by tail stock barrel. These may be used after drilling operation.

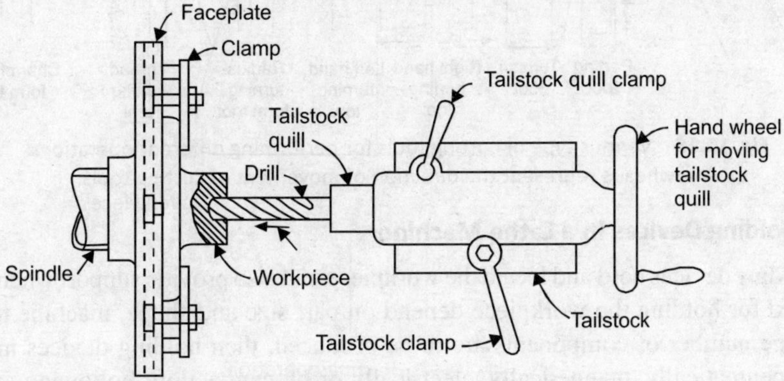

Fig. 10.14 Drilling operation on lathe machine.

8. **Boring.** Boring is always the operation of enlarging an already existing hole. Hole may be made by a drill or may be because of a core in a casting. One important purpose of boring is also to correct any eccentricity in the hole which may have resulted during drilling operation. Thus holes obtained after boring are more accurate and have good surface finish.

The boring operation is somewhat similar to turning operation and may be called as internal turning operation. The cutting forces experienced are somewhat higher as compared to external turning and tool is less rigid than turning tool. Therefore, the cutting parameters (depth of cut, feed, speed) used in boring operation are somewhat lower than those used for turning.

In boring operation work is held between lathe chuck or held on a face plate. Workpiece keeps on rotating and boring tools are fed parallel to the axis of rotation of workpiece. Boring tools may be held by tail stock barrel.

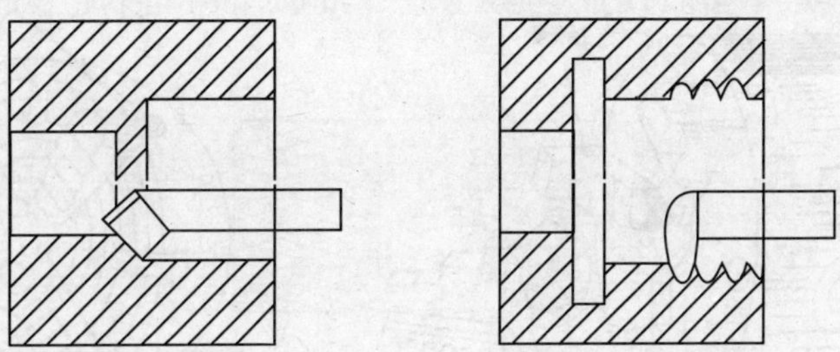

(a) Internal turning (b) Internal threading

Fig. 10.15

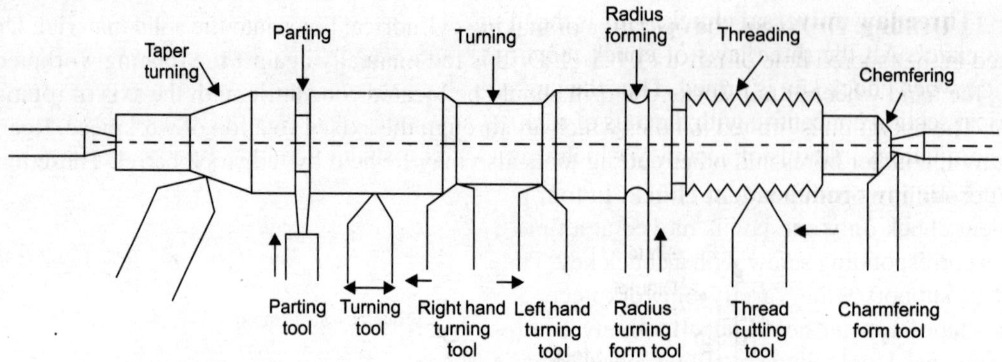

Fig. 10.16 Various type of cutting tools for performing different operations. Arrowheads represent the direction of movement of cutting tools.

10.4.4 Work Holding Devices in a Lathe Machine

Work holding devices hold and locate the workpiece and also provide support whenever necessary. Methods adopted for holding the workpiece depend on part size and shape, machine tool and quality produced. If large number of components are to be produced, then holding devices may be actuated mechanically, pneumatically, magnetically, electrically or by cam action. Following are the common work holding devices:

1. Holding workpiece in chuck.
2. Holding workpiece between centres.
3. Holding workpiece on a face plate.
4. Holding workpiece with help of mandrel.

1. **Holding the Workpiece in Chuck.** Chuck is the most common form of work holding devices. Jaws are provided in chuck to hold the workpiece. These jaws move in channels in the chuck body towards or away from centre with the help of the chuck key. The chuck body along with jaws and workpiece rotates while machining a workpiece. Chucks are made with central hole which allows the long workpieces or bars to extend through them into the hollow spindle of lathe. Following are two basic types of chucks.

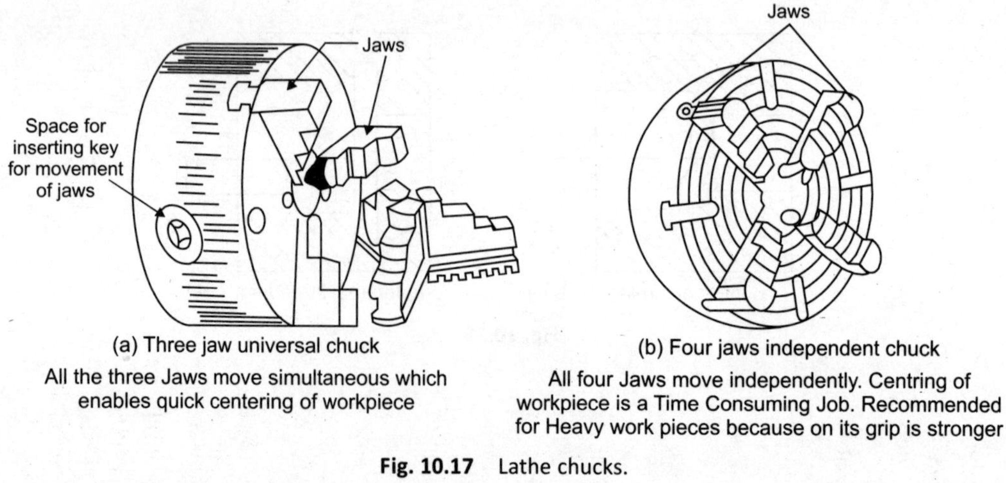

(a) Three jaw universal chuck
All the three Jaws move simultaneous which enables quick centering of workpiece

(b) Four jaws independent chuck
All four Jaws move independently. Centring of workpiece is a Time Consuming Job. Recommended for Heavy work pieces because on its grip is stronger

Fig. 10.17 Lathe chucks.

(*a*) **Three jaw universal chuck.** This is the most common type of chuck. It is also called as *self centering chuck*. All the three jaws of chuck move simultaneously equal distances towards or away from centre when chuck key is turned. Thus the jaws are able to locate centre axis of any cylindrical or symmetrical section concentric with the axis of rotation of spindle. This chuck enables quick centering but it has limitations in terms of accuracy which is gradually lost as a result of wear or misuse.

(*b*) **Four jaws independent chuck.** In four jaws independent chuck only one jaw is moved at a time by turning its corresponding screw with a chuck key. These are used to support wide variety of workpiece. Any irregular shapes can be held and effectively centred. These are also used for holding the cylindrical workpieces which are to be machined eccentric to the cylindrical surface, like machining of the component shown in Fig. 10.30.

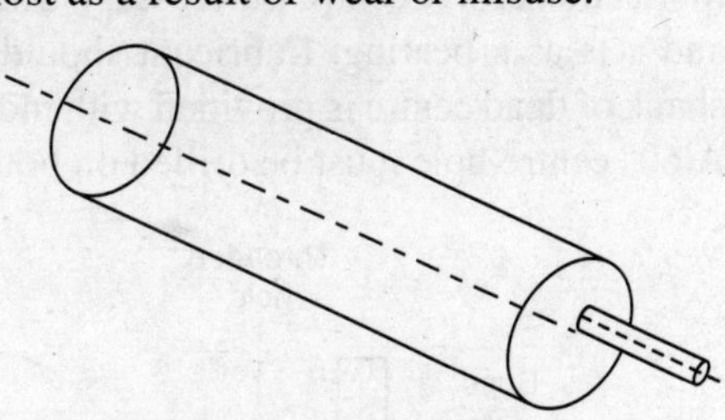

Fig. 10.18 Machining of a component which is not concentric.

Four jaws chucks are heavier and more rugged than three jaw chucks and they should be used for heavy work. Workpiece can be accurately adjusted with respect to spindle axis because each jaw can be individually moved. But it is of course time consuming process as the workpiece is centred by using dial indicator. The gripping force of four jaw chuck is more as compared to a three jaw chuck.

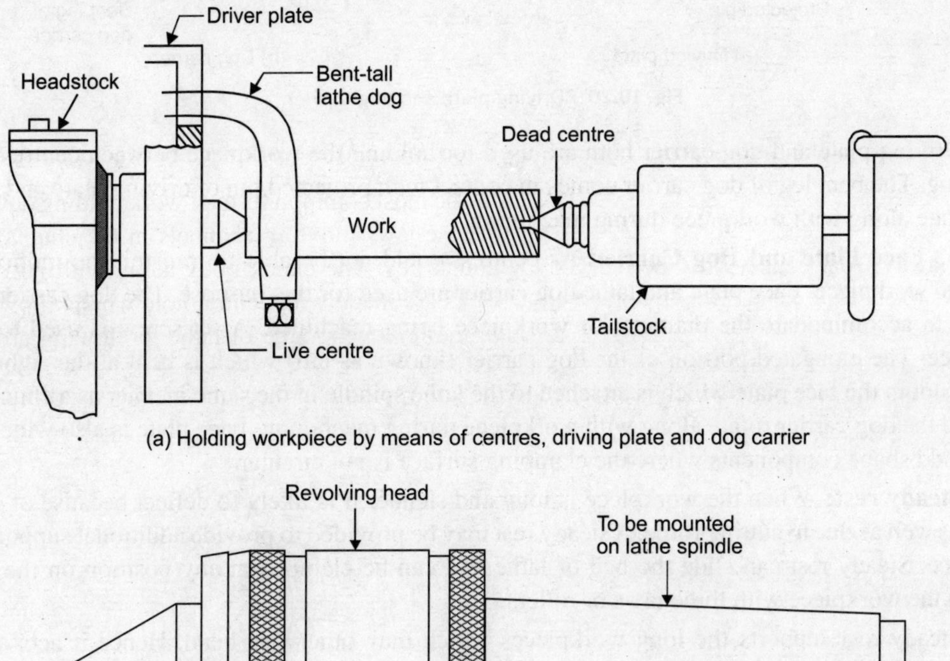

(a) Holding workpiece by means of centres, driving plate and dog carrier

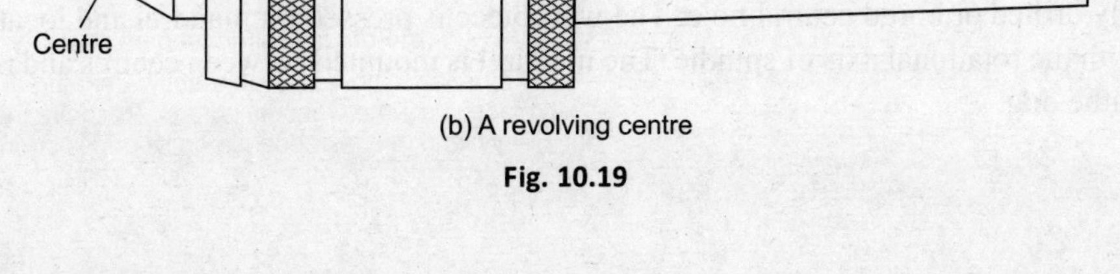

(b) A revolving centre

Fig. 10.19

2. **Holding by Means of Centres.** Workpieces which are relatively long are usually held between centres during machining. It is quite common way to support the workpiece in turning operation. Two lathe centres are used, one in spindle hole of the head stock and the other in barrel hole of tail stock. The centre which is in head stock spindle is termed as live centre and rotates with workpieces and hence no lubrication of live centre is necessary. The tail stock centre or dead centre does not rotate with workpieces and acts as a bearing. Lubricant should be provided and excessive pressure should be avoided. The shank of dead centre is provided with morse taper which fit into the tapered hole of spindle or tail stock. A 60° centre hole must be drilled on both ends of workpiece before mounting between centres.

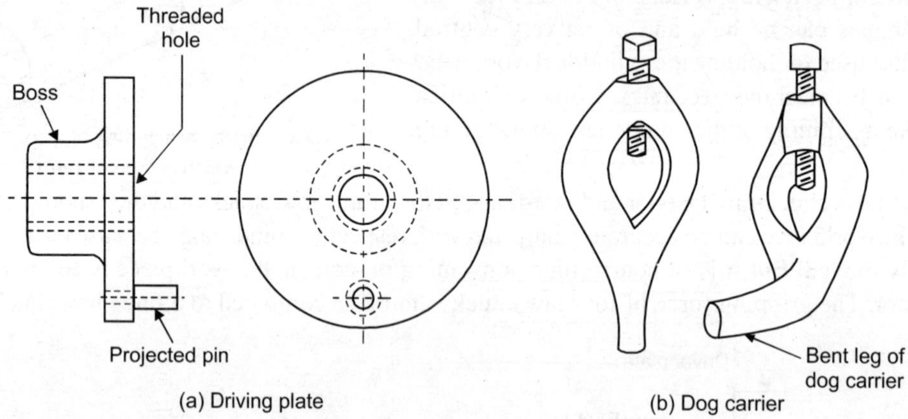

Fig. 10.20 Driving plate and dog Carrier.

Driving plate and dog carrier both are used for holding the workpiece between centres during machining. The bent leg of dog carrier comes in contact with projected pin of driving plate and both of these rotate along with workpiece during machining.

3. **Face Plate and Dog Carrier.** As centres would not be able to transmit the motion from spindle to workpiece. Face plate and lathe dog carrier are used for this purpose. The dog carrier has an opening to accommodate the diameter of workpiece being machined. A set screw is used to clamp workpiece. The elongated portion of the dog carrier (known as tail) which is bent at the right angle, enters a slot in the face plate which is attached to the lathe spindle in the same manner as a chuck. Face plate and the dog carrier rotate along with workpiece during machining. Face plate is also widely used to hold odd shape components where the clamping surface is not circular.

Steady rest: When the workpiece is long and slender it is likely to deflect because of its own weight as well as due to cutting force. A steady rest may be provided to provide additional support to the workpiece. Steady rest can slide the bed of lathe and can be clamped at any position on the bed. It supports the workpiece with three jaws or rollers.

Steady rest supports the long workpieces which may otherwise bend. Hence it acts as shaft bearing. It can slide on guideways.

4. **Holding Workpiece by Means of Mandrel.** Mandrel is used for holding workpiece which has previously drilled or bored central hole. The workpiece is pressed on mandrel and locates the hole concentric with the rotational axis of spindle. The mandrel is mounted between centres and rotated with the help of lathe dog.

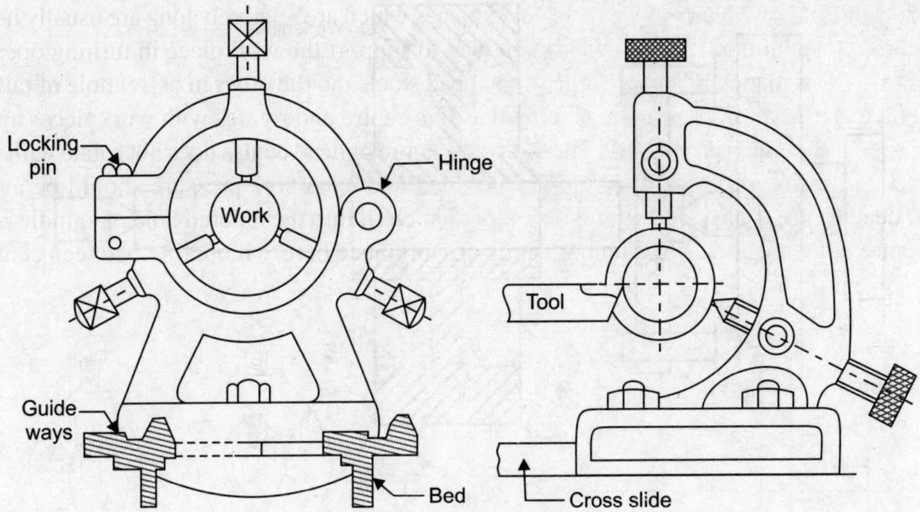

Fig. 10.21 Steady rest.

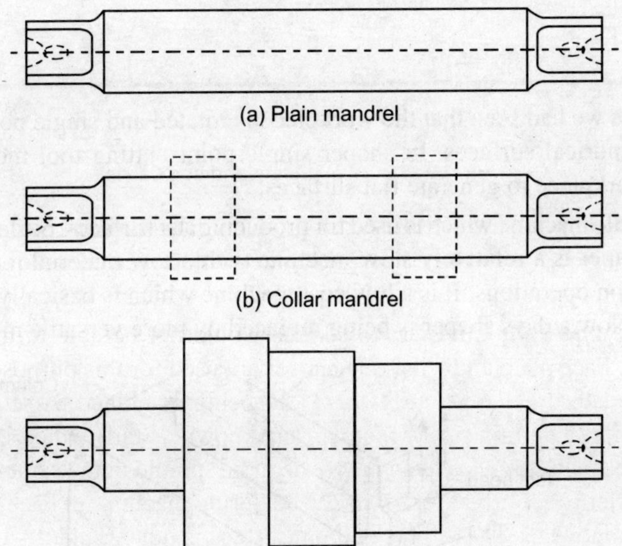

Fig. 10.22 Various types of mandrel.

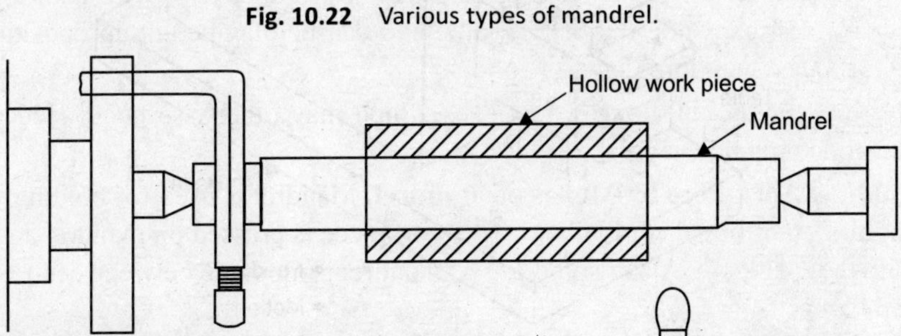

(a)

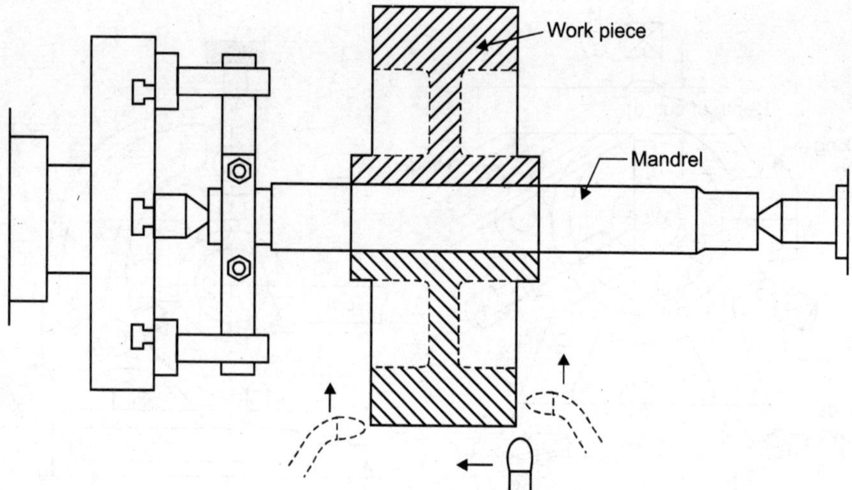

(b)

Fig. 10.23 Illustrate how hollow workpieces are held in mandrel during machining.

10.5 SHAPER

In lathe machines we had seen that the workpiece is rotated and single point cutting tool moves axially to generate cylindrical surfaces. In shaper single point cutting tool moves in a straight line motion in relation to workpiece to generate flat surfaces.

Shaper is a versatile machine which is used for producing flat surfaces. Surfaces may be horizontal, vertical or inclined. Shaper is a relatively slow machine with a low material removal capability. It is rarely used for production operations. It is a low cost machine which is basically used for initial rough machining operations. Now a days shaper is being replaced by more versatile milling machine.

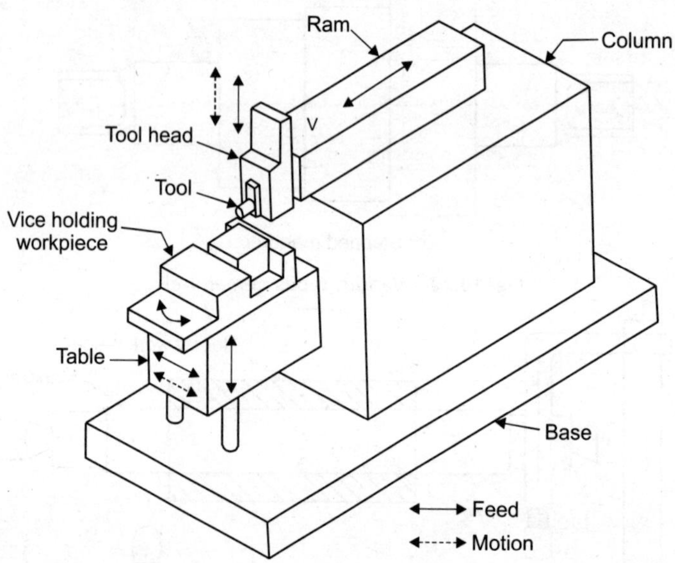

(a)

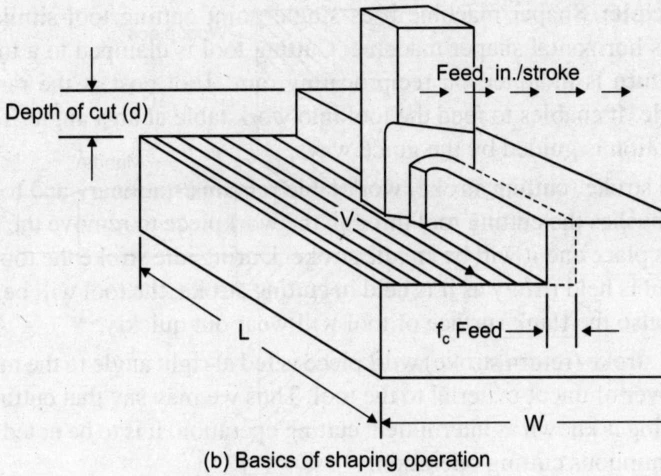

(b) Basics of shaping operation

Fig. 10.24 Shaper machine.

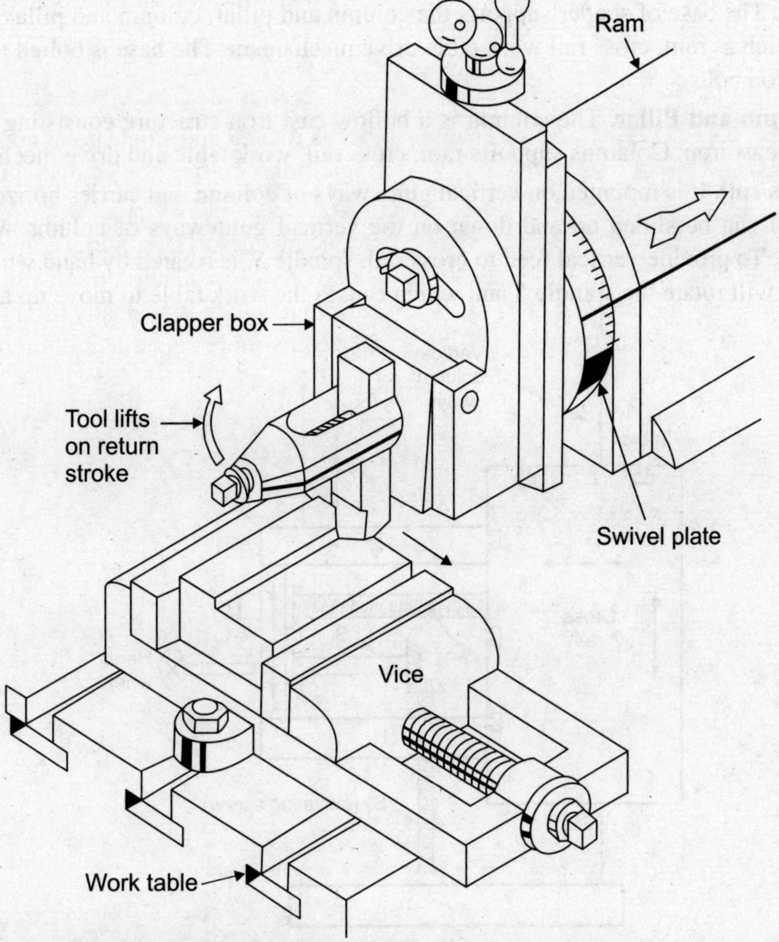

Fig. 10.25 Shaper machine tool lifting mechanism.

Working principle: Shaper machine uses single point cutting tool similar to that used in a lathe. Fig. 10.24 shows horizontal shaper machine. Cutting tool is clamped to a tool post mounted on clapper box which in turn is mounted on reciprocating ram. Tool post at the ram can be easy term rotated through an angle. It enables to feed the tool into work table at an a angle. The ram reciprocates horizontally and its motion is guided by the guide ways.

During forward stroke (cutting stroke) work table remains stationary and tool reciprocates. The ram in forward stroke pushes the cutting tool through the workpiece to remove the material. When ram returns no cutting takes place and it will be an idle stroke. During idle stroke the tool is lifted by clapper box arrangement. If tool is held firmly as it is held in cutting stroke, the tool will be rubbing the already machined surface and also the flank surface of tool will wear out quickly.

After every idle stroke (return stroke) workpiece is fed at right angle to the motion of cutting tool in order to provide a layer of uncut material to the tool. Thus we may say that cutting is not continuous and this type of machining is known as intermittent cutting operation. It is to be noted here that machining in case of lathe is a continuous cutting operation.

10.5.1 Construction Details of Shaper

1. **Base.** The base of shaper supports the column and pillar, column and pillar supports all the working parts such as ram, cross rail work table drive mechanism. The base is bolted to floor with the help of foundation bolts.

2. **Column and Pillar.** The column is a hollow cast iron structure consisting of two vertical walls cast from cast iron. Columns supports ram, cross rail, work table and drive mechanism.

3. **Cross rail.** It is mounted on vertical guideways of column and carries horizontal table slide ways. Cross rail can be slided up and down on the vertical guideways of column with the help of elevated screws. To provide vertical feed to cross rail, spindle *S*, is rotated by hand wheel provided on it. This, in turn, will rotate the spindle *S* and which causes the work table to move up and down.

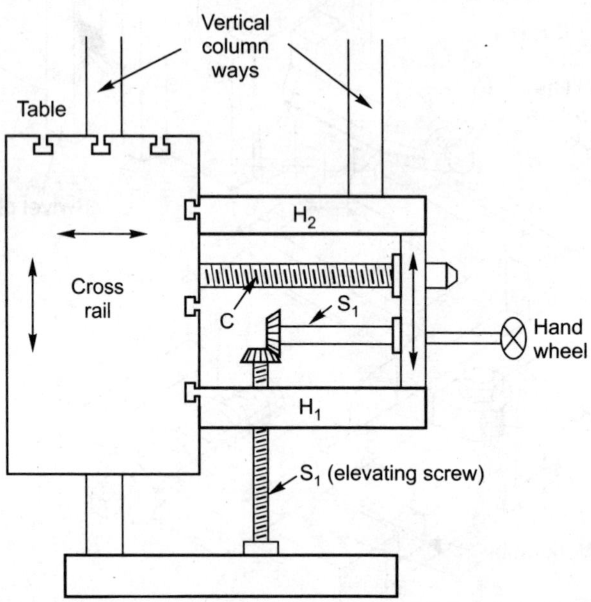

Fig. 10.26 Mechanism for up and down movement of cross rail and work table.

5. **Work Table.** Work table is bolted on the cross rail. The work table can be moved in horizontal as well as vertical direction. The horizontal movement of work table is performed by sliding it on horizontal slideways of cross rail, where as, the vertical movement of table is achieved by vertical movement of cross rail on vertical guideway of column with help of elevating screw and (please refer Fig. 10.26)

6. **Saddle.** Saddle is mounted on the cross rail and holds the work table. The horizontal cross-wise movement of saddle (by rotating the cross feed screw) causes the table to move in cross-wise direction.

7. **Ram.** The Ram is reciprocating member of shaper. The Ram is semicylindrical in shape and slides in the dovetail guideways provided on the top of column. The Ram is provided with internal ribs in order to provide strengths and to rigidity and reduce vibrations. The Ram contain stroke positioning mechanism. With the help of stroke positioning mechanism, stroke length of Ram may be increased or decreased. The Ram holds tool head at extreme forward end.

8. **Tool head.** Tool head is mounted at the extreme front end of the Ram. It consists of a vertical slide (tool slide), which can be moved vertically up and down by rotating the feed screw by means of tool feed handle. The feed screw rotates at its position in a circular threaded hole (nut) provided at the back of tool slide. In this way circular motion of the feed screw causes the tool slide to move up and down. At the back of tool slide there is a circular graduated plate (Fig.) called the swivel plate. This plate is bolted to the extreme front end of ram. This plate can be unbolted and swivelled to desire angle to make the tool head inclined for machining inclined surfaces. The inclination angle of tool can be read directly from the graduations provided on swivel plate.

In the front end of tool slide clapper box is mounted which carries clapper block. The clapper block, carrying the tool post, is hinged about the pin. During forward stroke (cutting stroke), the clapper is pressed against the clapper box due to cutting force and performs cutting. (Fig. 10.28). During backward stroke (idle stroke), it is lifted up. By this action, damage to the cutting edge of the tool and the work surface is prevented.

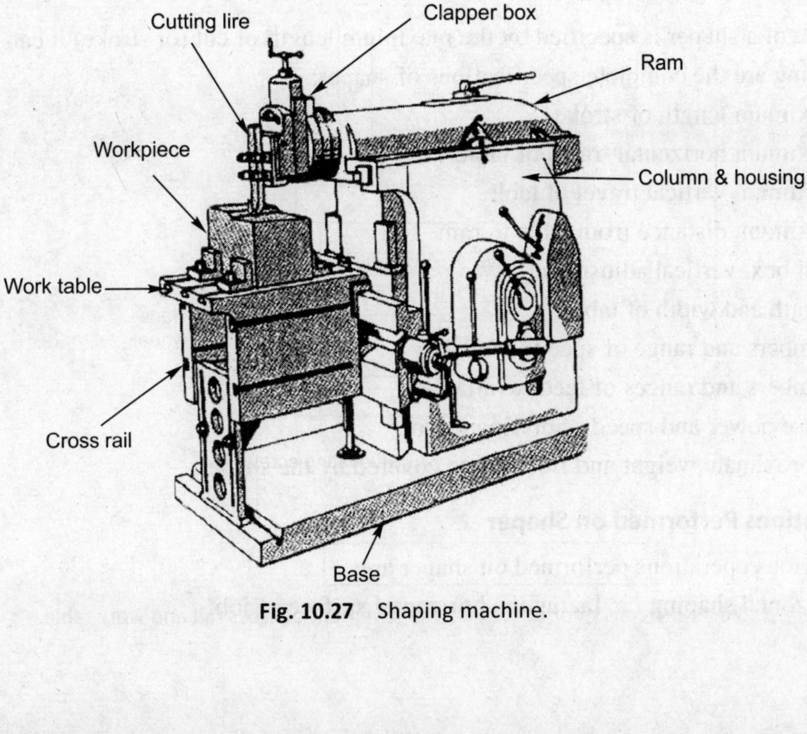

Fig. 10.27 Shaping machine.

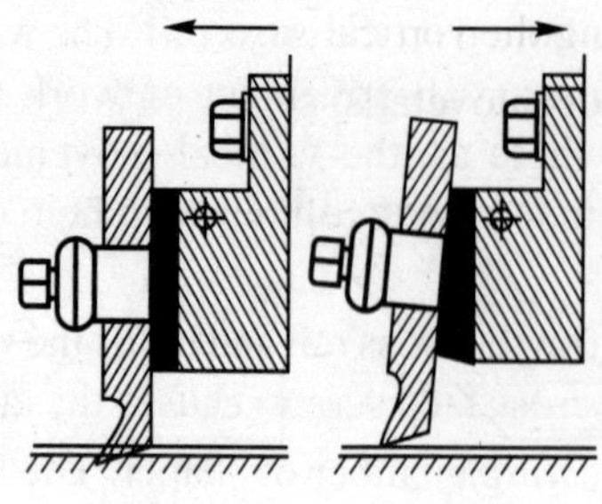

Fig. 10.28 (*a*) movement of tool and clapper during forward stroke (cutting stroke), (*b*) movement of tool and clapper during backward stroke (idle stroke).

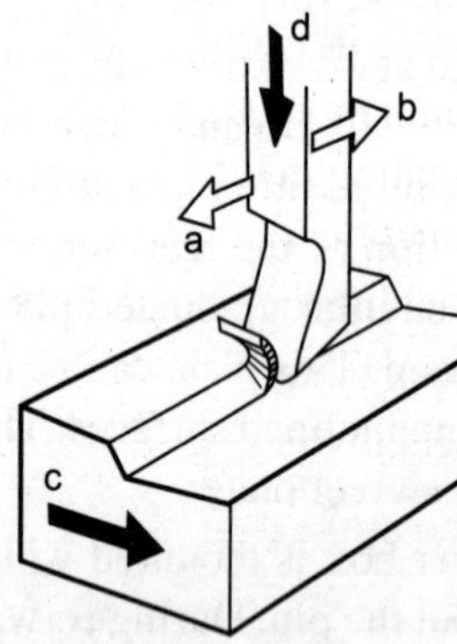

Fig. 10.29 Movements required for the horizontal shaping process. (*a*) Forward stroke, (*b*) backward stroke, (*c*) horizontal feed, (*d*) vertical adjustment.

10.5.2 Specifications of a Shaper

The size of a shaper is specified by the maximum length of cut (or stroke) it can make. Following are the complete specifications of shaper:

1. Maximum length of stroke
2. Maximum horizontal travel of table
3. Maximum vertical travel of table
4. Maximum distance from table to ram
5. Tool box, vertical adjustment
6. Length and width of table
7. Numbers and range of speeds available
8. Numbers and ranges of feeds available
9. Horse power and speed of driving motor
10. Approximate weight and floor space covered by the shaper.

10.5.3 Operations Performed on Shaper

The various operations performed on shaper are:

— horizontal shaping *i.e.* facing the horizontal surface of job.

— vertical shaping *i.e.* facing the vertical surface of job.

— shaping of grooves, T slots, dovetails, steps and angles.

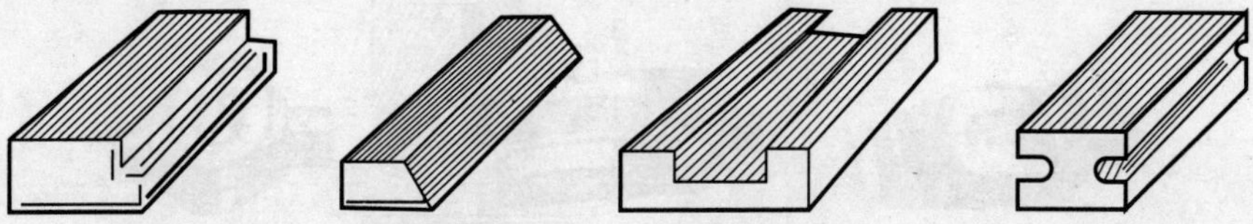

Fig. 10.30 Various surfaces machined by shaper (shaping machine).

10.5.4 Work Holding Devices

The primary function of work holding device is to hold and clamp rigidly the job to be machined.

In addition the clamping force should be such that it should not bend workpiece or cause pressure marks on the surface of workpiece. Depending upon the shape and size of workpiece the work holding devices are used to perform following functions:

1. To clamp the workpiece in machine vice.
2. To clamp the workpiece on work table.
3. To clamp the workpiece in specially designed fixture.

Various work holding devices used are:

1. Maching vice
2. Work table
3. Fixture
4. T bolts and clamps
5. Strip, Stop pin, and Toe dogs
6. Angle plate and wedge block
7. V block

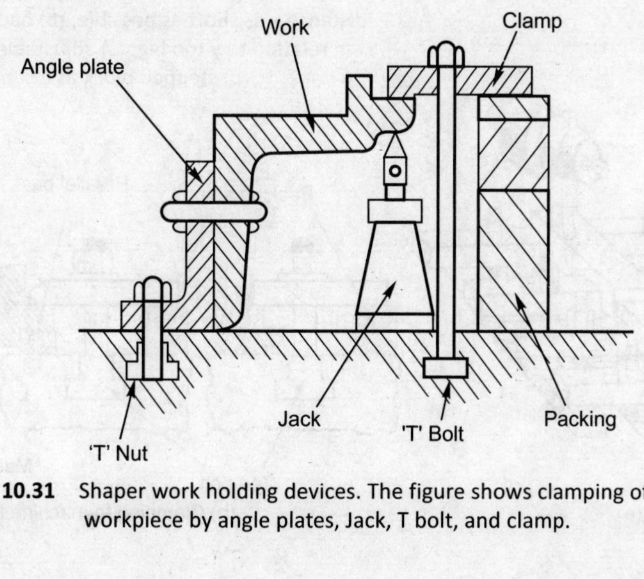

Fig. 10.31 Shaper work holding devices. The figure shows clamping of thin workpiece by angle plates, Jack, T bolt, and clamp.

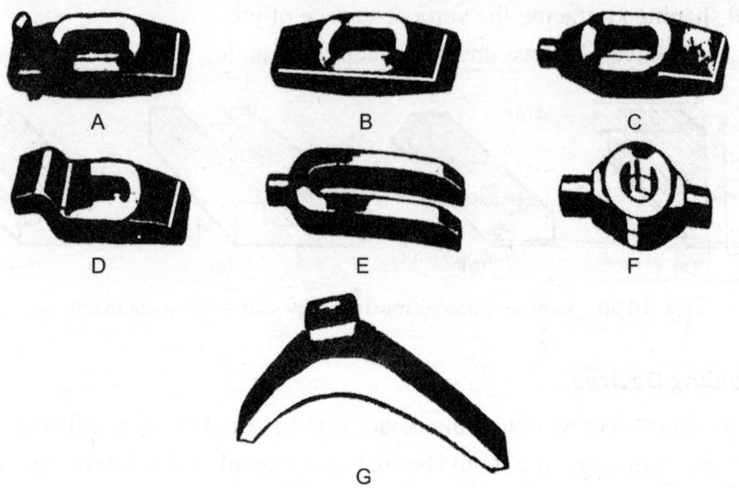

Fig. 10.32 *A*—Screw heel clamp, *B*—Plain clamp, *C*—Finger clamp, *D*—Gooseneck clamp, *E*—U-clamp, *F*—Double finger clamp, *G*—Universal adjustable clamp.

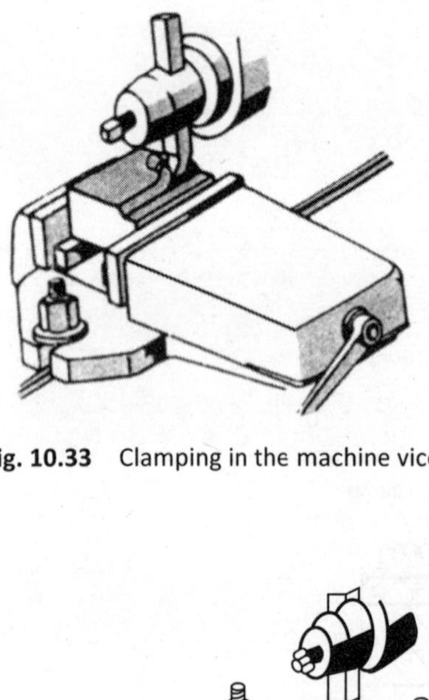

Fig. 10.33 Clamping in the machine vice.

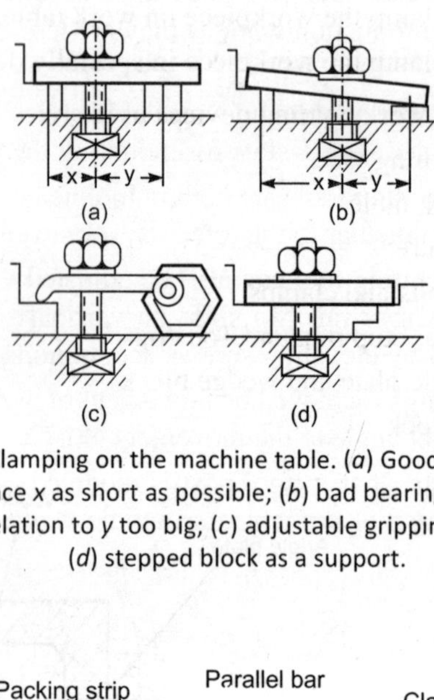

Fig. 10.34 Clamping on the machine table. (*a*) Good bearing surface, distance *x* as short as possible; (*b*) bad bearing surface, *x* in relation to *y* too big; (*c*) adjustable gripping device, (*d*) stepped block as a support.

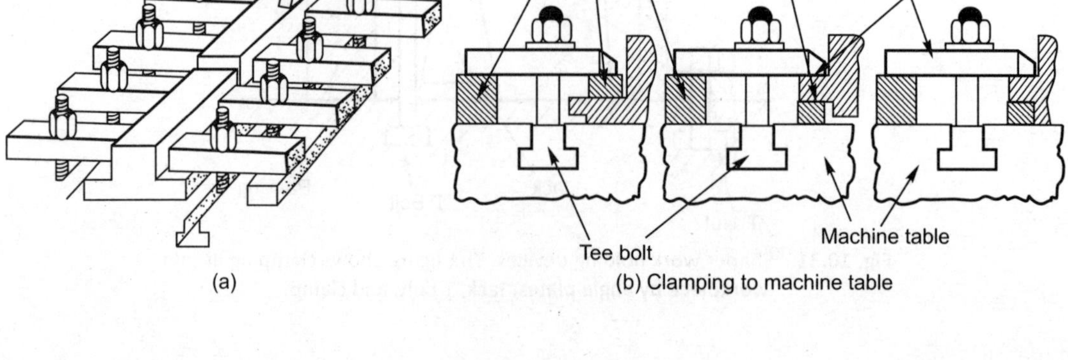

(a)

(b) Clamping to machine table

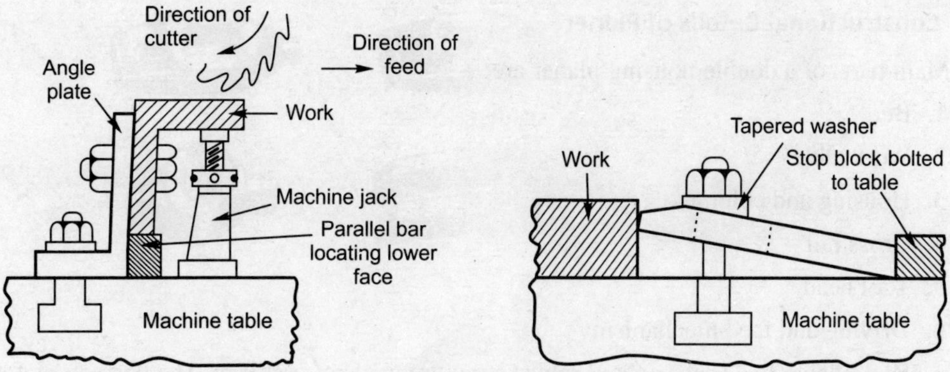

(c) Work bolted to angle plate (d) Clamping narrow work to a machine table

Fig. 10.35

10.6 PLANER

Working Principle. Planers are used for producing horizontal, vertical or inclined flat surfaces on the workpieces that are too large to be accommodated on a shaper or milling machine. Planers are rarely used for production work as they are less efficient than other machining processes like milling, grinding etc.

— In case of planer the work table slides in the guideways of machine bed. The work table is provided with T-slots for mounting the workpiece.

— The cutting tools are held in tool head and can slide horizontally on cross rail (in a direction perpendicular to the direction of movement of work table)

— The cutting tool can also be adjusted vertically with the up and down movement of cross rail. Cross rail can slide on vertical guideways of column. Cross rail may move up and down by elevating screws accommodated with in column.

— Cutting is achieved by movement of work table in guideways (motion X) and feeding the tool at right angle to the movement (Y).

— like the shaper, tool post is mounted on the clapper box to avoid interference between the cutting tool and table during idle stroke.

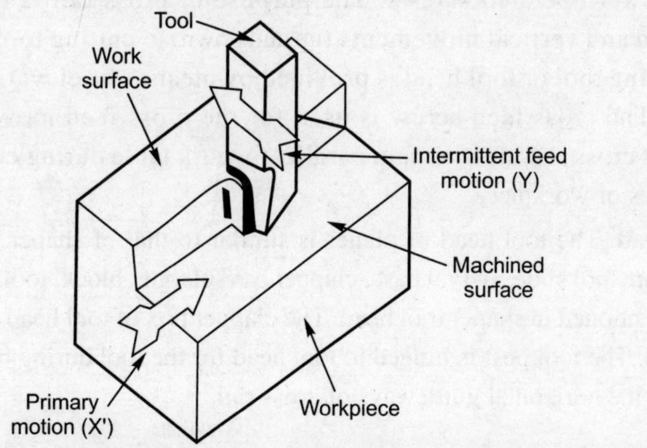

Fig. 10.36 Principle of operation of a planer.

10.6.1 Constructional Details of Planer

Main parts of a double housing planer are:

1. Bed
2. Work table
3. Housing and columns
4. Cross rail
5. Tool head
6. Driving and feed mechanism

1. **Bed.** Planer bed is strong and robust structure made of cast iron. The bed acts as foundation of machine and supports other parts of planer. The length of bed is slightly more than the length of stroke. For movement of work table on bed dovetail, flator V guideways are provided on the top of the bed. For lubrication of guideways, pressure lubrication is adopted. The hollow space with in the box like structure of bed houses the driving mechanisms of work table.

2. **Work table.** The work table holds and supports the workpiece. It slides on the guideways provided in the bed. The worktable is heavy rectangular shape and is made of cast iron. The top face of cast table is accurately finished to locate the workpiece. The T slots are provided on the table so that workpiece and work holding devices may be bolted upon it. Accurately machined holes are also there on top surface of worktable to support stop pins.

3. **Housing and columns.** The columns are vertical members and are there on both sides of bed in case of double housing planer and on one side in case of an open side planer. In the front of columns, accurately machined guideways are there along which cross rail may slide up and down. Columns enclose cross rail elevating screw, vertical and cross feed screws for vertical and horizontal movement of tool head. These screws may be operated either by hand or power. The column also enclose counter balancing weight for cross rail.

4. **Cross rail.** Cross rail is heavy box like construction between two columns. The cross rail slides on vertical guideways of column. The movement of cross rail may be controlled by hand or by power operated screws. The purpose of cross rail is to carry tool head and to provide cross feed and vertical movements (up and down) to cutting tool. Up and down (vertical) movement to cutting tool or tool head is provided by means of a elevating screw accommodated with in column. The cross feed screw is used for the cross feed movement of tool head. It is very essential that cross rail must remain parallel to work table during cutting to achieve accurate machined surfaces of workpiece.

5. **Tool head.** The tool head of planer is similar to that of shaper in construction. Important parts of tool head are tool slide, swivel base, clapper box, clapper block, tool post. The function of these parts is same as mentioned in shaper tool head. The clapper box of tool head carries a tool post in which cutting tool is held. The tool post is hinged to tool head for the tool during backward (idle) stroke. The tool head slides in the horizontal guideways of cross rail.

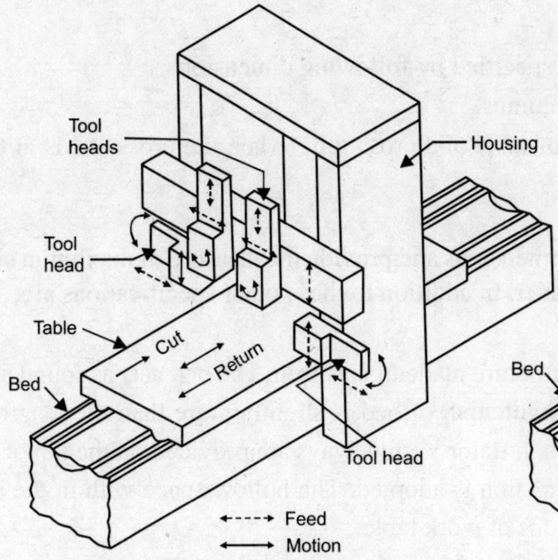

(a) Block diagram showing the basic components of a double-housing planer

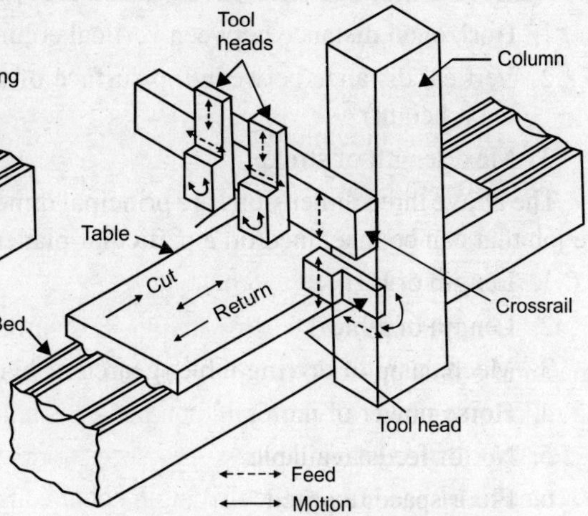

(b) Block diagram of an open-side planer

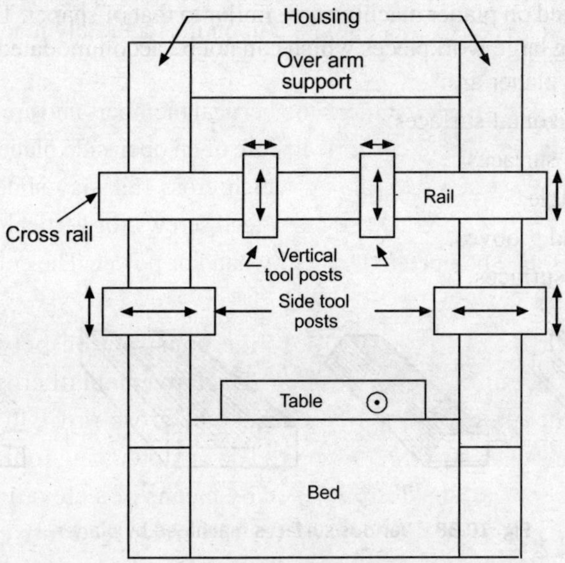

(c) Main parts of a planer

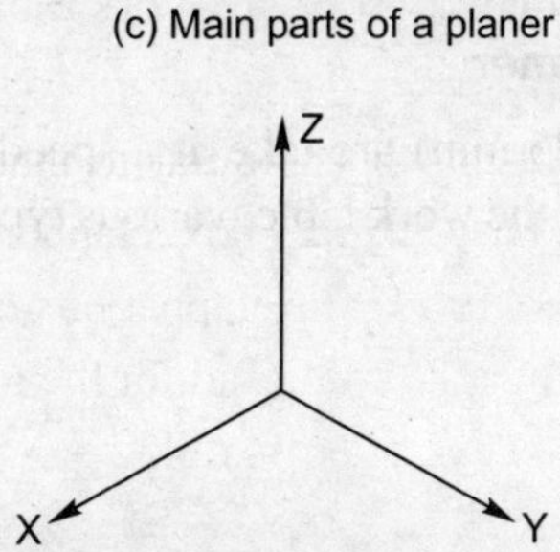

Fig. 10.37 Planer machine.

10.6.2 Specifications

Planers are made in different sizes and are specified by following dimensions:

1. Horizontal distance between vertical columns.
2. Vertical distance between top surface of table and cross rail, (when the cross rail is at the max. height).
3. Max. length of stroke.

The above three dimensions are principal dimensions and provide the measure of maximum size of the job that can be machined on a particular planer. In addition to these other specifications are:

1. Length of bed
2. Length of table
3. Mechanism of driving table (geared or hydraulic)
4. Horse power of motor
5. No. of feeds available
6. Floor space required
7. Net weight of the machine.

10.6.3 Operations Performed

Operations performed on planer machine are similar to that of shaper. The only difference is that planers are used for planing large workpieces which can not be accommodated by shaper. The common operations performed on a planer are:

1. Planing flat horizontal surfaces.
2. Planing vertical surfaces.
3. Planing at an angle.
4. Planing slots and grooves.
5. Planing curved surfaces.

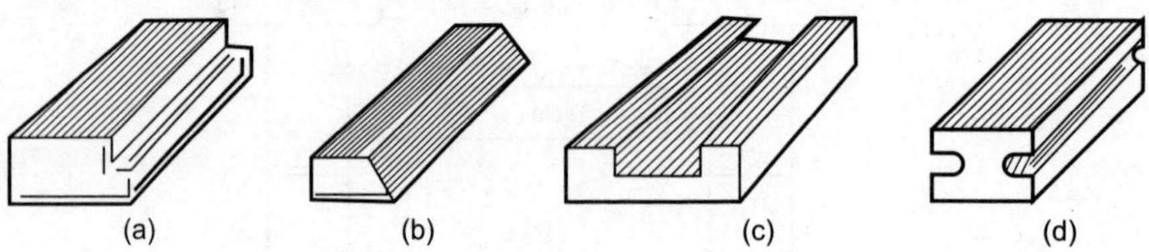

Fig. 10.38 Various surfaces machined by planer.

10.6.4 Work Holding Devices for Planer

Since heavy cuts (as much as 25 mm) are take at a speed of 20–30 m/minute on a planer, the workpiece must be properly fastened to the work table, various type of devices used for hold workpieces on planer worktable are:

1. Planer vice
2. V blocks
3. Angle plate
4. Parallelships

5. Cramps
6. Holding with dogs
7. T bolt, nut, washer, wedges and packings etc.

10.7 MILLING MACHINE

After lathe, the milling machine is most versatile of all machine tools. A large number of operations can be performed on a milling machines like cutting key way, slitting, sawing, slab and face milling, gear cutting and shaping irregular pieces. Alongwith good surface finish, milling machine provides a high rate of production and higher accuracy and, therefore, in many cases it has largely replaced shapers and slotters. But it is not very much suitable for larger jobs. Various components machined by milling process are shown in Fig. 10.39.

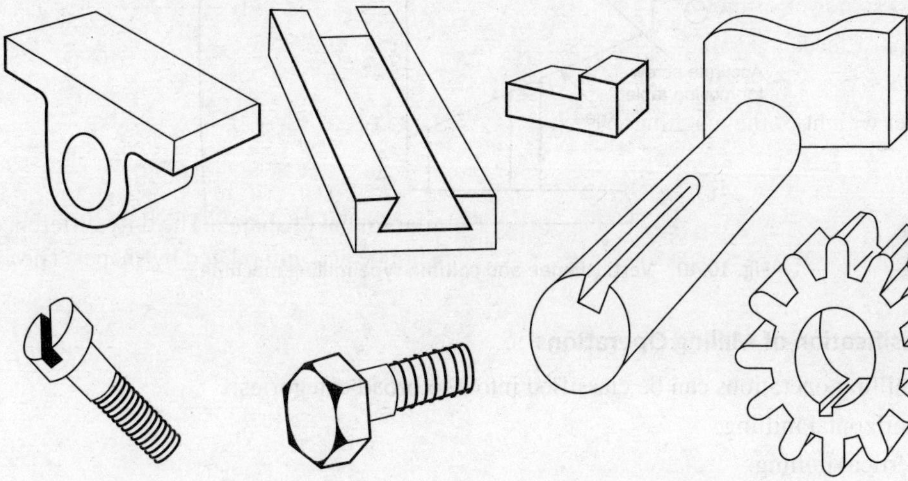

Fig. 10.39 Various components machined by milling process.

In milling machine metal is removed from workpiece with a rotating cutter. This rotating cutter is known as milling cutter. Milling cutter is a multipoint cutting tool (unlike lathe and shaper tools which are single point cutting tools). For cutting, workpiece is fed into the rotating milling cutter and metal is removed in form of chips. Following are the characteristics of milling operation:

- **Interrupted cutting.** Each of the cutting edge of milling cutter removes material for only a part of the rotation of the milling cutter. Therefore, the cutting has a time to cool before it removes materials. Thus the milling operation is much cooler compared to lathe operations and as a result we can have much larger material removal rate in milling.
- **Small size of chips.** As there are multiple cutting edges in the milling cutter the material removal rate is high even when the size of chips is small. Because of high removal rate, desired surface is generally obtained in a single pass unlike the turning process which requires a large number of cuts for producing finished surface.

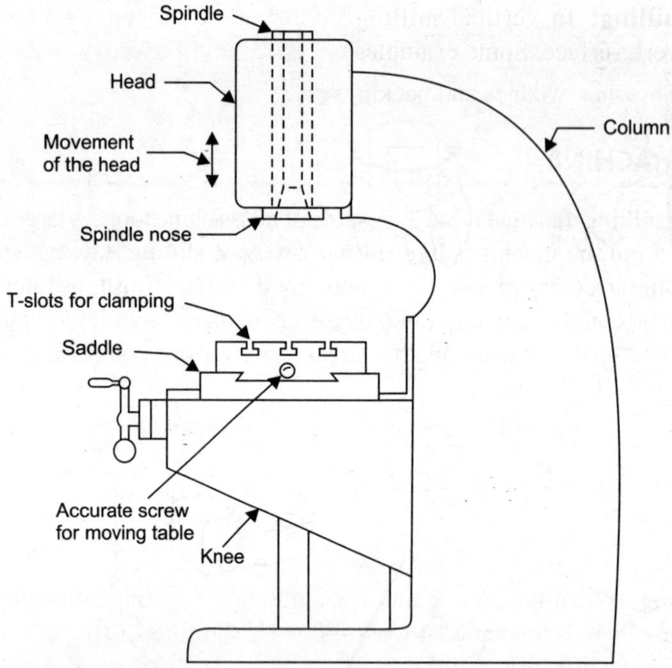

Fig. 10.40 Vertical knee and column type milling machine.

10.7.1 Classification of Milling Operations

The milling operations can be classified into two broad categories:

1. Horizontal milling
2. Vertical milling.

1. **Horizontal milling.** In horizontal milling the cutter axis is horizontal. Fig. 10.41 shows some horizontal milling operations.

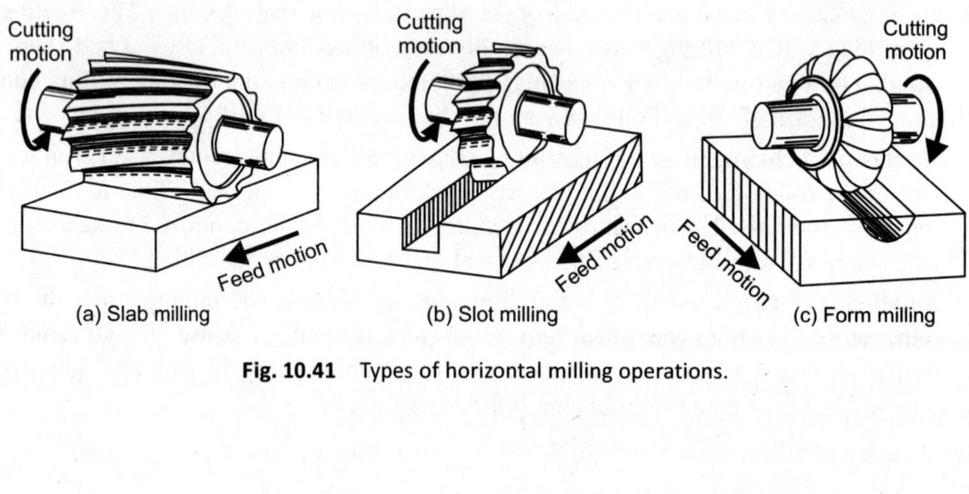

Fig. 10.41 Types of horizontal milling operations.

2. **Vertical milling:** In vertical milling operation cutter axis is vertical and generally perpendicular to the work surface. Some examples of vertical milling are shown in Fig. 10.42.

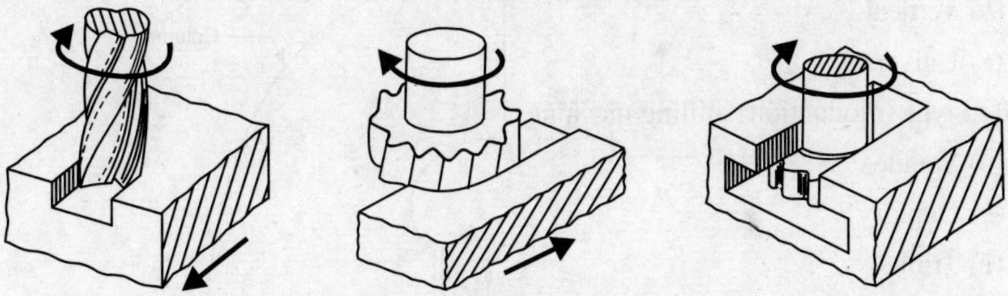

(a) Slot milling (b) Face milling (c) T-Slot milling

Fig. 10.42 Types of vertical milling operations.

Milling operations can also be classified as:

(*i*) Down milling

(*ii*) Up milling

(*i*) *Down Milling.* When the cutting and feed motion are in same direction, it is called down milling. In down milling there is tendency of workpiece being dragged into the cutter. However, down milling results in better surface finish and longer tool life. The process is good for thin and frail workpieces. When the cutting edges are helical the cutting operation is smoother and better surface finish is obtained; this is because of gradual engagement of cutting edge.

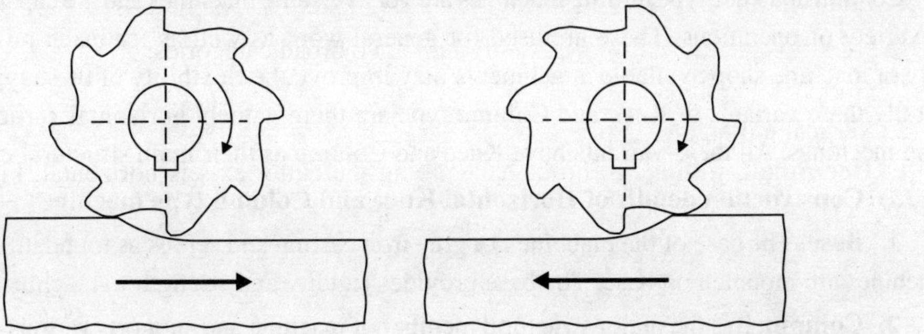

(a) Up milling (b) Down milling

Fig. 10.43 Up milling and down milling operations.

(*ii*) *Up Milling.* In up milling the cutting tool rotates in the opposite direction to the motion of feed. This tends to lift the workpiece from table. There is possibility that the cutting tool will rub the workpiece before starting the material removal. However, the process is safer and is commonly done.

10.7.2 Classification of Milling Machines and Constructional Details

Depending upon the type of milling cutter, shape and size of workpiece, and number of workpieces required. Milling machines are manufactured in various designs, shapes and sizes. Following is the broad classification of milling machines:

1. Knee and column type milling machines
   (*a*) Horizontal
   (*b*) Vertical
   (*c*) Universal
2. Bed type (production) milling machines
   (*a*) Simplex
   (*b*) Duplex
   (*c*) Triplex
3. Plano milling machines
4. Special types of milling machines
   (*a*) Rotary table
   (*b*) Drum type
   (*c*) Planetary type
   (*d*) Profile milling machine
   (*e*) Tracer controlled (copying) milling machine
   (*f*) Spline shaft milling machine
   (*g*) Keyway milling machine.

Column and knee type milling machines are very versatile machines and are capable of performing wide variety of operations. These are used for general work as well as for batch production. A large variety of machine shop available attachments may improve the flexibility of these types of machines. Generally three variants of Knee and Column type are there namely horizontal vertical and universal milling mechines. All these variants have Knee and Column as their main structural elements.

(*a*) **Construction details of Horizontal Knee and Column type machine**

1. **Base.** The base of the machine is a grey iron casting and serves as foundation. All other parts of machines are mounted on base. The base provides rigidity and strength to machine.

2. **Column.** It is the major structural member of machine and supports various parts and control mechanism.

The front vertical face of column has guideways in which Knee slides up and down (*Z*-axis movement). The top of the column supports the overarm which extends outward at the front of machine. The inner space of column houses driving motor and gear mechanism for transmitting power to spindle and work table.

3. **Knee.** The Knee is rigid grey iron casting that slides up and down on vertical guideways of column. The Knee can be moved up and down with the help of elevating screw. Knee mounts saddle which in turn mounts the work table. The knee moves up and down along Z direction whereas table and saddle moves in X and Y direction respectively. The movement of Knee is generally made manually while longitudinal and cross feed movements are made (by rotating hand wheel) manually as well as power driven.

4. **Saddle.** The saddle slides (cross feed or transverse motion) along *Y*-axis in the horizontal guideways provided in the Knee. A hand wheel and lever is provided in front of Knee for the manual and power feed of saddle respectively guideways are provided on the top surface of saddle for mounting the work table.

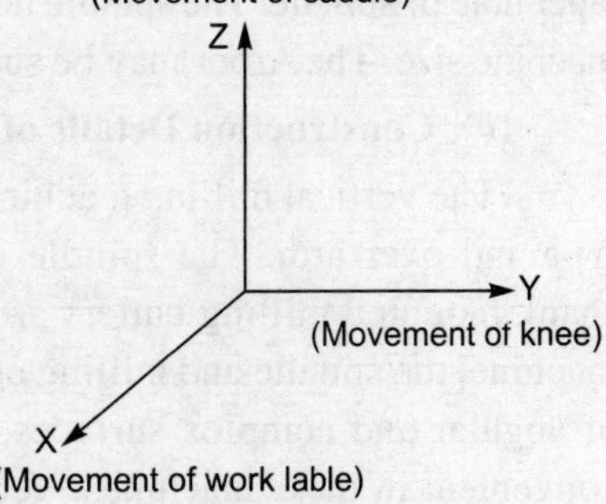

Fig. 10.43

5. **Work table.** The work table slide on the guideway provided on the saddle in the *X*-axis direction (longitudinal motion). The work table has *t* slots on its top surface for clamping the workpiece and fixtures on it. A lead screw under the table engages a nut on the saddle to move the table by hand or power. Sometimes hand wheel is provided on both sides of table for movement of table.

6. **Overarm.** The overarm is mounted on the top of column. The overarm is adjustable and can be tighted in any position on the guide ways provided on the top of column. The adjustment of overarm is required to facilitate bearing support to the milling cutter.

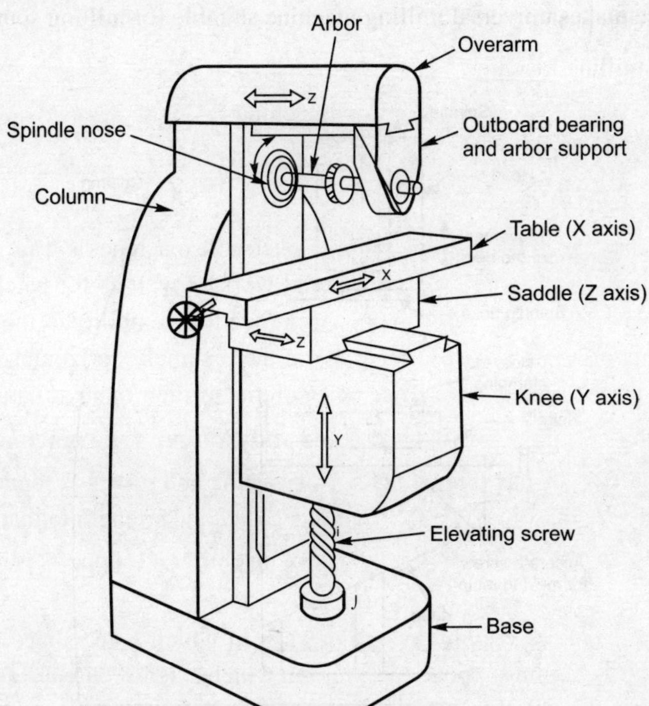

Fig. 10.44 Horizontal milling machine.

7. **Spindle.** Spindle is hollow shaft which revolves in bearing in the upper end of the column. It receives power from motor through belt and gear trains mounted with in the column. The front end of spindle is projected from column and has tapered holes for receiving standard arbors or shanks.

8. **Arbor.** Arbor is a shaft and may be considered an extension of machine spindle and are used for mounting of milling cutters. One end of Arbor has the taper end which fits into a corresponding

taper hole of spindle. The spindle nose has the standard morse taper of suitable size depending upon the machine size. The Arbor may be supported at farthest end by Arbor support.

(*b*) **Construction Details of Vertical Milling Machine (Knee and Column type)**

The vertical milling machine is very similar to horizontal milling machine except the spindle type and over arm. The spindle is perpendicular to work table and revolves along Y-axis. The shank mounted milling cutters are generally used in vertical milling machine. In vertical milling machine, the spindle and milling cutter assembly may be swivelled at an angle, allowing the milling of angular and complex surfaces. In addition, location of tool and setting of workplace is more convenient in these machines. Vertical milling machine may be regarded as more versatile due to presence of above features. Vertical milling machines are used to perform end milling and face milling operations.

(*c*) **Universal Milling Machine.** In Universal milling machine, table is provided with one more movement, *i.e.,* table may be swivelled horizontally along *X*-axis direction. For this purpose table is mounted on a circular base which in turn is more mounted on a saddle. The circular base has graduations in degrees. This feature makes universal milling machine suitable for milling spur gears, helical gears, worm gears and cams.

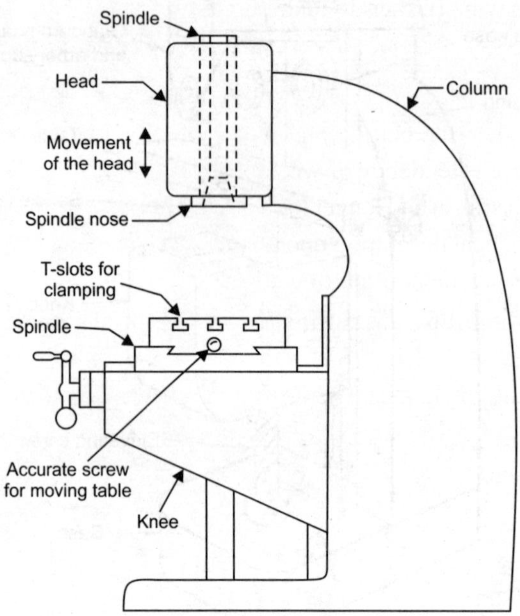

Fig. 10.45 Vertical knee and column type milling machine.

Classification of Milling Methods:

1. Depending upon the position of milling cutter

   Horizontal milling

   Vertical milling

2. Depending upon the method of mounting of milling cutter

   Arbor mounted milling cutter

   Shank mounted milling cutter

   Nose mounted milling cutter

3. Depending upon the rotation of milling cutter w.r.t. work table

   — Right hand rotation

   — Left hand rotation

4. Depending upon the construction details of milling cutter

   — Solid

   — Solid

   — Inserted tooth.

(*a*) **Horizontal milling (peripheral or slab).** In Horizontal milling axis of rotation of milling cutter is parallel to the surface of **work table.** The horizontal milling is usually done on the horizontal milling machines and milling cutter are mounted on arbor. The horizontal milling is used for flat as well as formed surface depending on the axial contour of milling cutter.

Horizontal milling may be further divided into two groups:

(*i*) Down milling

(*ii*) Up milling

(*i*) **Down milling.** When the cutting and feed motion are in same direction, it is called down milling. In down milling there is tendency of workpiece being dragged into the cutter. However, down milling results in better surface finish and longer tool life. The process is good for thin and frail workpieces. When the cutting edges are helical the cutting operation is smoother and better surface finish is obtained; this is because of gradual engagement of cutting edge.

(*ii*) **Up milling.** In up milling the cutting tool rotates in the opposite direction to the motion of feed. This tends to lift the workpiece from table. There is possibility that the cutting tool will rub the workpiece before starting the material removal. However, the process is safer and is commonly done.

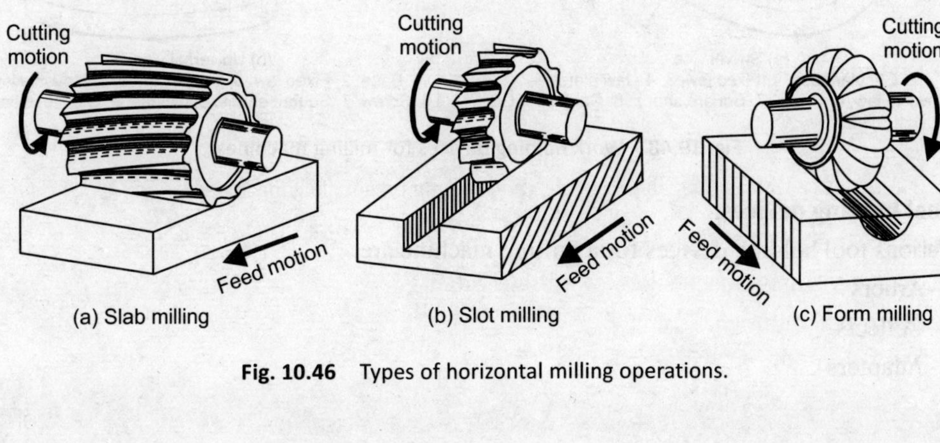

Fig. 10.46 Types of horizontal milling operations.

(*b*) **Vertical Milling** (end milling). In vertical milling, the milling cutter axis is perpendicular to the surface of work table. The vertical milling is performed on vertical milling machine and milling cutter cutters is mounted on the shanks which in turn are mounted in the spindle.

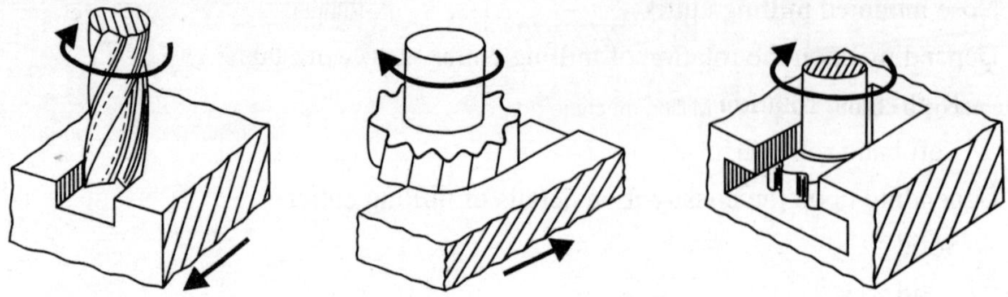

Fig. 10.47 Types of vertical milling operations.

10.7.3 Workholding Tool and Tool Holding Devices for Milling Machines

The work holding devices for mill machines are:

1. Vices – plan swivel vice and universal vice
2. T-bolt, clamp and angle plate (Please refer figures in with holding devices of shaper in (10.5.4).
3. Special fixtures

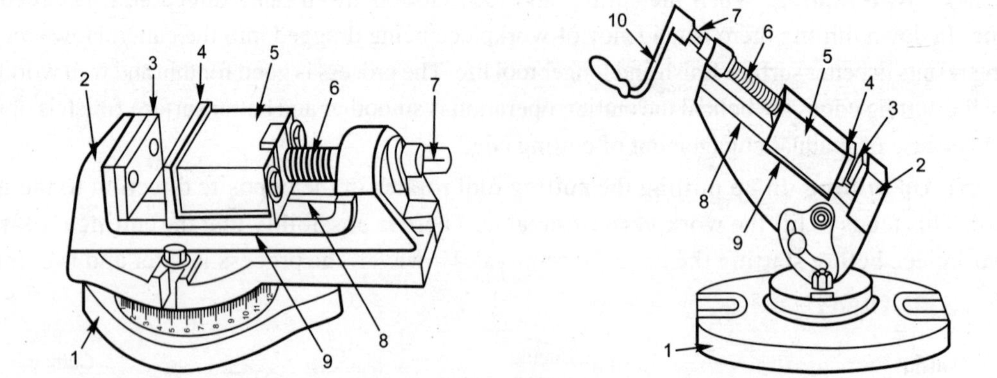

Fig. 10.48 Work holding devices for milling machines.

Tool holding devices:

Verious tool holding devices for a milling machine are:

—Arbors

—Collects

—Adapters

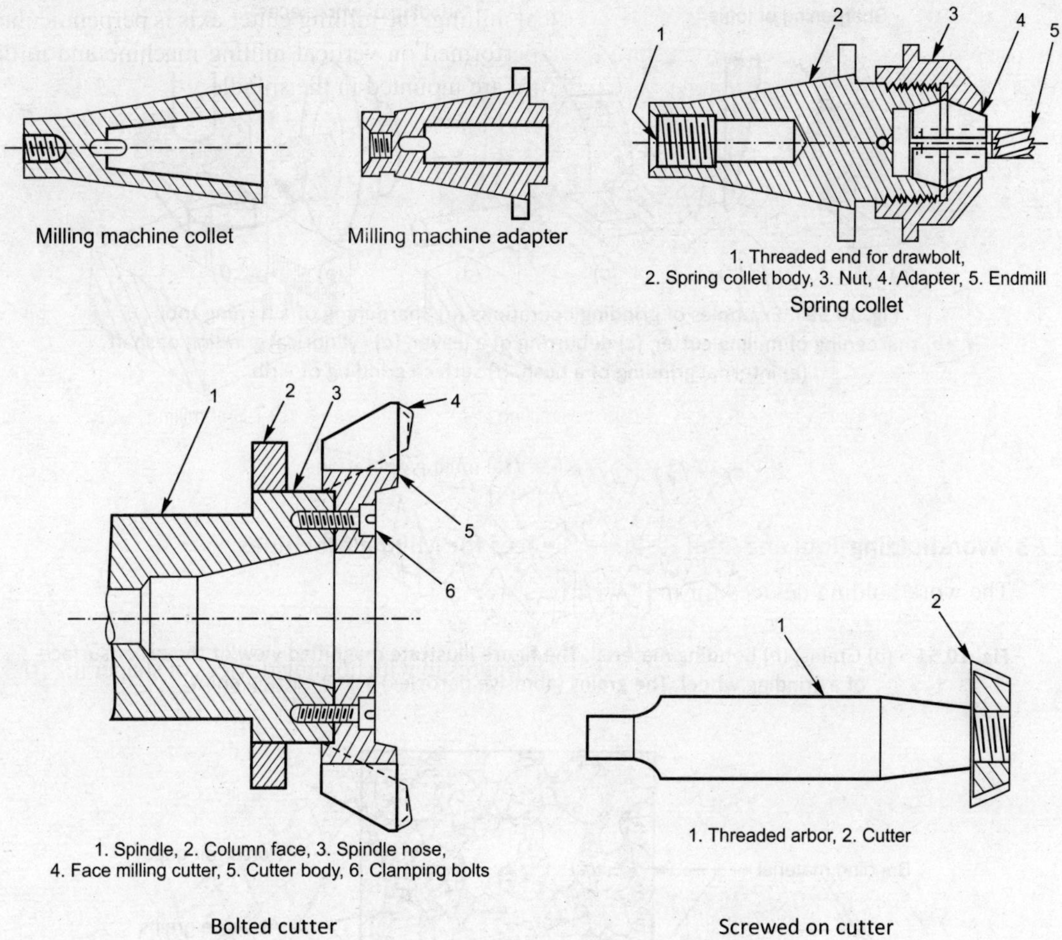

Fig. 10.49 Milling cutter holding devices.

10.8 GRINDING

Grinding process utilise very small abrasive grains to remove material from workpiece. In grinding process, very small amount of material is removed by means of a rotating abrasive wheel.

The cutting action of grinding wheel may be considered as similar to milling cutter. In a grinding wheel, many small abrasive particles (grains) are bonded together. Each abrasive particle acts as cutting point and removes very small amount of material through direct mechanical contact. Hence, grinding process may be considered as multipoint cutting process. Grinding is a finishing process and following are the salient feature of grinding process.

1. Materials which are too hard for machining by other machining processes such as hardened steel, tool and die steel can be machined by grinding process.
2. Close dimensional accuracy of the order of 0.3 to 0.5 micro meter is achieved.
3. High degree of surface finish ($Ra = 0.15$ to 1.25 μm) is achieved,

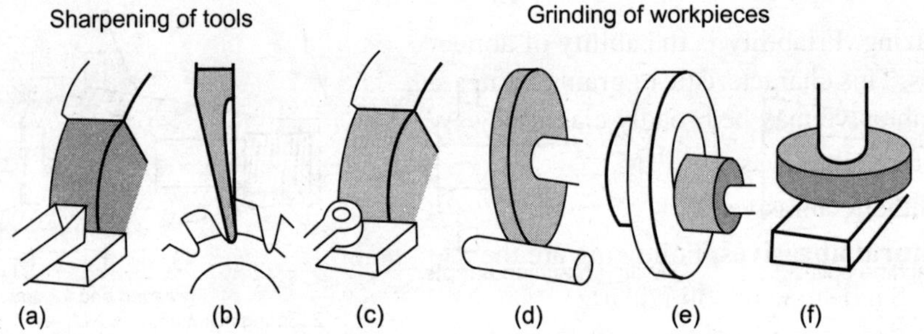

Fig. 10.50 Examples of grinding operations.(*a*) Sharpening of a turning tool, (*b*) sharpening of milling cutter, (*c*) deburring of a leaver, (*d*) cylindrical grinding of shaft, (*e*) internal grinding of a bush, (*f*) surface grinding of a rib.

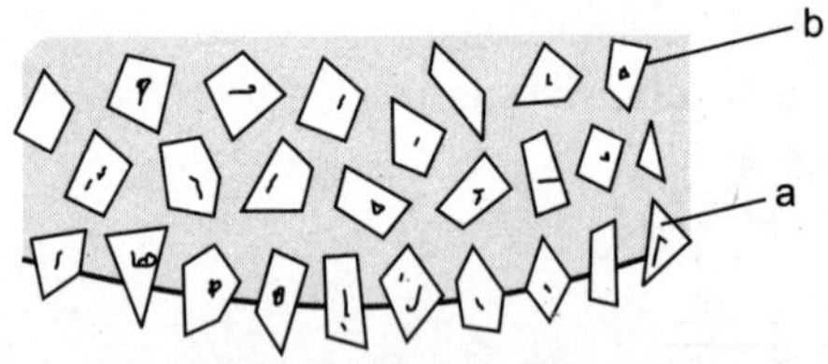

Fig. 10.51 (*a*) Grains, (*b*) bonding material. The figure Illustrate magnified view of the order surface of a grinding wheel. The grains (abrasive particles) act as cutting point.

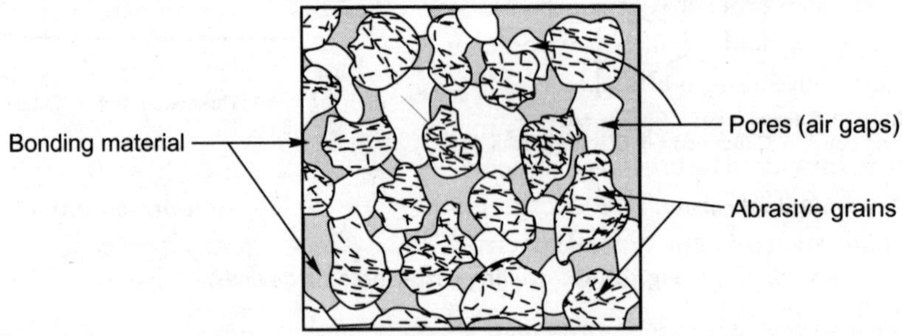

Fig. 10.51A. Illustrate the magnified inside view of a grinding wheel showing abrasive grains, bonding material and air gaps.

10.8.1 Composition of Grinding Wheel

The grinding wheel consists of hard and sharp edged abrasive grains which are embedded together in a bonding material (Grinding wheel). The characteristics of the grinding wheel as well as grinding process depends upon a number of variable such as type of abrasive grain, size and shape of abrasive grain, type of bonding material, balancing of grinding wheel and truing & dress of grinding wheels etc. These variables are described below.

10.8.2 Type of Abrasive

The abrasives are tiny sharp edge grains which should be hard and tough so that they will be able to act as cutting edges for sufficiently long time. Two other significant properties of abrasive grains are attrition and friability. Attrition is abrasive wear action of grains resulting in dull edges, grit flattening and wheel glazing. Friability is the ability of abrasive grain to fracture in small pieces when force on them increases. This characteristic of grains creates self sharpened and new sharp edges during grinding process. The abrasive may be broadly classified as:

1. Natural abrasives
2. Synthetic abrasive

1. **Natural abrasives.** Following are the natural abrasives:
   1. Sand stone or solid quartz
   2. Emery (50–60% crystalline and iron oxide)
   3. Corundum (70–90% crystalline and iron oxide)
   4. Diamonds

The natural abrasive are obtained directly from mines. The natural abrasive are soft and have impurities as a result there performance is not consistent. Almost all the natural abrasives except diamond are obsolete. In most cases synthetic abrasives are used for manufacturing grinding wheels.

2. **Synthetic abrasives:**
   (*a*) Aluminium Oxide, (Al_2O_3)
   (*b*) Silicon Carbide (SiC)
   (*c*) Cubic Boron Carbides (CBN)
   (*d*) Diamond

(*a*) **Aluminium Oxide (Al_2O_3).** Aluminium oxide is obtained by fusing the Aluminium ore (bauxite) in electric arc furnace. Aluminium oxide is a most widely used artificial abrasive. Although aluminium oxide is softer then silicon carbide, it is considerable tougher. It is used to grind steel, ferrous alloys, and high strength alloys.

(*b*) **Silicon Carbide.** Silicon carbide is made by charging silica sand, petroleum coke, salt and saw dust in electric furnace. The crystals of silicon carbide are very hard and rather brittle which restricts its use. Its grinding application include ductile metals such as Aluminium, brass and stainless steel as well as brittle metals such as some type of cast irons and ceramics. It is not used effectively for grinding steels because of chemical reaction between carbon (C) in silicon carbide and the iron (Fe) in steel.

(*c*) **Cubic boron nitride.** It is produced by combination of intensive heat and pressure in the presence of a catalyst. It is next to diamond in hardness. It maintains its hardness and chemical integrity even at higher temperature (approx. 1400°C) as compared to diamond which may begin to revert to plain carbon dioxide at 650°C. It is also quite expensive (10 to 20 times) then cost of conventional abrasives such as aluminium oxide. Its application includes mostly hard materials such as hardened tool steels and aero space alloys. It is less reactive with ferrous, nickle base and cobalt base super alloys.

(*d*) **Diamond.** Diamond is hardest of all materials. Diamond abrasives are either natural (garnets) or synthetic. The natural diamond which are not suitable for gemes are used specifically for abrasive purposes.

Synthetic diamonds are more friable and thus cut faster and have low co-efficient of thermal expansion. Diamond abrasive wheels are extensively used for sharpening carbide and ceramic cutting tools. These can also used for truing and dressing other type of abrasive wheels.

Grain Size and Structure

As we know that in grinding process, each abrasive grain may be considered as tiny cutting tool. The size of abrasive grain has a vital role for providing surface finish. The fine grain would take very small depth of cut and hence better surface finish is obtained. The size of grain, usually called as grit is denoted by a number. This number indicates the number of openings per square inch. A number 60 grit would pass through a standard sieve having 60 openings per inch but it would not pass through one having 80 openings per inch. Therefore larger the number, smaller (finer) is the grain size and vice versa. Fine grains varies from grain size 500 to 600 where as coarse grain varies from 6 to 8. Coarse grain are good for higher material removal rate where as fine grains are good for high surface finish.

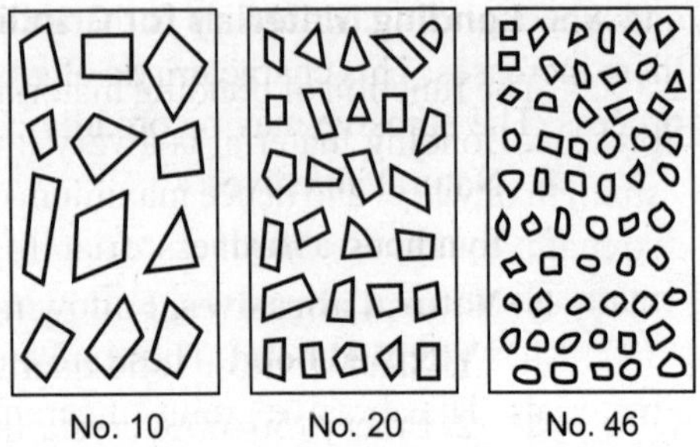

Fig. 10.52 Examples of various grain sizes (grits) (drawn exaggerated).

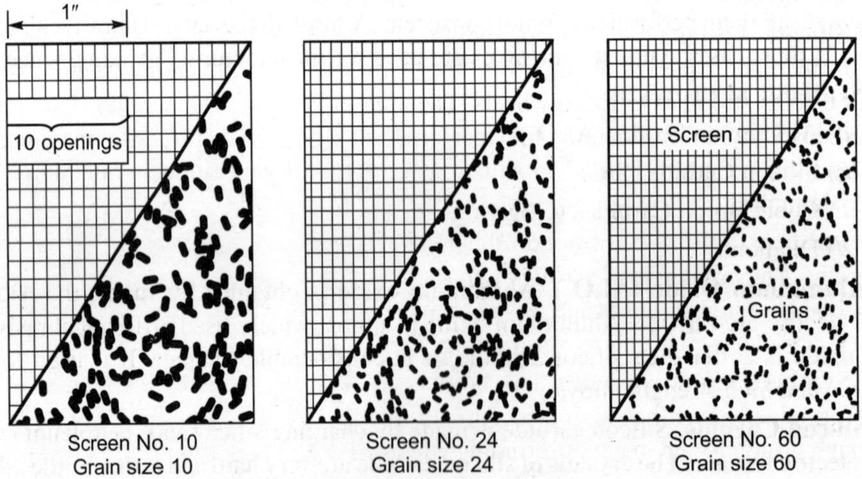

Fig. 10.53 Screen number and grain sizes. As the screen no. (no. of openings per unit inch) increases the grain size decreases.

Structure of Grinding Wheel. The spacing of abrasive grains in grinding wheel is called structure. Closely spaced grains means dense structure whereas widely spaced grains are open structure. Open structured grinding wheel allows chips to collect but have fewer cutting edges per unit area. Open structure are used for high material removal rate and produce rough surface finish. Dense structures are used for better surface finish and dimensional control.

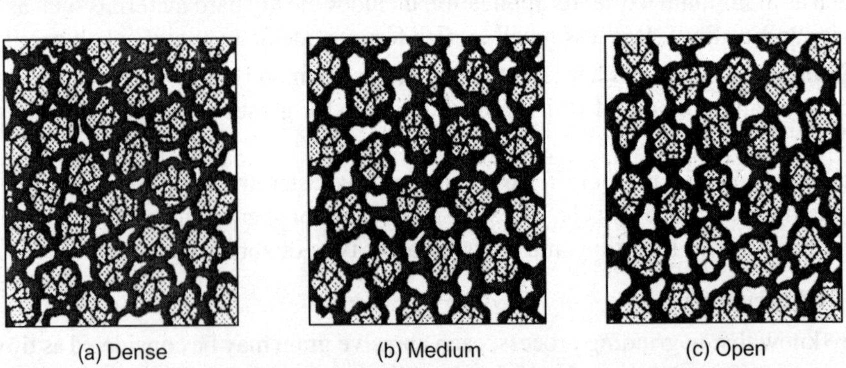

Fig. 10.54 Grinding wheel structure.

10.8.3 Bonding Materials for Grinding Wheels

The function of bonding material is to keep the abrasive grains together during grinding (cutting) process. Bonding material is a very important factor in selection of grinding wheel. It determines the strength of wheel and hence maximum operating speed. Desirable properties of bonding material include strength, toughness, hardness, friability and temperature resistance etc. The commonly used bonding materials are:

1. **Vitrified bond.** These are most widely used bonds. The bond is composed of clay and ceramic materials. The abrasive grains of ceramic material are mixed with wet clay so that each grain is coated with clay. They are made by pressing drying and firing in a kiln at about 1250°C. It results in a bonding material becoming hard and strong. Vitrified grinding wheels are porous, rigid, resistant to elevated temperatures, and are not affected by grinding fluid (water and oil). However these wheels are sensitive against shock and blow.

2. **Silicate.** In this type of bond, sodium silicate ($NaSiO_3$) is used as bonding material. The wheel are formed and baked at about 500°C for a day or more. On heating, sodium silicate hardens. However it is to as strong as vitrified wheels. The abrasive grains in silicate bond are more brittle and are released quickly and consequently they perform cutting at lower surface temperature than vitrified wheels. These are generally limited to situations where heat generation is minimum. They are also sensitive to dampness.

3. **Rubber.** In this type of bond rubber, sulphur and other vulcanizing agents are mixed with abrasive grains. The mixture is then rolled into sheets. This is the most flexible bond and can resist side thrust. The bond has high strength and less porocity. These are used as a bonding material in thin cut off wheel. These are also used in centreless grinding. These are affected by dampness and alkaline solution.

4. **Synthetic resin (resinoid).** This bond is made of various thermosetting resin materials such as phenol formaldehyde. It has high strength and are more elastic than vitrified bond. It is generally used for rough grinding, and in cut off grinding wheels. These wheels produce finer finish than vitrified wheel.

5. **Shellac.** These are made by mixing abrasive grains with shellac, pressing and rolling in desired shape and baking for several hours at **300°F**. The wheels are relatively strong and have some elasticity. They have limited applications and a generally used for high surface finish operations like rolls, camshaft and cutlery grinding.

6. **Metal.** Metallic bonds (generally bronze) are the common bond materials for diamond and cubic boron nitride grinding wheel. Powder metallurgy techniques are used to make the abrasive periphery of the wheel. The wheels are made of high thermal conductive material like Cu or Al alloys. The grains are firmly held in shock resistant bond and can not be knocked out until fully utilised.

10.8.4 Grinding Grade

The grinding grade indicates how strongly abrasive grains are held in grinding wheel. In other words, grinding grade indicates wheel strength to retain abrasive grains during grinding (cutting). It depends upon two factors (*i*) the strength of bonding material and amount of bonding material connecting the grains. If the strength amount of bonding agent is high, it will hold the grains firmly and large force is required to detach the grain from wheel. These are called hard wheels.

— Hard wheels are generally used for grinding soft materials while soft wheels are used for grinding hard materials.

During grinding it is strongly recommended that worn out grains should detach from the wheel to expose new sharp edges. Therefore

— While grinding hard material with soft wheel, the grain are likely to become dull quickly thereby increasing grinding force which tend to detach dull abrasive grains. This keeps the grinding wheel in sharp condition.

— While grinding soft material with hard wheel, the grinding wheel will be able to hold the grain for longer period and results in high material removal rate.

10.8.5 Grinding Wheel Specification

The figure 10.55 represents standard marking system for grinding wheels as per ANSI standard B74.13–1970. In the following example A grinding wheel marked 30 A46M6V23 is specified:

Sequence :	Prefix	1	2	3	4	5	Suffix
	Manfacturer symbol	Abrasive type	Grain size	Grade	Structure	Bond type	Manufacture's record
	30	**A**	**46**	**M**	**6**	**V**	**23**

Prefix	Abrasive type	Abrasive (Grain) Size: Coarse	Medium	Fine	Very fine	Grade: Soft	Medium	Hard	Structure: Dense to open	Bond type	Manufacture's record
Manufactuer's symbol indicating exact kind of abrasive (optional)	A – Aluminium oxide	8	30	70	220	A E	I M	Q V	1 9	B – Resinold	Manufactuer's private marking to identify wheel (optional)
		10	36	80	240	B F	J N	R W	2 10	BF – Resinold reinforced	
		12	46	90	280	C G	K O	S X	3 11		
		14	54	100	320	D H	L P	T Y	4 12	E – Shellac	
	C – Silicon carbide	16	60	120	400			U Z	5 13	O – Oxychloride	
		20		150	500				6 14	R – Rubber	
		24		180	600				7 15	RF – Rubber reinforced	
									8 16	S – Silicate	
									etc.	V – Vetrifed	
									(Optional)		

Fig. 10.55 Standard marking system for grinding wheels (ANSI Standard B74.13-1970).

Table: Grinding wheel hardness for different work materials

Workpiece material	*Wheel hardness*			
	Cylindrical grinding	*Surface grinding*	*Internal grinding*	*Deburring*
Steel up to 80 kg/mm^2	L, M, N	K, L	K, L	
Steel up to 140 kg/mm^2	K	K, J	J	O, P, Q, R
Steel more than 140 kg/mm^2	J	I, J	I	
Light alloys	J	I, K	I	
Cast iron	K	J	J	
Bronze, brass and copper	L, M	J, K	J	

10.8.6 Grinding Wheel Shapes

1. Straight
2. Cylinderical
3. One side recessed
4. Two side recessed
5. Tapered
6. Straight cup
7. Flaring cup
8. Dish

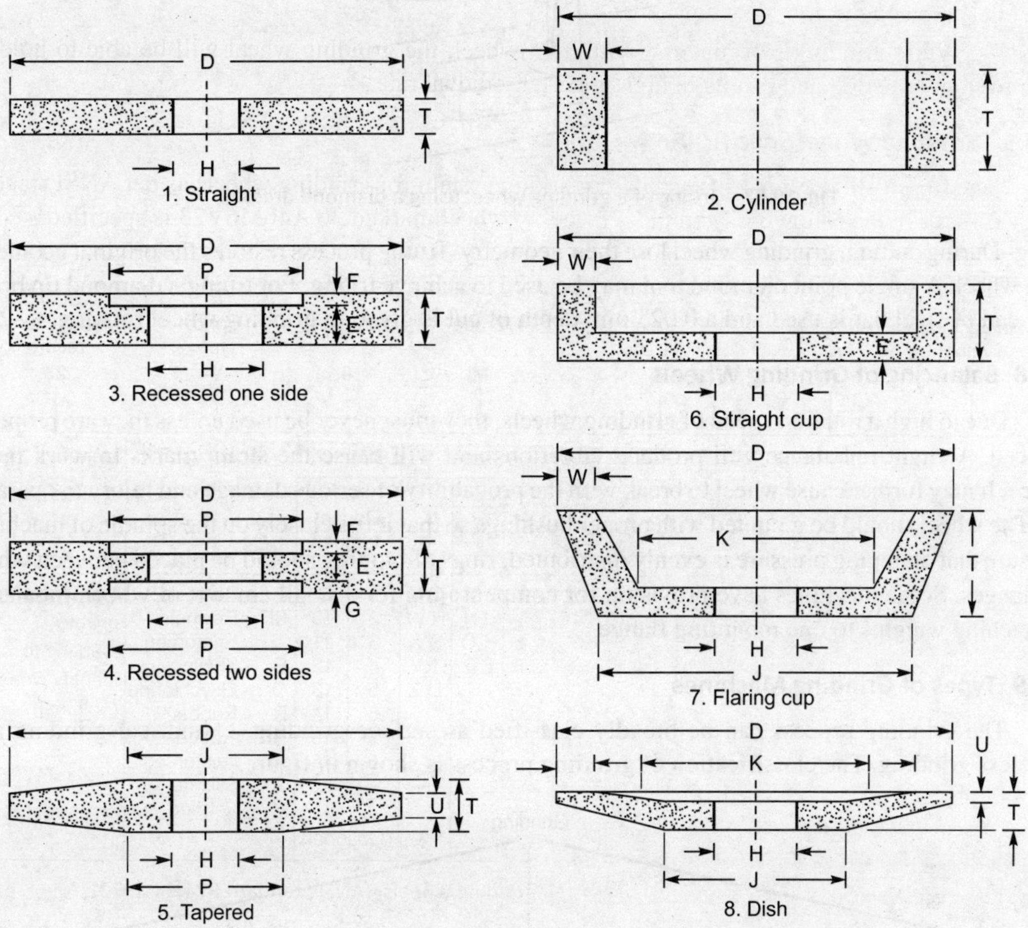

Fig. 10.56 Standard grinding-wheel shapes which are commonly used.

10.8.7 Truing and Dressing

As the grinding wheel is used, the sharp abrasive grains become worn and dull. In addition, some metal chips are embedded in the spaces between the grains with the resulting condition known as loaded wheel. Loading of wheel is generally caused while cutting soft and ductile materials. The loaded wheel reduces the efficiency of grinding wheel and also affects surface finish of surface.

To restore the grinding efficiency. It is necessary to resharpen abrasive grain and to clean the spaces between grains. It is known as dressing.

Dressing: Dressing can be done using metal stick, abrasive stick, abrasive wheel and single point diamonds etc. These sticks when come in contact with grinding wheel face will sharpen the wheel by removing the dull grain and exposing the sharpen grains.

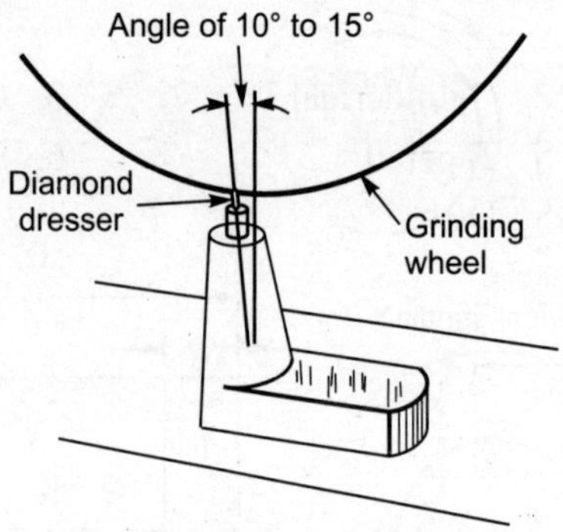

Fig. 10.57 Truing of a grinding wheel using a diamond dresser.

Truing–During cutting, grinding wheel lose their geometry. Truing process restores the original geometry of the wheel. A single point diamond tool may be used to achieve truing. For truing a diamond tip braze at the end of steel bar is used and a 0.025 mm depth of cut is given to grinding wheel rotating.

10.8.8 Balancing of Grinding Wheels

Due to high a rotation speed of grinding wheels, they must never be used unless they are properly balanced. A slight imbalance will produce vibrations and will cause the strain marks in work piece surface. It may further cause wheel to break with the probability of serious damage and injury to operator also. The wheel should be mounted with proper bushings so that it fits closely on the spindle of machine. To assure that clamping pressure is evenly distributed, rings of blotting should be placed between wheel and flanges. Some machines have provision for compensating for a small amount of wheel imbalance by attaching weights to one mounting flange.

10.8.9 Types of Grinding Machines

The grinding process can be broadly classified as surface grinding, cylindrical grinding and centreless grinding. The classification of grinding process is shown in figure.

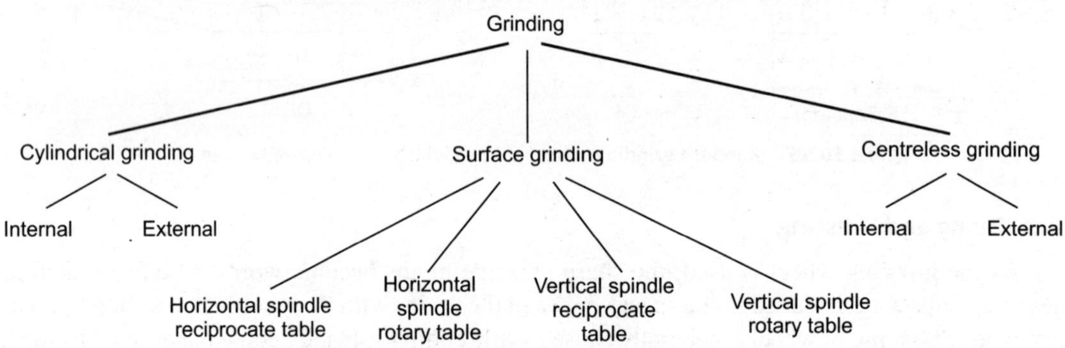

Fig. 10.58 Classification of grinding process.

Grinding machines can be broadly classified:

1. Surface grinding machines
2. Cylindrical grinding machines
3. Centreless grinding machines.

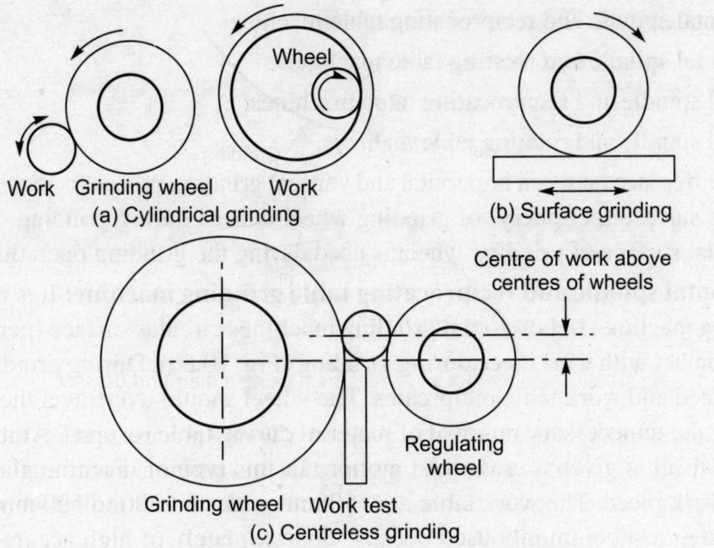

Fig. 10.59 Various grinding operations.

Table: Grinding Machines

Type of Machine	*Type of surface*	*Specific Types or Features*
Cylindrical External	External surface on rotating, usually cylindrical parts	work rotated between centers Centreless Chucking Tool post Crankshaft, cam, etc.
Internal	Internal diameters of holes	Chucking Planetary (work stationary) Centerless
Surface Conventional	Flat surfaces	Reciprocating table or rotating table Horizontal or vertical spindle
Creep feed	Deep slots, profiles in hard steels, carbides and ceramics using CBN and diamond	Rigid, chatter-free, creep feed rate Continuous dressing Heaving coolant flows NC or CNC control Variable speed wheel
Tool grinders	Tool angles and geometries	Universal Special
Other	Special or any of the above	Disk, contour, thread, flexible shaft, swing frame, snag, pedestal, bench

1. **Surface grinding machines.** These are used to grind flat surface. These machine are similar to milling machine in construction as well as in motion. There are four basic types of surface grinding machines.

(*a*) Horizontal spindle and reciprocating table machines

(*b*) Horizontal spindle and rotating table machines.

(*c*) Vertical spindle and reciprocating table machines.

(*d*) Vertical spindle and rotating table mahines.

The basic difference between horizontal and vertical grinding process is that in horizontal spindle machines, circular surface (periphery) of grinding wheel is used during grinding. Whereas in vertical spindle machine flat surface of grinding wheel is used during the grinding operation.

(*a*) **Horizontal spindle and reciprocating table grinding machine:** It is most common types of surface grinding machine. In horizontal grinding machines circular surface (periphery) of grinding wheel comes in contact with workpiece during grinding (Fig. 10.60). During grinding, grinding wheel is provided with feed and work table reciprcates. The wheel should overtravel the work piece at both the end to prevent the unnecessary removal of material during table reversal. At the end of each table motion grinding wheel is given a cross feed motion. In this type of machine, the width of wheel is less than that of work piece. The work table are 150 mm wide and 300 to 500 mm long. This type of surface grinders are most commonly used because approximately of high accuracy and fine surface finishes.

(*b*) **Horizontal spindle rolating table grinding machine.** In these machines, spindle traverses horizontally from edge of table to centre of table. The infeed is provided by moving the rotary work table into the grinding wheel. These type of machines are not used to a greater extent.

(*c*) **Vertical spindle reciprocating table grinding machine.** In these machines, grinding wheel is cylindrical and flat surface comes in contact with workpiece during grinding. The workpiece is feed by reciprocating motion of the work table. The diameter of grinding wheel is higher than the surface to be ground and as a result no traverse feed is requied.These machines produces very flat surface and also remove large amount of material. They are used for high production work.

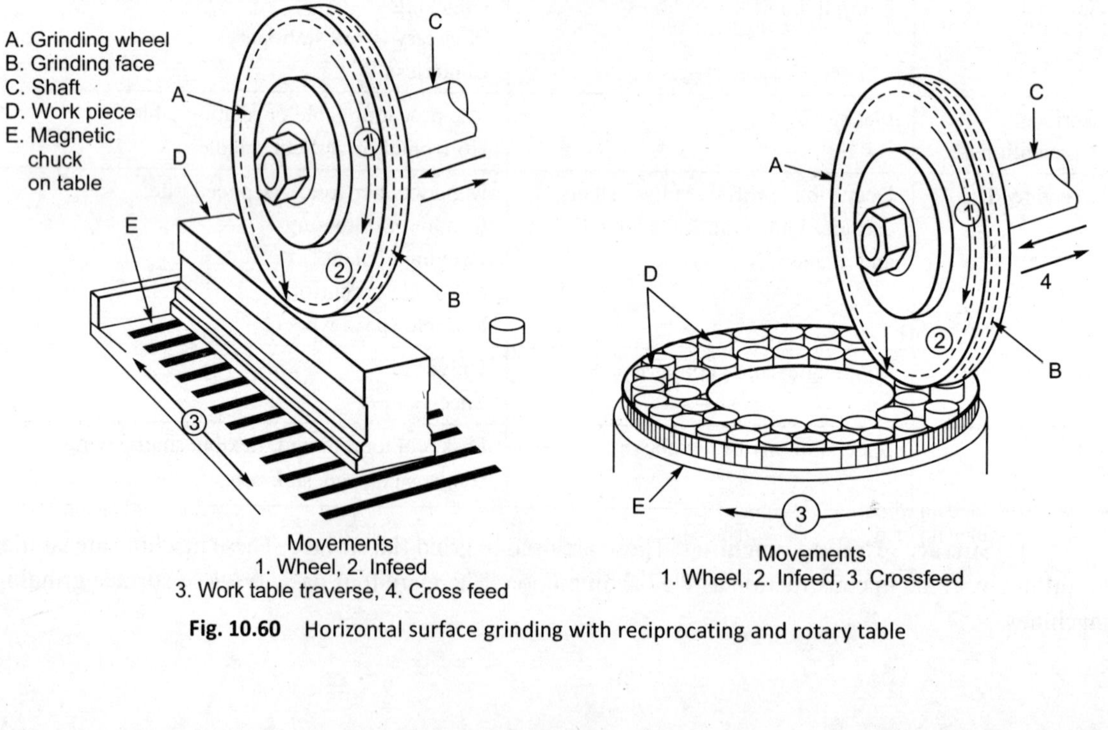

Fig. 10.60 Horizontal surface grinding with reciprocating and rotary table

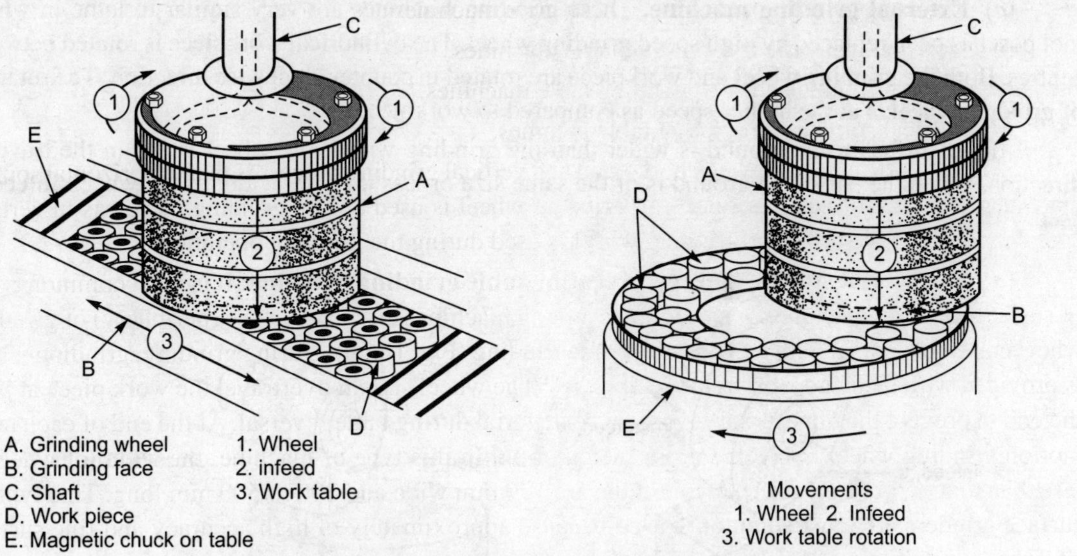

Fig. 10.61 Vertical surface grinding with reciprocating and rotary table.

(*d*) **Vertical Spindle rotating table machine.** In this case also flat surface of grinding wheel cuts rather than periphery. These are ususally machines of higher capacity. Here, both the workpiece and grinding wheel rotate and feed into each other. These machines have higher material removal rate and are used for production work. The work pieces are held on magnetic chuck or special fixture attached to the table.

2. **Cylindrical grinding (centre grinding)**

Cylinder grinding process is used for grinding cylindrical surfaces. It is used for producing both external and internal surfaces. Two types of cylindrical grinding operations are external cylindrical grinding and internal cylindrical grinding.

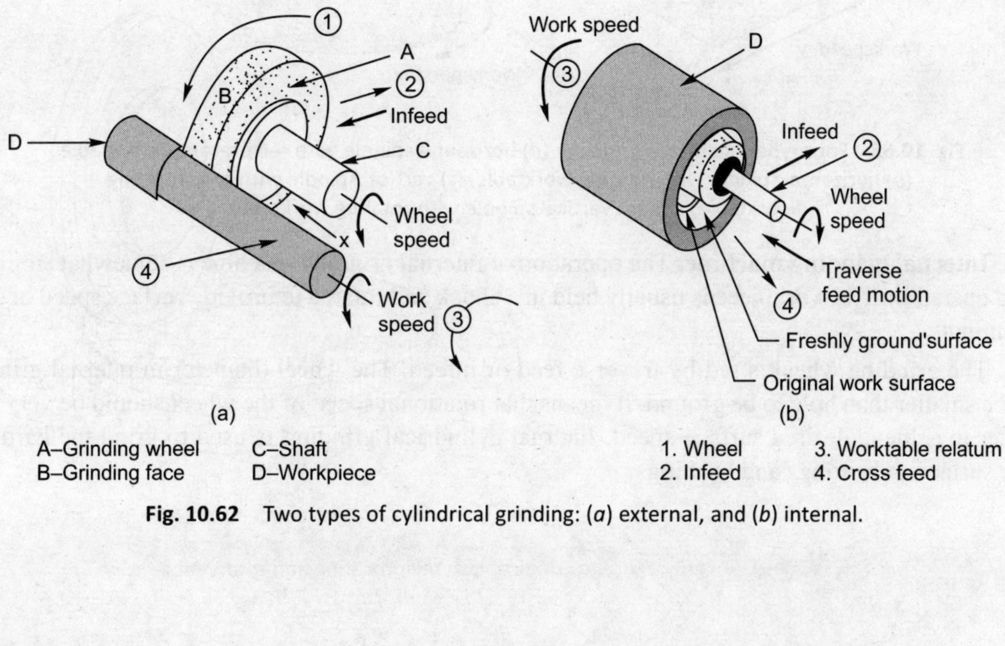

Fig. 10.62 Two types of cylindrical grinding: (*a*) external, and (*b*) internal.

(*a*) **External grinding machine.** These grinding machines are very similar to lathe in which tool post has been replaced by high speed grinding wheel. The cylindrical workpiece is rotated between centres. Both the grinding wheel and workpiece are rotated in counter clockwise direction. The rotation of grinding wheel is at the higher speed as compared to workpiece.

If the surface to be ground is wider than the grinding wheel, the wheel is fed in the traverse direction. When the part to be ground is of the same size or less than the width of the wheel, infeed is used.

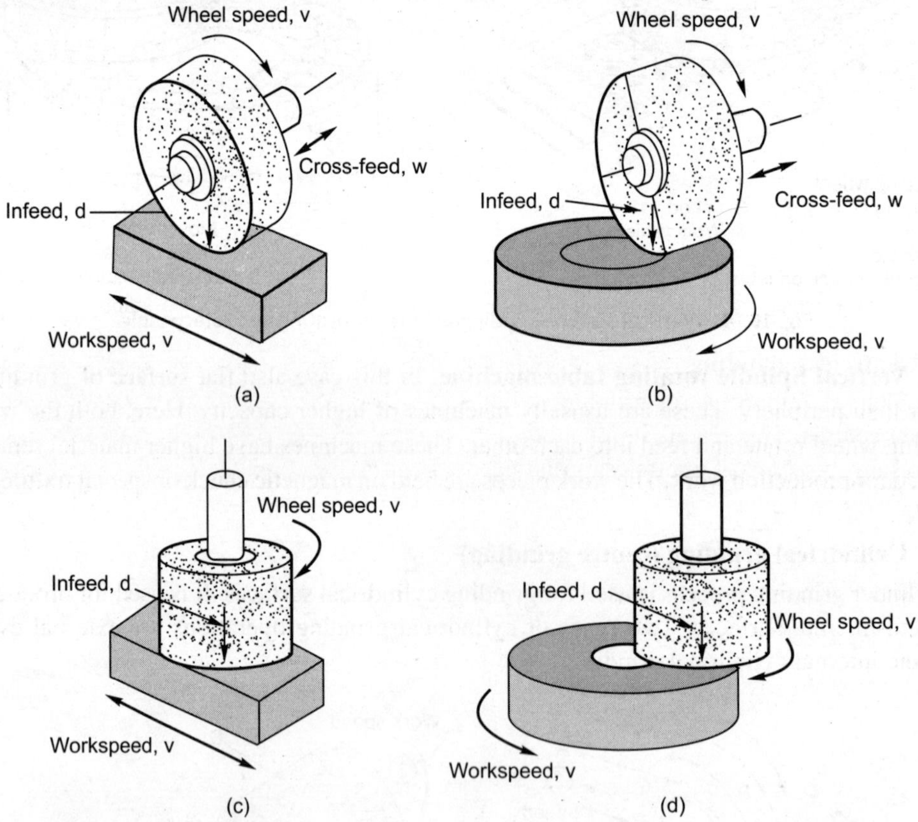

Fig. 10.63 Four types of surface grinding: (*a*) horizontal spindle with reciprocating worktable, (*b*) horizontal spindle with rotating worktable, (*c*) vertical spindle with reciprocating worktable, and (*d*) vertical spindle with rotating worktable.

Internal grinding machine. The operation of internal grinding machine is somewhat similar to boring operation. The workpiece is usually held in a chuck and rotated to provide surface speed of 20 to 60 m/minute.

The grinding wheel is fed by traverse feed or infeed. The wheel diameter in internal grinding must be smaller than hole to be ground. It means that rotational speed of the wheel should be very high in order to achieve desired surface speed. Internal cylindrical grinding is used to grind the hardened inside surface of bearings and bushings.

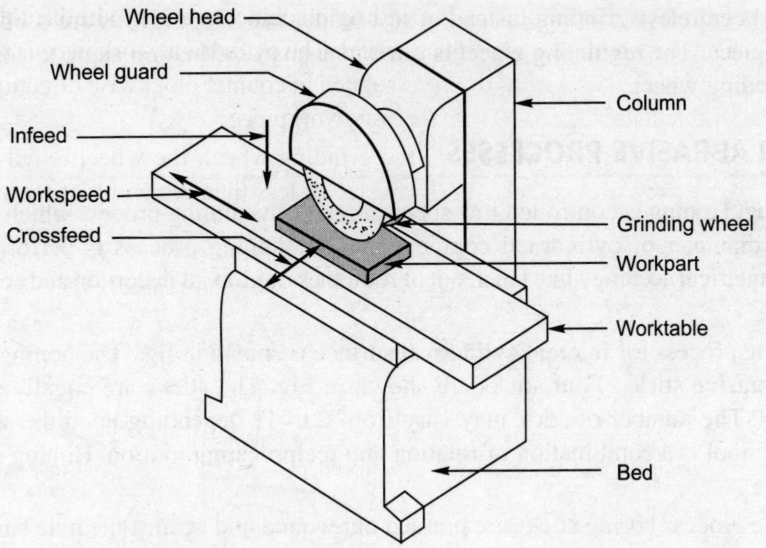

Fig. 10.64 Surface grinder with horizontal spindle and reciprocating worktable.

3. **Centreless grinding**

It is alternative process for grinding cylindrical workpieces. In this case the workpiece is not held between centres. This results in reduction of workpiece setting time, hence used for production work. There are two type of centreless grinding processes namely external centreless grinding and internal centreless grinding.

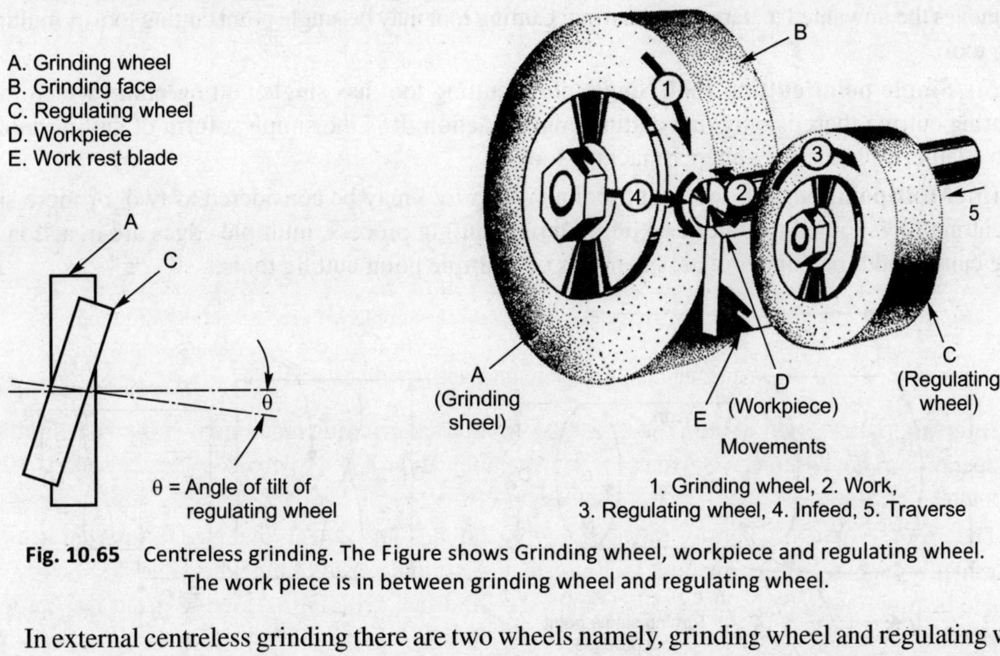

Fig. 10.65 Centreless grinding. The Figure shows Grinding wheel, workpiece and regulating wheel. The work piece is in between grinding wheel and regulating wheel.

In external centreless grinding there are two wheels namely, grinding wheel and regulating wheel. The workpiece are supported by rest blade (as shown in figure) and feed in between two wheels as shown in Fig. (10.65). The regulating wheel rotates at much lower speed and is inclined at a slight angle to control the through feed of workpiece.

In Internal centreless grinding instead of rest blade, two support rolls are used to maintain the position of workpiece. The regulating wheel is tilted at a small inclination angle to control the feed of work against grinding wheel.

10.9 OTHER ABRASIVE PROCESSES

(*a*) **Honing.** Honing is controlled low speed abrasive machining process which is used most for finishing inside diameter of cylindrical components. The honing process is performed to achieve, correction of geometrical accuracy like taper, out of roundness and axial distortion and achieve dimension accuracy.

The honing process for internal cylindrical surface is shown in fig. The honing tool consists of set of bonded abrasive sticks. Four sticks are shown in Fig. The sticks are equally spaced over the periphery of tool. The number of stick may vary from 2 to 12 depending upon the size of hole. The motion of honing tool is a combination of rotation and reciprocating motion. Honing speeds are 15 to 150 metre/minute.

During the process boring sticks are pressed out wound and against the hole surface to produce desired abrasive cutting action. The abrasive grain used range from 80 to 600 grit.

10.10 MISCELLANEOUS TOPICS

10.10.1 Cutting Tools

We had seen in previous paragraph that in material removal processes unwanted material is removed in form of chips from raw material by an harder tool. This tool is called cutting tool. Cutting tool removes the unwanted material by shearing. Cutting tool may be single point cutting tool or multipoint cutting tool.

(*a*) **Single point cutting tool.** Single point cutting tool has single cutting point, which means that during cutting there is only one cutting edge in action. It is the simplest form of cutting tool and used in a lathe, a planner or a shaper machine tool.

(*b*) **Multipoint cutting tool.** Multipoint cutting tool may be considered as two or more single point cutting tools joined together as a unit. During cutting process, multiple edges are in action. The milling cutter and broaching tool are examples of multiple point cutting tools.

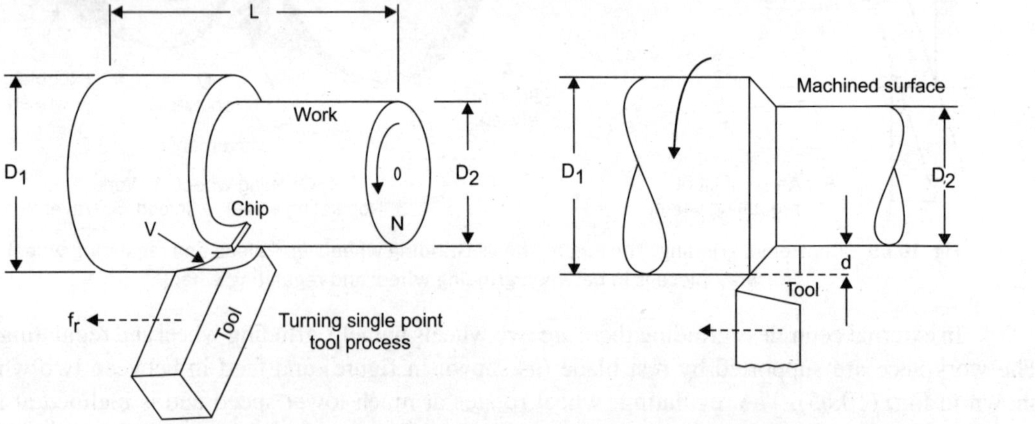

Fig. 10.66 Single point cutting tool (in cylindrical turning operation).

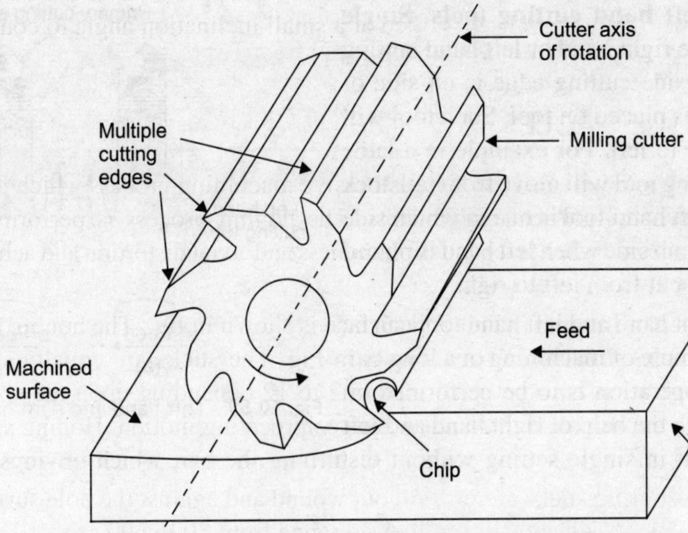

Fig. 10.67 Multipoint cutting tool (in milling operation).

Following are the terms associated with single point cutting tool.

(*a*) **Shank.** Shank is the main body of tool.

(*b*) **Flank.** The adjacent surface below the cutting edges are called flank of the tool.

(*c*) **Face.** The surface on which chip slides after cutting is called face of the tool.

(*d*) **Nose.** It is the point where side cutting edge and end cutting edge intersect.

(*e*) **Cutting edge.** It is the edge on the face of the tool. Cutting edge consists of side cutting edge and end cutting edge. Side cutting edge is formed by intersection of face and side flank. End cutting edge is formed by intersection of the face and the end flank.

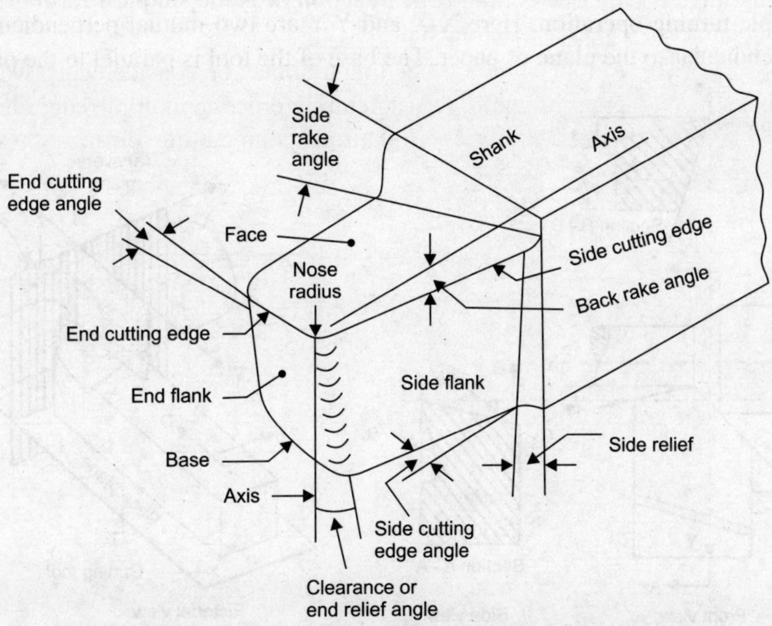

Fig. 10.68 Terms associated with single point cutting tool.

(*f*) **Right and left hand cutting tools.** Single point cutting tool may be right hand or left hand cutting tool. In right hand tool side cutting edge is on side of thumb when right hand is placed on tool. Such tool will cut when fed from right to left. For example in a lathe machine right hand cutting tool will move from tailstock towards head stock. A left hand tool is one in which side cutting edge is on the thumb side when left hand is placed on tool. Such a tool will cut from left to right.

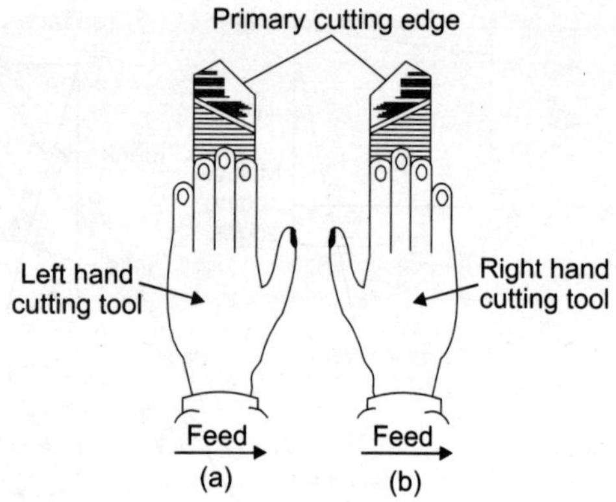

Fig. 10.69 Left hand and right hand cutting tools.

The utility of right hand and left hand tool can be explained by taking example of machining of a long bar. Suppose some cutting operation is to be performed on both ends of the bar. With the help of right hand and left hand tools we can cut it in single setting without disturbing the bar, which obviously means more accurate machining.

10.10.2 Nomenclature of Single Point Cutting Tool

The nomenclature or designation of cutting tools means the designation of the geometry of the cutting part of a tool Fig. 10.70 and Fig. 10.71.

The two systems for nomenclature of tool are:

1. American Standard Association System (ASA) or American National Standard Institute.
2. Orthogonal Rake System.

ASA system. In ASA system the angles of tool face are defined in two mutually perpendicular planes (orthogonal planes). One plane is perpendicular to the axis of cutting tool whereas other is parallel to the axis of cutting tool. Both planes are perpendicular to the base of cutting tool. Fig. 10.70 shows tool geometry and various angles of right hand single point cutting tools. Fig. 10.70 shows ASA system for simple turning operation. Here X-X and Y-Y are two mutual perpendicular planes. Both planes are perpendicular to the plane of paper. The base of the tool is parallel to the plane of paper.

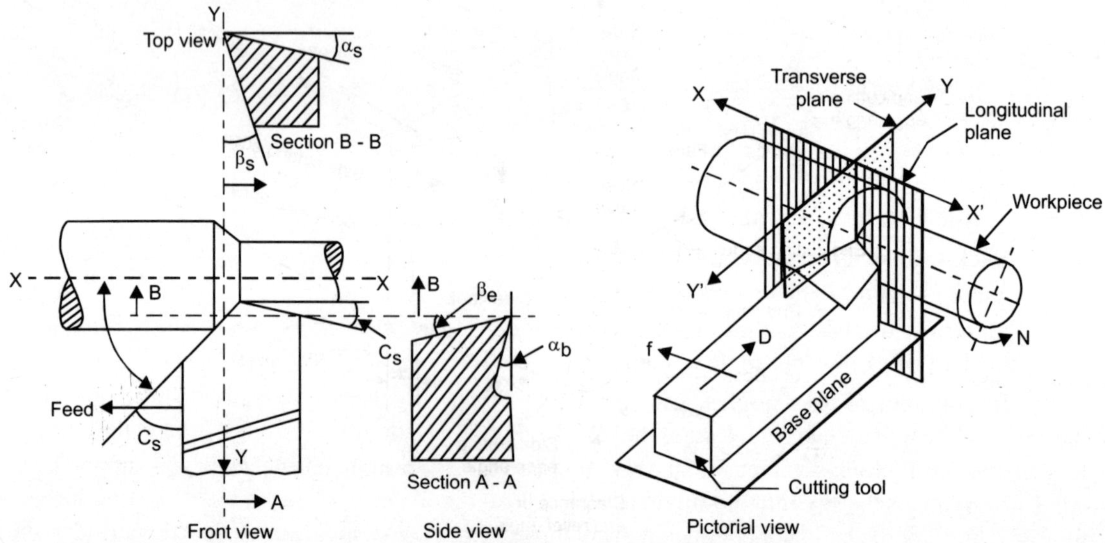

Fig. 10.70 Workpiece, cutting tool and reference planes in ASA (American Standard Association System).

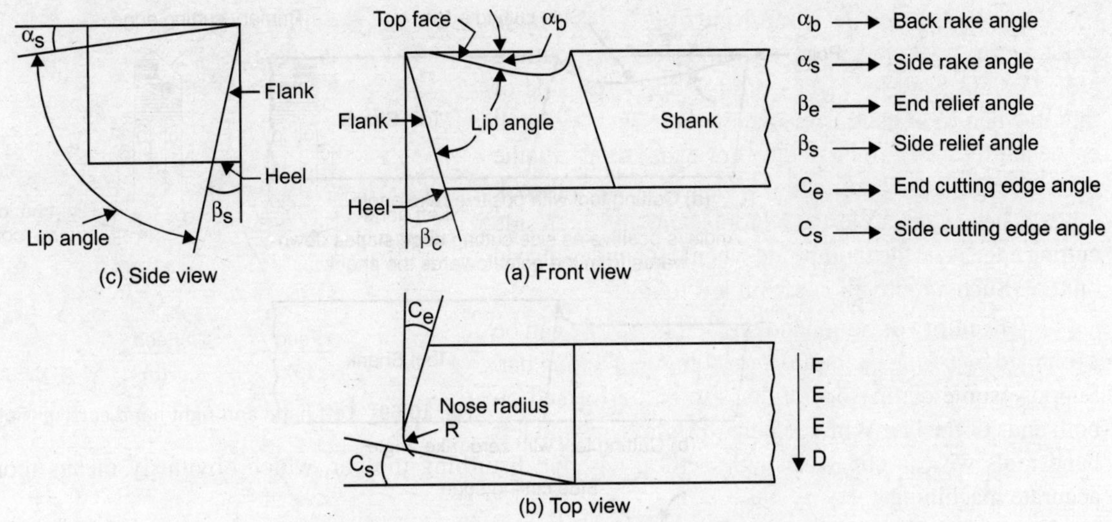

Fig. 10.71 Various angles in single point cutting tool.

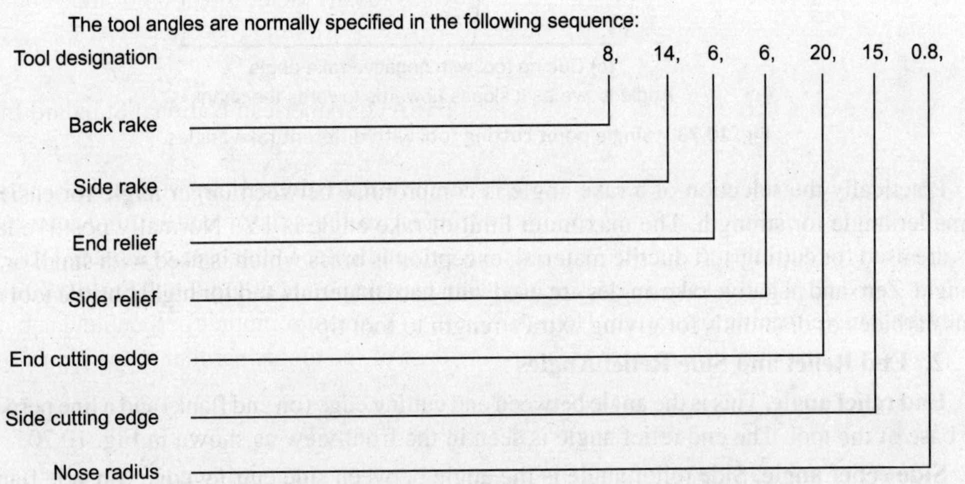

Fig. 10.72 Sequence for specification of tool angles in ASA.

Following are the various angles in a single point cutting tool:

1. Back Rake (Top rake) and Side Rake Angles

Back rake or (Top rake) angle. It is the angle between side cutting edge (on the tool face) and a plane parallel to the shank of tool. This angle is seen in front view.

Side rake angle. It is the angle between end cutting edge (on the tool face) and a plane parallel to the shank of the tool. This angle is seen in side view.

Importance of rake angles. Rake angle may be positive, zero, or negative. This angle specifies the ease with which metal is cut. Larger the rake angle smaller the cutting angle, better is the cutting and less are the cutting forces. Increasing the rake angle reduces the amount of metal which supports cutting edge. This reduces the strength of tool tip as well as heat dissipaation through the tool. When strength of tool is reduced and heat is not conducted away the tool may wear out or fail in shorter periods of time.

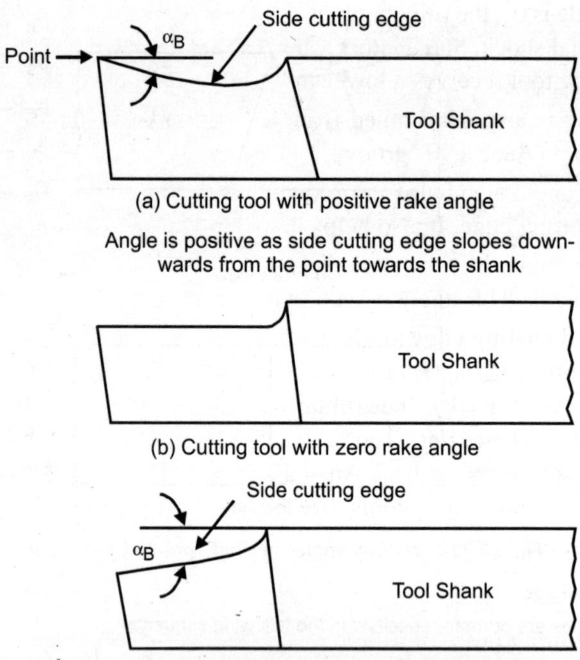

(a) Cutting tool with positive rake angle
Angle is positive as side cutting edge slopes downwards from the point towards the shank

(b) Cutting tool with zero rake angle

(c) Cutting tool with negative rake angle
Angle is –ve as it slopes upwards towards the shank

Fig. 10.73 Single point cutting tool with different rake angles.

Practically the selection of a rake angle is compromise between larger angle for easier cutting and smaller angle for strength. The maximum limit of rake angle is 15°. Normally positive large rake angles are used for cutting soft ductile material, exception is brass which is used with small or negative rake angle. Zero and negative rake angles are used with hard materials and for highly brittle tool materials such as carbides or diamonds for giving extra strength to tool tip.

2. End Relief and Side Relief Angles

End relief angle. This is the angle between end cutting edge (on end flank) and a line perpendicular to the base of the tool. The end relief angle is seen in the front view as shown in Fig. 10.70.

Side relief angle. Side relief angle is the angle between side cutting edge (on side flank) and a line perpendicular to the base of the tool. The angle is seen in side view.

Importance of relief angles. The relief angles allow the tool to cut without rubbing on the workpiece. These angles reduce the cutting force and hence power consumed is less. These angles mainly effect tool life and surface finish of workpiece. Relief angles in range of 5 to 10° are normally used for high speed steel. For carbides lower values of relief angles are used to give added strength to the tool. Too large relief angles reduce the strength of tool tip and also there is less material at cutting edge to conduct away the heat.

3. Side Cutting and End Cutting Edge Angles

Side cutting edge angle. It is the angle between side cutting edge and the plane parallel to the plane of tool shank. This angle is seen in top view.

Importance of side cutting edge angle. Side cutting edge angle enables the tool to contact the workpiece first behind the tip at the start of cut and thus helps in protecting tool tip. This angle may vary from 0° to 90°. When angle is 0°, the full length of cutting edge is in contact with workpiece at once and tool receives a strong initial shock. Satisfactory value of side cutting edge angle is 15° to 30°. With 30° side cutting edge angle the tool receives a lower initial shock from workpiece.

The value of angle is also determined from shape of workpiece. For example 0° side cutting edge angle is desirable to produce a 90° groove.

Side cutting edge angle also helps in producing thinner and wider chips which distribute heat produced over more of cutting edge. It also helps in avoiding Built Up Edges (BUE).

End cutting edge angle. It is the angle between end cutting edge and a plane perpendicular to the base of the tool and shank. This angle is seen in top view.

Importance of end cutting edge angle. End cutting edge angle provides a clearance to the non cutting portion of end cutting edge so that there is no rubbing action between the machined surface and non cutting portion of end cutting edge. The rubbing on the tool tends to create vibration and which in turn may cause noise (called chattering). Too large cutting edge angle reduces the metal that supports the cutting edge and conducts away the heat. An angle of about 8° to 15° has been found satisfactory in most cases like boring tools and turning tools. Cut off and necking tools often have zero value of end cutting edge angle.

10.10.3 Cutting Parameters

For all the metal cutting processes three major cutting parameters are speed, feed and depth of cut. These parameters have an important influence upon metal cutting. Let us have the example of turning operation of lathe machine to explain these parameters.

1. **Cutting Speed.** It is the surface speed or rate at which workpiece moves with respect to cutting tool. It is measured in meter/second. In lathe machine cutting speed is the motion which relates the peripheral velocity of rotating workpiece with respect to stationary cutting tool. Cutting speed is expressed by formula $V = \frac{\pi DN}{60}$ = meters/second, where D is the diameter of rotating workpiece, N is the revolution per minute.

Selection of cutting speed depends on:

(*i*) Work material
(*ii*) Cutting tool material
(*iii*) Depth of cut and feed
(*iv*) Tool life and
(*v*) Condition of machine tool and cutting tool.

2. **Feed.** Feed is the relatively small movement per cycle of the cutting tool relative to the workpiece in a direction which is usually normal to cutting speed direction.

- For lathe turning operation it is rate at which cutting tool advances into the workpiece per revolution. So for turning operation it is expressed in mm/revolution.
- For a shaper it is the rate at which cutting tool advances along the workpiece during one stroke. So for shaping it is expressed in mm/stroke.

The factors which affect the feed are:

(*a*) Smoothness and surface finish required
(*b*) Depth of the cut
(*c*) Tool life
(*d*) Power available, and condition of the machine tool.

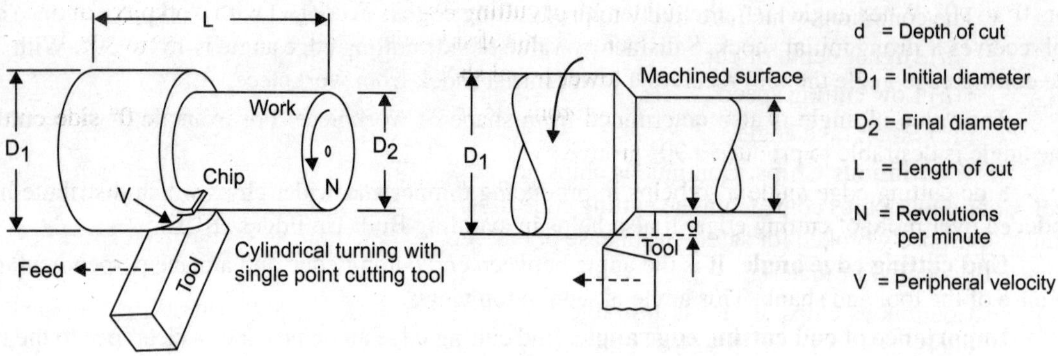

$$\text{DOC (Depth of cut)} = \frac{D_1 - D_2}{2} = d,\ V = \frac{\pi DN}{60}$$

Fig. 10.74 Various cutting parameters: Speed, feed and depth of cut in turning operation.

3. **Depth of Cut.** Depth of cut is the thickness of layer of metal removed in one cut or pass measured in a direction perpendicular to the direction of feed motion. It can be also defined as distance through which tool plunges into the surface of workpiece. It is difference between (D_1) initial diameter of workpiece and (D_2) diameter of workpiece after single cut or pass.

$$\text{Depth of cut} = \frac{D_1 - D_2}{2} = d$$

d = Thickness of layer of metal removed.

Depth of cut depends on

(*i*) Type of operation (whether rough or finishing operation)

(*ii*) Tool life

(*iii*) Power required.

10.10.4 Type of Chips

The chips produced during machining may vary both in shape and size. The shape and size of chip may depend upon the various factors like material of workpiece, tool geometry, cutting conditions (speed, feed, depth of cut), cutting fluid etc. Chip formation in machining may be broadly classified as

1. Discontinuous chips
2. Continuous chips
3. Continuous chip with BUE (Built up edges)

1. **Discontinuous Chips.** These chips are formed while cutting sufficiently brittle material such as cast iron and bronze. During machining metal ahead of tool is fractured into fairly small pieces.

- These types of chips are convenient to collect, handle and dispose off.
- As the cutting edge smoothern over the irregularities in workpiece a fairly good surface finish is obtained and tool life is also reasonably good.
- In case of ductile material they cause poor surface finish and low tool life. Discontinuous chips in case of ductile materials are indication of poor cutting condition.

➢ The conditions which are likely to produce discontinuous chips are:

(*a*) Higher depth of cut

(*b*) Low cutting speed

(*c*) Small rake angle.

2. **Continuous Chips.** Continuous chips are produced while machining ductile materials like steel. Continuous chips come from the cutting edge of cutting tool as single piece and will remain as it is unless purposely broken for safety or for ease of handling. Production of continuous chips is possible because of ductility of material.

➢ Continuous chips are most desirable since cutting is smooth and surface finish obtained is good.

➢ It helps in having higher tool life and lower power consumption.

➢ However, these chips are difficult to handle and dispose off as chip coils in a helix and curl around the workpiece. Chip breakers may be used to overcome this difficulty. Function of chip breakers is to reduce the radius of curvature of chip and thus break it.

➢ Conditions which promote the formation of continuous chips are:

(*a*) High cutting speeds

(*b*) Small chip thickness

(*c*) Large negative rake angle

(*d*) Less friction between chip tool interface through.

3. **Continuous Chips With Built Up Edges (BUE).** These types of chips are formed while machining ductile materials which have a high coefficient of friction. As the tool starts the cut some of the material because of high coefficient of friction builds up ahead of cutting edge. Some portion of the workpiece may even get welded on the tool point and thus is known as a built up edge (BUE).

➢ As some particle of chip will adhere to the tool tip and cutting is carried by this built up edge and not by actual tool tip. The life of cutting tool increases while cutting with BUE.

➢ BUE is not harmful while rough machining but surface smoothness is not as good as with the continuous type chips.

➢ The factors which reduce the BUE are:

(*a*) Higher cutting speed

(*b*) Increased rake angle

(*c*) Small chip thickness.

➢ In case of many ductile materials it is not possible to eliminate BUE entirely.

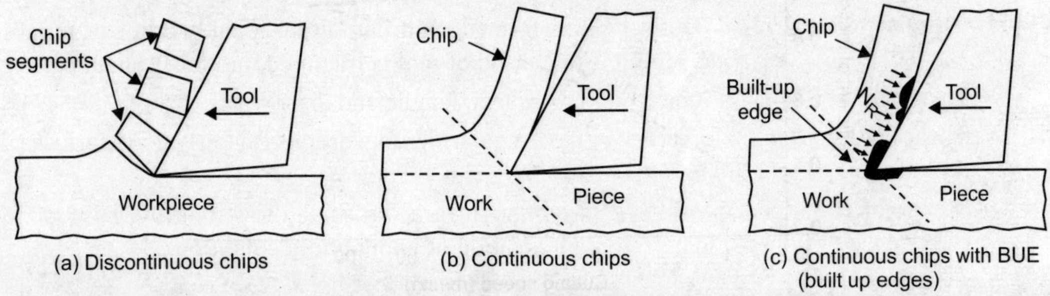

Fig. 10.75 Types of chips in machining processes.

10.10.5 Cutting Fluids or Coolants

Cutting fluids are frequently used in almost every machining process. Cutting fluid primarily acts as coolant and secondly as a lubricant. For some machining processes a cutting fluid which is predominantly a lubricant is desirable. For other machining processes a cutting fluid which is predominantly a coolant is desirable. Following are the two major functions of a cutting fluid.

1. Cooling action

(*a*) Cutting fluids cool down the tool workpiece zone by carrying away some of the generated heat.

(*b*) Removal of heat causes the reduction in temperature which greatly helps in retaining the hardness of tool and thus extending the tool life.

(*c*) Removal of heat reduces the thermal distortion of workpiece and permits better dimensional control.

2. Lubricating action

(*a*) Cutting fluids act as lubricant and reduces the co-efficient of friction at tool chip interface. Reduction in friction lowers the machining force and energy consumed in machining.

(*b*) Lubricating action prevents the formation of BUE (Built up edges) and surface finish is improved.

(*c*) Cutting fluids wash away the chips and clears the machining zone.

(*d*) Cutting fluid protects the both work and machine from corrosion.

Desirable characteristics of cutting fluids are

(*i*) It should have high specific heat and thermal conductivity.

(*ii*) It should have low viscosity.

(*iii*) It should have high flash temperature.

(*iv*) It should be non poisonous and non corrosive.

(*v*) It should be inexpensive and easily available.

Types of Cutting Fluids

Cutting fluids are mainly of two types:

1. **Water based cutting fluids.** Water very effectively works as coolant as it has high specific heat but causes the corrosion of work and tool and is not effective as a lubricant. Soda ash and trisodium phosphate are added to water to prevent corrosion.

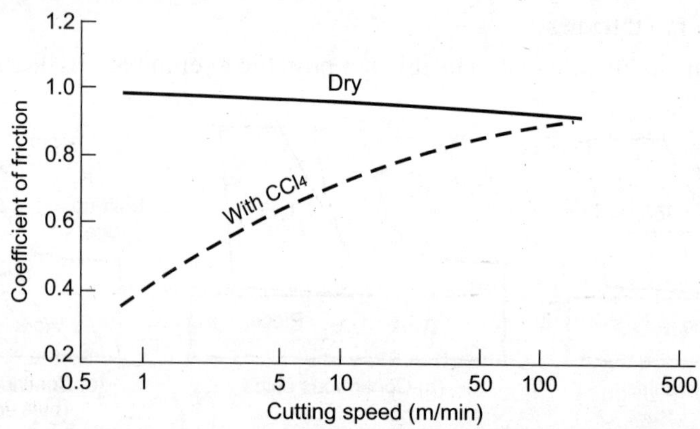

Fig. 10.76 Relationship between coefficient of friction speed when machining copper.

2. **Mineral oil based fluids.** Oil acts as less effective coolant but do not cause corrosion and have some lubricating properties. Sulphur and some times cutting oils are added to prevent the formation of BUE (Built up edges) on the cutting tool. Various additives are added to both of these types to achieve desirable properties.

The effectiveness of a cutting fluid also depends on the technique of application. It has been observed that when cutting zone is flooded coolant faces resistance to enter the cutting zone (Chip tool inter face) and effectiveness is thus not high, whereas if coolant is forcibly injected into cutting zone the fluid acts in more effective manner.

10.10.6 Tool Wear

During any machining process the cutting tool is subjected to severe cutting conditions like:

- Very high stresses
- High temperature
- High stress gradients
- High temperature gradients and
- Metal to metal contact with work and chip.

Because of these conditions the tool will start giving unsatisfactory performance after some time. Unsatisfactory performance may involve loss of dimensional accuracy, increased vibrations, increased surface roughness and increased power requirements. The unsatisfactory performance is due to tool wear and its continued use. When tool wears out it may be either reconditioned by grinding or may be replaced. Both cases will result in loss of production due to machine down time and/or loss of money because of replacing the worn out tool. Thus we may say that study of tool wear is important from stand point of performance and economics.

Tool wear depends upon large number of factors like:

(*i*) Type and hardness of tool
(*ii*) Condition and materials of workpiece
(*iii*) Feed, depth of cut, tool geometry and tools temperature
(*iv*) Cutting fluid, etc.

Tool wear may be classified as:

1. Flank wear
2. Crater wear
3. Localised wear such as rounding of cutting edge etc.

1. Flank Wear

- Flank wear is usually caused due to abrasion by hard particles and shearing of micro welds between tool and workpiece.
- Flank wear modifies the tool geometry and changes cutting parameters like depth of cut.

2. Crater Wear. Crater wear is caused due to severe abrasion and high temperature in tool chip interface reaching the softening or melting.

- Crater wear does not always extend to tool tip, but may end at a distance from tool tip. It increases the cutting forces, modifies the tool geometry and softens the tool tip. Crater wear is more common in cutting ductile materials which produce continuous chips.

3. Localised Wear. The reason for rounding of cutting edges may be because of one or more reasons discussed above.

10.11 MACHINE SHOP EXERCISE

Aim: To manufacture the job on centre lathe as per the given drawing.

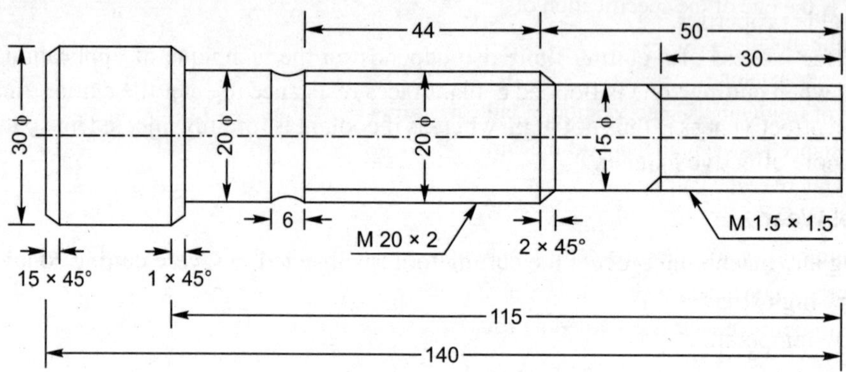

Fig. 10.77 Shape of the lathe job

Method: Set the speed, feeds and depth of cut for different operations as recommended. Following sequence of operation is followed:

1. Load the job on three jaw chuck of Lathe machine from one end.
2. By Facing tool face the end of the job.
3. By Turning tool turn the ϕ 30 mm × 25 mm.
4. By Knurling tool complete the Knurling of ϕ 30 × 25 mm end.
5. By chamfering tool chamfer the dimension 1.5 mm × 45°.
6. Unload the job again load it on the machine from the machine end.
7. Face the end of the job for a length of 140 mm, i.e., total length by Facing Tool.
8. Drill at the centre of job by fixing centre drill on the tail stock and then support this end of the job by centre.
9. Then by turning tool turn ϕ 20 × 50 mm.
10. Then again by Turning tool turn ϕ 15 × 50 mm.
11. By Grooving tool complete the grooving of ϕ 15 × 6 mm.
12. By chamfering tool chamfer 2 mm × 45°.
13. Cut the thread M 20 × 2 mm by threading tool.
14. By threading tool cut the thread M 15 × 1.5 mm.
15. Chamfering 1 mm × 45° by chamfering tool.
16. Unload the job from the machine and deburr it.

MULTIPLE CHOICE QUESTIONS

1. The operations performed on lathe machine are

(*a*) metal forming processes (*b*) material removal processes

(*c*) metal joining processes (*d*) any of above

2. The type of long used in lathe machine during cylindrical turning process is

(*a*) single point culting boil (*b*) multipoint cutting boil

(*c*) both single point and multipoint cutting (*d*) none of above

3. Swing is the one of the specification of

(*a*) lathe machine (*b*) shaper machine

(*c*) planer machine (*d*) grinding machine

4. Which of following is specification of lathe machine

(*a*) distance between line and dead centre (*b*) swing over the bed

(*c*) swing over the carriage (*d*) all of above

5. In lathe machine

(*a*) cutting movement is provided by rotating workpiece and feed movement is given by tool and carriage

(*b*) cutting movement is provided by reciprocating tool and feed movement is given by work lathe

(*c*) cutting movement is provided by rotating cutting tool and feed movement is also given by cutting tool

(*d*) cutting movement is provided by reciprocating worktable & feed movement is given by cutting tool.

6. In drilling machine

(*a*) cutting movement is provided by reciprocating tool and feed movement is given by work table

(*b*) cutting movement is provided by rotating cutting tool and feed movement is also given by cutting tool

(*c*) cutting movement is provided by reciprocating work tool and feed movement is given by cutting tool

7. In shaper machine

(*a*) cutting movement is provided by rotating cutting tool and feed movement is also given by cutting tool

(*b*) cutting movement is provided by recriprocating tool and feed movement is given by work table

(*c*) cutting movement is provided by reciprocating work table and feed movement is given by cutting tool

(*d*) cutting movement is given by rotating workpiece and feed movement is given by tool and carriage.

8. In planner machine

(*a*) cutting movement is provided by reciprocating tool and feed movement is given by work table

(*b*) cutting movement is provided by rotating workpiece and feed movement is given by tool and carriage

(*c*) cutting movement is provided by reciprocating work table and feed movement is given by cutting tool

(*d*) cutting movement is provided by rotating cutting tool and feed movement is also given by cutting tool

9. In horizontal milling machine

(*a*) cutting movement is provided by rotating workpiece and feed movement is given by tool

(*b*) cutting movement is provided by rotating tool and feed movement is given by work table

(*c*) cutting movement is provided by reciprocating work table and feed movement is given by cutting tool

(*d*) cutting movement is provided by reciprocating tool and feed movement is given by work table

10. Which of the following is the micro finishing operation

(*a*) reaming (*b*) milling

(*c*) honning (*d*) drilling

11. Which of following operation is also known as internal turning operation

(*a*) milling (*b*) boring

(*c*) drilling (*d*) tapping

12. Which of following operations produces chips of very small size

(*a*) boring (*b*) drilling

(*c*) milling (*d*) grinding

13. Which of following can be considered as chipless operation

(*a*) reaming (*b*) milling

(*c*) knurling (*d*) boring

14. The cutting tool materials are

(*a*) high speed steel (*b*) cemented carbide

(*c*) cubic boron nitride (*d*) all of above

15. Inserts are usually made of

(*a*) high speed steel (*b*) high carbon steel

(*c*) cemented carbide (*d*) ceramics

16. The chuck is used in lathe to

(*a*) hold the work piece (*b*) hold the cutting tool

(*c*) to hold both work piece and cutting tool

(*d*) none of above

17. The type of lathe chuck used for eccentric turning and operations performed on heavy and irregular shape work pieces are

(*a*) 3 jaw universal chuck (*b*) magnetic chuck

(*c*) 4 jaw independent chuck (*d*) any of above

18. Capstain lathe are used for

(*a*) mass production

(*b*) copying the shape of work peice

(*c*) for mating tools and dies

(*d*) for applications which cannot be handled on conventional lathe machine

19. Lathe centre on tail stock is known as

(*a*) live centre (*b*) dead centre

(*c*) both live and dead centre (*d*) none of above

20. The operations like taper turning, facing, thread cutting and grooving are performed on

(*a*) lathe machine (*b*) shaper machine

(*c*) planner machine (*d*) milling machine

21. The live centre of lathe machine is held in

(*a*) head stock (lathe spindle) (*b*) tool post

(*c*) tail stock (*d*) carriage

22. In shaper machine material is removed

(*a*) during forward stroke (*b*) during return stroke

(*c*) both the forward and return stroke (*d*) none of above

23. Slotter can also be considered as

(*a*) vertical shaper (*b*) planner

(*c*) horizontal shaper (*d*) lathe machine

24. The point angle of twist drill is
(*a*) 18° (*b*) 118°
(*c*) 45° (*d*) 90°

25. The size of shaper is specified by
(*a*) maximum length of stroke (length of cut) (*b*) horse power and speed of driving motor
(*c*) weight of shaper (*d*) floor space covered by shaper

26. In down milling operation
(*a*) cutting motion and feed motion are in same direction
(*b*) cutting motion and feed motion use in opposite direction
(*c*) cutting motion and feed motion may be either in same or opposite direction
(*d*) none of above

27. In up milling operation
(*a*) cutting motion and feed motion are in same direction
(*b*) cutting motion and feed are in opposite direction
(*c*) cutting motion and feed motion may be either in same or opposite direction
(*d*) none of above

28. The function of cutting fluid is to
(*a*) remove heat from tool work piece zone (*b*) lubricate tool work piece zone
(*c*) to protect both work piece and machine from corrosion
(*d*) all of above

29. Lathe bed is made of
(*a*) cast iron (*b*) wrought iron
(*c*) medium carbon steel (*d*) mild steel

30. The shape is used for machining
(*a*) horizontal surface (*b*) angular surfaces
(*c*) vertical surfaces (*d*) all of above

31. Discontinuous chips are produced when
(*a*) cutting tool with low rake angle are used (*b*) brittle materials are machined
(*c*) depth of cut is higher (*d*) all of above

32. Continuous chips are produced because of
(*a*) ductile materials are machined (*b*) chip thickness is small
(*c*) high cutting speeds are used (*d*) all of above

33. The grinding operation is a
(*a*) metal forming operation (*b*) surface finishing operation
(*c*) metal cutting operation (*d*) none of above

34. The material most commonly used as abrasive in grinding process is
(*a*) diamond (*b*) silicon carbide
(*c*) magnesium oxide (*d*) aluminium oxide

35. Grinding wheel marking system should specify
(*a*) abrasive used (*b*) grain size and grade
(*c*) structure and bond (*d*) all of above

FILL IN THE BLANKS

1. The machining processes in which material is removed by harder material (tool) through direct mechanical contact are known as ....
2. The cutting tool used in lathe, shaper and planner is .... cutting tool.
3. The cutting tool used in milling machine is .... cutting tool.
4. .... is a machine tool in which cutting movement is provided by rotating work piece and feed movement is provided by tool and carriage.
5. In drilling machine cutting movement is provided by .... and feed movement is given by ....
6. In milling machine cutting movement is provided by .... and feed movement is provided by ....
7. Grinding is a .... operation.
8. In cylinderical turning process diameter of work piece .... and length of work piece ....
9. .... is the operation of generating flat surfaces perpendicular to the rotational axis of spindle. The feed is given by cross slide.
10. .... is the operation of generating external cylinderical surfaces.
11. Boring is also known as .... process.
12. The lathe chuck in which all the three jaws more simultanious for holding work piece is known as ....
13. The work pieces which are too large to hold on shaper are machined on ....
14. Slotter is known as ....
15. The shaper is specified by ....
16. In shaper cutting is performed in .... stroke.
17. Thread cutting operation can be performed on machine tool.
18. In shaper cutting movement is provided by .... and feed movement is given by ....
19. In milling operation, when cutting and feed motion are in same direction it is known as ....
20. In milling operation, when cutting and feed motion are in opposite direction it is known as ....
21. The point angle of twist drill is ....
22. The grinding wheel can be considered as .... tool.
23. The most widely used abrasive material is ....
24. In grinding process the material is removed by ....
25. Lapping, honning and grinding are .... operations.

VIVA VOCE QUESTIONS

1. What are machine tools used in machine shop? Name a few.
2. Name five parts of lathe machine.
3. What is taper turning operation?
4. What is difference between Lathe and a Milling machine?
5. What is the function of Ram in a shaper?
6. What is the principle of Drilling M/C?
7. What is countersinking operation?
8. What is the difference between up and the down milling?
9. Name few cutters used in milling machine.
10. What is a twist drill, in which machine it is used?

INDEX

A

B

C

T

U

V

W

Win Prizes

ATTENTION READERS

Win Prizes

We request for your frank assessment regarding various aspects of the book as given below :

WORKSHOP TECHNOLOGY

By: Virender Narula

Please Fill up the Sheet in Neat and Clean Writing

(i) What topics of your syllabus, important for your examination point of view, are not covered in this book ?

....

....

....

....

(ii) What are the chapters wherein the treatment of the subject matter is not systematic or organised or updated ?

....

....

....

....

(iii) Have you come across misprints/mistakes/factual inaccuracies in the book? Please specify the chapters and the page numbers.

....

....

....

(iv) What other books on the same subject have you found/heard better than the present book? Please specify in terms of price and quality.

....

....

....

....

(v) Further suggestions and comments for the improvement of this book.

....

....

....

....

(PS: If need be please attach a separate sheet to write your views)

OTHER DETAILS

(i) Who recommended you this book ? (Please tick (✓) in the box near the option relevant to you.

☐ Teacher ☐ Friends ☐ Bookseller

....

....

....

(ii) Name of the recommending teacher, his designation and address :

....

....

....

(iii) Name and address of the bookseller you purchased the book from :

....

....

....

(iv) Name and address of your institution (Please mention the University or Board as the case may be)

....

....

....

(v) Your name and complete postal address :

....

....

....

(vi) Course you are enrolled in :

....

....

....

The best assessment will be awarded each month: The award will be in the form of our publications as decided by the Editorial Board.

Please mail the filled up coupon at your earliest to :

S.K. KATARIA & SONS®
4885/109, Prakash Mahal, Dr. Subhash Bhargav Lane,
Opposite Delhi Medical Association, Daryaganj, New Delhi-110002 (India)
+91-11-23243489, +91-11-23269324, +91-11-43551243 Telefax: 011-23243489
Email: katariabooks@yahoo.com; katariabook@gmail.com
Website : www.skkatariaandsons.com, www.katsonbooks.com